# Industrial Ecology and Environmental Management

Volume 5

Industrial Ecology and Environmental Management have been recognized as two powerful disciplines for solving the global challenges such as the depletion of non-renewable resources, global warming and energy/water shortage.

This book series includes informative books with comprehensive texts focusing on both theory and practice of industrial ecology and environmental management, it aims to present the multi-disciplinary frontiers of methods, models and technologies in industrial ecology and environmental management for scientists, engineers and researcher from academia, industry, government and institutions. It aims to build a platform for the worldwide scientists, engineers and researcher to share the frontiers of the methods, models and technologies in industrial ecology and environmental management for achieving a more sustainable future.

The books series will include both the theoretical tools and the practical applications of industrial ecology and environmental management for sustainable energy, industry and society.

Xianlong Ge

# Urban Transportation Energy Supply Network

## Research on the Negative Effects Assessment and Load Reduction Mechanism

Springer

Xianlong Ge
Economics and Management
Chongqing Jiaotong University
Chongqing, China

ISSN 2730-6070 ISSN 2730-6089 (electronic)
Industrial Ecology and Environmental Management
ISBN 978-3-031-64618-8 ISBN 978-3-031-64619-5 (eBook)
https://doi.org/10.1007/978-3-031-64619-5



This Springer imprint is published by the registered company Springer Nature Switzerland AG
The registered company address is: Gewerbestrasse 11, 6330 Cham, Switzerland

# Preamble

Traffic Energy Supply Network (TESN) is a transportation equipment energy supply network composed of gas stations, filling stations, charging piles, and other facilities and is an important infrastructure to support the flow of urban people and goods. According to the "China Gas Station Industry Market Outlook and Investment Strategic Planning Analysis Report" and the "China Statistical Yearbook" released data, by the end of 2022, there are 107,600 gas stations, 12,000 gas stations, and 111,000 charging stations in China and 417 million car ownership. Vehicle energy demand is large, random, unreasonable layout of energy stations, insufficient energy supply, low efficiency of resource use, resulting in vehicle energy difficulties or long queues, resulting in urban traffic congestion, environmental pollution, and waste of time and other negative effects, the huge urban transportation energy supply network operation, the negative effect caused by the economic loss is obviously inestimable. Therefore, in order to respond to the national policy requirements, it is urgent to achieve the goal of "double carbon" and reduce the negative effect of TESN. The main work of this book is as follows:

1. Urban transportation energy supply network operation status and negative effect formation mechanism analysis
   By collecting a large number of urban transportation energy supply network data, sorting out the urban transportation energy supply structure, network structure, urban transportation energy supply station layout and network distribution and future development trend, summed up the current situation of supply and demand of China's urban transportation energy supply network. The paper analyzes the balance of urban energy supply and demand from the macro and middle perspectives, summarizes a series of negative impacts caused by the current contradiction of urban transportation energy supply and demand, and defines the concept of the negative effect of urban transportation energy supply network, and puts forward the specific definition of the negative effect. According to whether the negative effect can be quantified, the expression forms of different negative effects are summarized and discussed in detail, and specific measurement methods are put forward for each negative effect of supplementary energy.

2. Research on evaluation methods of negative effects of urban transportation energy supply network
   Based on the operation status of the urban transportation energy supply network, firstly, starting from the negative effect evaluation of a single station, the negative effect evaluation method of a single recharge station is proposed, and the specific evaluation index system is constructed. On this basis, the evaluation system of the negative effect of urban transportation energy supply network is designed. Considering the characteristics of the surge of travel demand in holidays, the evaluation methods of the negative effects of transportation energy supply network in working days and holidays are established respectively. Finally, taking Chongqing regional road network as a case study, a case study on the evaluation of the negative effects of supplementary energy stations and networks is carried out.
3. Research on negative effect reduction of urban transportation energy supply network
   In the study of the negative effect reduction of urban transportation energy supply network, the negative effect reduction mechanism of TESN is constructed from three aspects: supply station layout optimization, station cooperation game, and supplementary energy scheduling. In the optimization of energy station layout, on the basis of the existing layout, the minimization of the negative effect of energy supplement network is taken as the optimization goal, and the layout optimization of urban energy supply station is studied. Through the comparison of different site selection schemes, the results show that the optimization of site layout can reduce the queuing time of vehicle energy supplement, and then reduce the negative effect of energy supplement. In the multi-agent cooperative game of energy supply station, through the application of cooperative game theory, considering the cooperation between transportation and energy supply stations, Shapely correction value is used for distribution. Numerical examples show that the cooperation between energy supplement stations can not only improve the revenue of the station but also achieve the balance between supply and demand of transportation and energy. Considering the cooperation among the government, the transportation energy supply station and the vehicle, the cost of the three can be reduced to 1/3 of the original through Shapley value distribution, which shows the effectiveness of the tripartite cooperation and the role of the government in reducing the negative effect of energy supplement. In the replenishment vehicle scheduling, the incentive scheduling method is used to set a number of incentive levels, balance the demand and supply of replenishment energy in the road network, and through building a double-layer optimization model, it is concluded that the incentive scheme can effectively reduce the travel and replenishment cost of the entire road network.

4. Evaluation of the negative effect of urban transportation energy supply network and empirical research on the mechanism of reducing the burden
   The AnyLogic simulation software was used to simulate and verify the urban transportation energy supply network, simulate the actual operating status of the energy supply network, analyze the key points of the energy replenishment bottleneck accumulated by negative effects, and verify the feasibility of reducing the negative effects of the urban transportation energy supply network by adopting traffic section restriction, station service rate adjustment, different station prices, and other reducing strategies.

December 2022 Xianlong Ge

# About This Book

This book focuses on the evaluation and burden reduction of the negative effects of urban transportation energy supply networks. By analyzing the current situation of supply and demand of urban transportation energy supply network in China, this paper defines the concept of the negative effect of urban transportation energy supply network and puts forward the specific definition and measurement of the negative effect. The load reduction mechanism of the urban transportation energy supply network is constructed from three aspects: transportation energy station layout optimization, station cooperation and energy supplement scheduling, and the simulation is verified by AnyLogic simulation software.

This book is suitable for senior undergraduates and postgraduates majoring in management science and engineering, transportation economics, transportation planning and management, operations research, system science, system engineering, etc., it is also a reference for university teachers and engineers.

# Contents

# Chapter 1
# Introduction

## 1.1 Research Background and Significance

### *1.1.1 Research Background*

TESN is a transportation equipment energy supply network composed of gas stations, gas stations, charging pile, and other facilities, which is the lifeline to ensure the development of urban transportation, and also an important infrastructure to support the flow of urban personnel and goods. The basic function of TESN is to provide energy supply for transportation equipment. The large-scale randomness of vehicle energy replenishment demand, improper site selection, insufficient energy supply, and competition of supply stations cause difficulties in vehicle energy replenishment or long queues, which affects traffic and then causes traffic congestion, environmental pollution, and waste of time. Obviously, these are not the expected consequences when TESN is designed. It is defined here as the negative effect of TESN.

The implementation of the strategy has further accelerated the development of transport and provided a solid guarantee for the realization of a better life for the people and the promotion of common prosperity. As shown in Fig. 1.1, there are serious problems of "energy replenishment difficulties" and "long queues" in many cities, which adversely affect the development of transport undertakings. In addition, due to the lack of overall layout planning of supplementary energy stations, some remote stations have been idle for a long time, resulting in waste of energy supply resources of supplementary energy stations.

According to the Traffic Management Bureau of the Ministry of Public Security, by the end of 2021, there will be 302 million vehicles in China, of which 10.01 million are new energy vehicles, accounting for 3.23 percent of the total number of vehicles. Seventy-nine cities, led by Beijing, Chengdu, and Chongqing, have more than 1 million cars, 35 cities have more than 2 million, and 20 cities have more than 3 million. Car ownership has led to a rapid increase in the number of urban transportation energy supply stations, which has brought a lot of negative

 
X. Ge, *Urban Transportation Energy Supply Network*, Industrial Ecology and Environmental Management 5, https://doi.org/10.1007/978-3-031-64619-5_1

**Fig. 1.1** Example diagram of "Difficulty in replenishing energy" and "long queue" at urban transportation energy supply stations

impacts on the urban transportation system. For example, with the arrival of the peak demand for vehicle energy replenishment, energy replenishment vehicles overload the use of urban energy supply stations, which makes the energy supply stations have negative effects on the urban transportation system such as energy consumption, carbon emissions, and environmental safety, and cause disturbances to the normal traffic flow on the main road. While increasing the risk of traffic safety, it also increases the pressure on social economy and environment.

In October 2021, The State Council issued the Action Plan for Peaking Carbon emissions before 2030, proposing to "ensure that the growth of carbon emissions in the transport sector is kept within a reasonable range", and "the concept of green and low carbon is integrated into the whole process of planning, construction, operation, and maintenance of transport infrastructure, so as to reduce energy consumption and carbon emissions in the whole life cycle". From 2015 to the end of 2021, the number of new energy vehicles in the country has increased from 420,000 to 10.01 million. However, the configuration of the domestic charging pile system still needs to be improved. As shown in Fig. 1.2, although the ratio between the number of new energy vehicles and the number of charging piles in China from 2017 to 2020 shows a gradual downward trend, it still cannot meet the development demand of new energy vehicles, resulting in a large negative effect of energy replenishment of electric vehicles in the process of energy replenishment, especially during the holiday peak period. The long time of waiting and queuing for energy replenishing brings psychological anxiety and time delay to drivers. In addition, the high load of the power grid is also easy to cause security risks. The "National Day Big Data Report 2021" shows that about 2.32 million passengers entered Chongqing during the National Day golden period, entering the top ten in the country, and the phenomenon of waiting for 4 h for electric vehicle charging even appeared after 1 h of charging. The root cause of this phenomenon is the spatio-temporal mismatch between vehicle energy demand and site and the charging characteristics of electric vehicles. This makes the resource part of the urban transportation energy supply station appear too

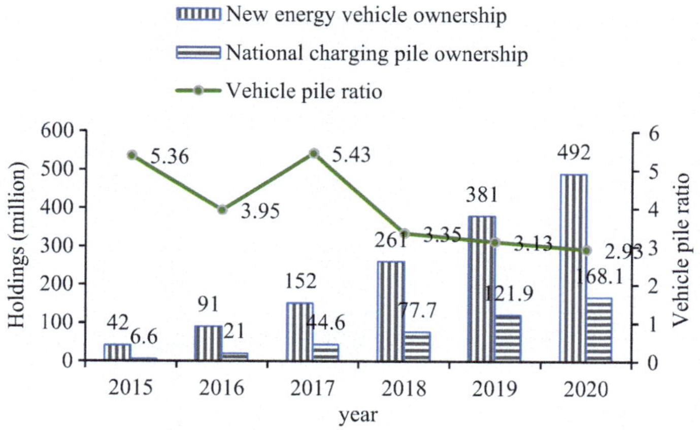


**Fig. 1.2** The number of new energy vehicles and charging piles in China

high load and some stations appear idle, resulting in unreasonable allocation of urban energy resources. In particular, the surge in travel demand during holidays, along with the increase in the number of travel vehicles, the pressure of urban transportation energy supply system to replenish energy increases, aggravates the negative effect of the energy replenishment process, brings hidden dangers to urban traffic, but also brings huge economic losses, affecting social development.

In the process of urban development, the flow trend of the urban road network is not unchanged, and the carrying capacity of the energy supply stations that have been built will also change in the development and change of the transportation network. This will inevitably cause a mismatch between the demand for energy replenishment of traffic vehicles and the supply capacity of some energy supply stations in the direction of time and space, resulting in excessive load of some stations. The phenomenon is that no one cares about some stations.

### *1.1.2 Research Significance*

Transportation and energy supply network is not only an important guarantee to support the flow of urban personnel and goods but also an essential foundation for the social activities and economic operation of the whole city. The volume and scale of urban transportation energy supply network are huge, and the economic losses caused by their negative effects are huge. The research on the negative effect of urban transportation energy supply network involves site layout, supply and demand balance, route selection, station queuing, road network planning, and so on. In this book, by analyzing the causes of the negative effect, we use the mechanism of site layout optimization and vehicle guidance to reduce the negative effect. The significance of the research is mainly reflected in:

#### 1.1.2.1 Enriching the Theoretical Research of Urban Transportation Energy Supply Network

This paper studies the optimization of urban transportation energy supply network from the perspective of negative effects, reveals the causes and rules of large-scale distributed dynamic random energy replenishment demand and queuing at supply stations in cities, and puts forward research methods and models of negative effects of urban transportation energy supply network with popularization and application value. Design optimization models and methods of urban transportation energy supply network based on the minimization of negative effects at different levels and angles to enrich the theoretical research of urban transportation energy supply network.

#### 1.1.2.2 One of the Concrete Realization Ways of National Smart Transportation and Green Transportation

The research on large-scale distributed random replenishment demand is helpful to improve the overall planning of urban transportation energy supply network, and the performance of urban transportation energy supply network in China is investigated and analyzed from the perspective of the negative effects of urban transportation energy supply network. On the basis of ensuring the demand for transportation energy supplement, resource optimization and energy saving and emission reduction of urban transportation energy supply network are helpful to alleviate urban traffic congestion, promote the upgrade of intelligent transportation, and the construction of smart city.

#### 1.1.2.3 Improving the Travel Experience of Transportation Participants

This book constructs the burden reduction mechanism of urban transportation energy supply network from three aspects: site layout optimization, site cooperation game, and vehicle energy replenishment scheduling, proposes load reduction strategies from various perspectives, and verifies the effectiveness of load reduction strategies in AnyLogic simulation software. A variety of strategies reduce the negative effect of energy supplement stations, help drivers quickly complete vehicle energy supplement, reduce energy supplement costs such as queuing time, and greatly eliminate drivers' mileage anxiety, which is conducive to healthy and green travel and improve travel experience.

## 1.2 Analysis of Domestic and Foreign Research Status

### *1.2.1 Analysis of Research Status of Negative Effect Evaluation of Urban Transportation Energy Supply Network*

Urban transportation energy supply network is a transportation equipment energy supply network composed of refueling stations, charging piles, and other facilities. Due to the multi-mode, multi-cycle, and random characteristics of large-scale vehicle energy replenishment demand, it leads to the imbalance of station resource utilization, traffic congestion, energy consumption, carbon emissions, and other negative effects. At home and abroad, some scholars have studied the negative effect assessment of energy supply network from multiple perspectives, using different methods.

#### 1.2.1.1 Research Status of Influencing Factors of Negative Effect of Energy Supply in Urban Transportation

The construction of urban transportation energy supply stations and the occurrence of energy demand will have various impacts on the surrounding environment, and many scholars have done research on many aspects. On the basis of the negative effect of the station, the negative effect of the vehicle driving on the network path is added.

The negative effect of station operation mainly comes from the time delay and carbon emission (Watabe et al. 2019). Most of the studies on time delay refer to the queuing delay (Xiao et al. 2020). For example, Cen et al. proposed a Mixed User Equilibrium (MUE) model, which considers the replenishing behavior (Cen et al. 2018). Lucas et al. propose a method to evaluate energy use and $CO_2$ emissions for the entire life cycle of energy supply sites for EVs and fossil fuel vehicles from construction, maintenance, and abandonment, and applied to a case study in Portugal (Lucas et al. 2012). Queuing at energy supply stations is the direct cause of most of the negative effects. Therefore, in order to quantitatively assess the negative effects of transportation energy supply networks, it is necessary to study queuing delays at stations. Considering the competition between different charging stations, Dupont et al. constructed a non-cooperative queuing model based on user preferences to verify the system equilibrium under anarchic cost (Dupont et al. 2023). Jung et al. proposed an SDIRQ model, which was based on the two-layer optimization of simulation and introduced the complexity (Jung et al. 2014). De Filippo et al. estimated the energy use of electric buses operating on different routes, including mixed driving in urban and suburban areas, and combined it with the simulation model to study the type and number of chargers deployed, as well as the effect (Filippo et al. 2014). In addition, many people use computer simulation to solve the queuing problem. Mohamed et al. developed a real-time simulation system to assess the impact of charging demand on the operation of the distribution network by generating a load profile of the EV transportation network and assigning it to each EV (Mohamed et al. 2017). Johnson and Yang et al. respectively conducted simulation analysis (Johnson et al. 2013; Yang et al. 2013). The comprehensive evaluation research of the supply station includes the evaluation of service operation and internal layout. For example, Cahyono evaluated the spatial service level of gas stations in Surabaya (Cahyono 2016).

The oil and gas fuel used at the station is flammable and explosive, causing great damage to employees, consumers, and surrounding facilities, especially in densely populated areas. It is mainly divided into fire protection facilities and equipment assessment and explosion safety distance study. For example, Chin et al. (1999) put forward a regional assessment model based on travel time extraction of online map to predict the demand for fire equipment at gas stations. Liu et al. (2020) proposed a regional evaluation model of firefighting equipment demand for gas stations based on online map travel time. With the development of the economy, the demand for energy is increasing (Havranek and Kokes 2015). Case studies have shown that fire and explosion accidents occurred many times in the past at the storage tank yard, causing property damage and casualties (Tauseef et al. 2018). In recent years, there have been many fire and explosion accidents at gas stations, which have seriously threatened people's lives and property safety. In the area of large gas stations or oil storage tanks, there may even be a domino effect caused by fire (Li et al. 2017). To better assess the risks, some studies use traditional quantitative risk analysis methods such as fault trees (Mohammadfam and Zarei 2015). However, the risk of gas station operations cannot be accurately simulated, and therefore the rigorous risk analysis of the vehicles in demand, as well as the risks of the site during the operation process, are ignored. Therefore, Zarei et al. applied Bayesian networks to the integrated dynamic safety risk modeling of natural gas stations, and the application of this method provides a risk-based investigation to identify the risk levels (Zarei et al. 2017). Luo MA uses the quantitative risk analysis method of Bayesian network to simulate the process of gas explosion to reduce the loss caused by gas station accidents (Ma and Huang 2019).

Filling stations usually store a large amount of oil and gas fuel and will be replenished regularly. Urban filling stations have large crude oil revenues and large residential activity populations around them, and these environmental impacts cannot be ignored. How to minimize the environmental impact at the edge of the city station and the environment at the edge of the station is a problem that managers need to think about. At present, the main environmental impact research is the impact of air pollution. Kathuria compared the change (Kathuria 2004). Narain and Krupnick also used monthly time series data from 1990 to 2005 to estimate the impact (Narain and Krupnick 2007). On the other hand, there is groundwater pollution at petrol stations. Lesage et al. distinguished the sources of petroleum hydrocarbons in groundwater and used principal component analysis to study the sources of hydrocarbons in soils (Lesage et al. 1997).

The impact of traffic efficiency mainly includes the disturbance effect of energy replenishing vehicles on normal vehicles running on the main road and the accompanying environmental impact caused by vehicle energy replenishing behavior. There are relatively few studies on this aspect. However, considering that this scenario is basically similar to the scenario of vehicle intersection at a common intersection, relevant studies are also of important research reference significance. For example, Vaziri uses the Lanekilometer duration index (LKDIF) to identify traffic congestion in California (Vaziri 2002). Sun et al. comprehensively considered various factors affecting traffic conditions at intersections and built a comprehensive evaluation index system to identify the risk degree of traffic congestion and predict its impact on future traffic flow (Sun et al. 2020). Pandian et al. studied the relationship between traffic, vehicle, and intersection characteristics and vehicle exhaust emission (Pandian et al. 2009).

It can be seen that the research on the negative impact factors of urban transportation energy supply mainly focuses on four aspects: site service, safety, environment, and traffic impact. However, the main problem is that the research directions of various scholars are relatively scattered, and few scholars have conducted comprehensive evaluation of the negative effects of urban traffic stations. The problem of site negative effect is composed of many factors, not just a single aspect. Therefore, constructing the comprehensive evaluation of urban transportation energy supply stations is the source of the negative effect, which can play a basic role in the analysis of network negative effect.

#### 1.2.1.2 Research Status of Urban Transportation Energy Supply Station Evaluation System

The selection of evaluation model for urban transportation energy supply network is related to the accuracy and reliability of negative effect evaluation. There are many methods to analyze index weight at home and abroad, which are mainly divided into subjective weighting methods and objective weighting methods. Subjective weighting methods mainly include Delphi method, analytic hierarchy process, cloud model, and so on, among which the analytic hierarchy process is the most widely used. Objective weighting methods mainly include the coefficient of variation method, principal component analysis method, and entropy weight method, among which the entropy weight method is the most widely used. For example, Chen established a fuzzy comprehensive evaluation model based on entropy weight and conducted an empirical study (Chen 2020). According to different evaluation contents, different qualitative and quantitative indexes may be included in the evaluation indicators. Of course, some scholars also use the method of combining subjective and objective weights.

Domestic and foreign scholars are more diversified in the construction and application of evaluation models. According to the specific types of index systems that need to be evaluated in their research, appropriate evaluation models are established, which mainly include principal component analysis, gray theory, TOPSIS, neural network, data envelope, cloud model, matter-element model, fuzzy comprehensive evaluation, etc. Among them, gray theory, neural network, cloud model, fuzzy comprehensive evaluation model are commonly used. For example, Du et al. use a fuzzy comprehensive evaluation model of traffic road safety influencing factors in cold areas based on attribute recognition theory (Du et al. 2018). Dong et al. used the running speed to evaluate the parking decision-making behavior at urban road intersections (Dong and Zhou 2020). Chen's empirical research on traffic operation efficiency also uses the fuzzy comprehensive evaluation method (Chen 2020). Each evaluation model has its own advantages and disadvantages, so it is necessary to use a suitable evaluation model according to the actual situation of their own research.

Although domestic and foreign scholars have considered the impact of all aspects and the evaluation of operation level in the evaluation of transportation energy supply stations, there are few comprehensive negative effects evaluation of energy supply stations at present. Comprehensive negative effect evaluation is conducive to the analysis and observation of the overall state of the station and its impact on the surrounding area and can compare the correlation of multiple factors, providing a reference for the overall layout optimization and coverage adjustment of the urban transportation energy supply system.

#### 1.2.1.3 Research Status of Urban Transportation Energy Supply Network Evaluation

When the vehicle has a demand for supplementary energy and drives to the supplementary energy station for supplementary energy activities, the increase of supplementary vehicles at the station will have a negative effect, especially when the service force of the transportation energy supply station cannot meet the demand. The overflow of vehicles will affect the road vehicles, resulting in the congestion of the main road and increasing the energy consumption and carbon emissions of vehicles on the road. And the competition between stations will also cause the idle waste of local site resources. In the process of studying the layout and location of filling stations, Zheng and other scholars found that vehicles in demand mainly complete the energy replenishment process based on the principle of the nearest home or the minimum detour distance in the journey, which may cause negative effects (Zheng et al. 2014).

Road traffic impact analysis, also known as road traffic impact assessment, refers to the systematic analysis of the impact degree and impact area of urban traffic development projects on road traffic services, so as to formulate measures to ensure that the quality of road traffic services is not reduced or adjust development plan measures, so as to reduce the impact of development projects on road traffic load (Kristensson et al. 2004). Domestic urban traffic impact assessment mainly adopts the traditional four-step method provided by Beckmann et al. Most of them follow the research process (Beckmann et al. 1956). Urban traffic congestion has become one of the major challenges faced by most transportation networks, which not only increases traffic delays, reduces mobility, and brings many inconveniences to travelers but also has negative impacts on the environment and economy, increasing air pollution and energy consumption. Therefore, more and more attention has been paid to the research of traffic congestion related issues, especially in the field of traffic congestion status identification and diffusion law analysis. Traffic congestion has not only the nature of the road section but also has strong temporal and spatial characteristics (Karlaftis et al. 2011). The existing traffic congestion identification research usually takes the threshold value of a certain flow parameter as the criterion to establish the identification model. When the actual value exceeds the threshold, traffic congestion is considered to have occurred. Some scholars have proposed evaluation indicators such as speed, traffic density, vehicle queue length, and traffic flow to identify the current traffic state. In the study of congestion diffusion law, the congestion index of queue length is often quantified according to the volatility, hysteresis, and propagation characteristics of congestion diffusion. For example, Ermagun and Levinson systems reviewed the studies on predicting short-term traffic conditions by using the spatial correlation between road segments. Spatial information improved the accuracy of prediction, especially under (Ermagun and Levinson 2018). In recent years, more and more attention has been paid to machine learning methods, which outperform purely statistical methods such as historical averages and exponential smoothing. However, there is no guarantee that machine learning methods are superior to advanced statistical methods such as spatio-temporal autoregressive comprehensive moving average. There is a big gap between the real spatial dependence of the traffic link in the actual network and the research network. In addition, these literatures also provide different evaluation methods and indicators for each influence object by studying the type of road and selecting a comprehensive evaluation index according to the weight calculation of each indicator.

As an important part of carbon dioxide emissions, carbon emissions from transportation have become an important concern (Dhakal 2009). Vafa-Arani has studied the development mode and regulation policy of urban low-carbon transportation from the perspective of system science and system engineering (Vafa-Arani et al. 2014). Traffic congestion increases energy consumption, further adding additional carbon emissions. Carbon emissions are generally studied in terms of vehicle carbon emissions and road network carbon emissions. As for vehicle carbon emission models, the models adopted internationally are mainly based on energy statistics, emission coefficient, and emission rate. Liu et al. adopted the system dynamics method to construct the carbon emission model (Liu et al. 2015). The calculation methods to determine the total carbon emissions of vehicles include measurement (Wallace et al. 2012), simulation (Karkatsoulis et al. 2017), and formula derivation. In practical case studies, many scholars have identified and continuously optimized vehicle operating parameters, such as speed, acceleration, power, and upwind area (Ottosen et al. 2015), for measuring the vehicle's overall emission status. The research on road network carbon emission mainly focuses on two aspects: road section environment and intersection environment. With the continuous development of information technology, the field gradually pays attention to the combination and application of information technology. For example, Shan et al. used a real-time database to analyze emission factors (Shan et al. 2018). Based on the combination of instantaneous emission data and second-by-second activity data, Lai et al. calculated the delay correction coefficient under different congestion levels and then established a delay correction model for estimating bus emissions (Lai et al. 2013). Mangones used a tool that combines a traffic model with emission factors in COPERTIV to study the impact of increasing road capacity on traffic-related emissions in Bogo (Mangones et al. 2020).

Existing research at home and abroad shows that the study of urban transportation energy supply network mainly focuses on the impact of traffic congestion at supply stations and the calculation of vehicle and road network emission models. Although these studies have laid a certain foundation for the study of the negative effects of urban transport energy supply network, there are still some deficiencies. First of all, most of the existing studies have analyzed the negative effects of supply stations from the perspective of the energy economy, without quantitative research on the impact of negative effects. Secondly, the existing station research only considers a single gas station or filling station and does not consider the stations with multiple service types, which is not in line with the actual situation. Finally, although there is a lot of literature on traffic congestion and carbon emissions at present. There is very little literature that systematically combines the two aspects, and there is almost no literature on traffic congestion and carbon emissions caused by supply stations.

## *1.2.2 Analysis of the Status Quo of Research on Site Layout Optimization of Urban Transportation Energy Supply Network*

The problem of site selection is one of the classic problems of operations research. The problem is to determine the number and location of site selection objects in the network by determining the site selection object, site selection target area, cost function, constraint conditions, etc., aiming at the minimum total cost or the optimal total service level or the maximum social benefit, so as to rationally plan the network structure (Wang et al. 2023a; Weninger and Wolsey 2023). At present, the commonly used site selection methods are divided into point demand site selection and cutoff site selection. Point demand site selection models include: P-median model, P-center model, coverage model, and so on. Traffic demand is to establish a model according to the maximum traffic flow of the site service and intercept as many demand flow paths as possible.

### 1.2.2.1 Point Demand Location

The point demand location strategy is a static deterministic location problem that assumes that the needs are located in different places, such as residential areas, workplaces, etc. Demand measurement or demand density is usually derived from population data, topographic data, cadastral data, survey data, etc. P-median model is based on the demand point, considers the distance from the demand point to the site, and establishes the model with the goal of minimizing the total cost of users. The limitations of this model are reflected in the assumption that the service capacity of each site is unlimited, and the site will not produce queues (Hakimi 1964). With the continuous expansion of demand, the model is difficult to meet the large-scale needs of site selection. Mirheli and Hajibabai considered the size of investment and established a site selection model function (Mirheli and Hajibabai 2023). The above literatures are designed to minimize the weighted average distance from each demand point to the facility. The obvious advantage of this method is that it is easy to find the median point of site selection and can provide services fairly. However, there are also some problems, such as insufficient consideration of service accessibility in emergency, and the emergency demand cannot be well met based on the median model alone. The P-center model starts from the station and finds the model with the smallest total travel cost to the point of demand. The alternative points of service stations in the model can meet all the needs. The P-central problem in the road network was first proposed by Hakimi, and later scholars Kariv and Hakimi proved that the problem is NP-hard (Kariv and Hakimi 1979). Therefore, some scholars later put forward solving methods, such as regional local search method and heuristic algorithm, to solve this problem. Under the premise of covering the requirements of the specified range, this method minimizes the farthest distance between each demand point and the facility. On the premise of ensuring service coverage, the robustness of service quality is improved. However, it is necessary to guarantee a certain range of services, which involves the distribution of demand points of each facility, leading to more complicated problems.

Coverage model is one of the important models in the study of location selection, which can be divided into set coverage, maximum coverage, and multiple repeated coverage model. In the set coverage problem, under the premise that the goal of site selection is to meet the needs, the number of site selection facilities is assumed to be unlimited, and a facility with the least number of facilities to meet the needs of the minimum total cost is sought. Set coverage is generally based on the service radius of the site selection (Pelegrín and Xu 2023). Some scholars have improved the set coverage model according to the research needs. Considering the uncertainty and difference in demand, Chen et al. (2023a) established the facility coverage location model. According to the quantitative evaluation of the urgency of the needs, Vazifeh et al. built a site selection model to minimize the risk of demand points and construction costs to ensure that the needs were met (Vazifeh et al. 2019). Guan et al. divided the coverage threshold based on regional population characteristics and the coverage quality of demand points and considered the time series characteristics and the types and quantities of stored materials to establish a multi-objective set coverage model (Guan et al. 2021). Based on traffic scenarios, Sebastian et al. built a facility location model with expected coverage with the goal of maximizing the expected coverage in emergency situations through multi-regional cooperative vehicle scheduling strategies (Rodriguez et al. 2020). The purpose of this site selection strategy is to minimize facility construction/operation costs while covering the needs of the specified area. Reduce costs by ensuring service coverage, so as to achieve a trade-off between benefits and service needs. However, it is necessary to guarantee a certain range of services, which involves the distribution of demand points of each facility, leading to more complicated problems.

The maximum coverage model is developed from the set coverage model, which aims to cover as many demand points as possible under the constraint of travel costs (Ghaffarinasab et al. 2022). Toregas et al. used the fewest stations to cover as many demands as possible and established a collection coverage location model. However, limited by resources, it is difficult for this model to cover all demands (Toregas et al. 1971). In view of this, Church et al. built the maximum coverage siting model under a limited number of sites to maximize the coverage (Church and ReVelle 1974). On the basis of the previous work, Kinay established a full coverage model, taking the O-D path into account within the coverage range, and found out the optimal location (Knay et al. 2021). Regarding the improvement of the maximum coverage location algorithm, Zhang Zhe established an improved maximized coverage model considering partial coverage through the grid discretization requirements to reduce the errors caused by the discretization solution of continuous problems. Yi and Liu et al. based on the characteristics of travel starting point-destination and social dimension, an improved Geographic PageRank (MGPR) model of electric vehicle charging demand was established, and the real charging data was input into the Maximum Coverage Localization Problem (CMCLP) model to optimize the spatial layout of public charging stations (Yi et al. 2022). Based on the historical flood event data, Hossein et al. predicted the areas where disasters might occur through the machine learning multilayer perceptron model and compared the prediction results with the data of weather stations and observation inventory events to build the minimum impedance and maximum coverage location model (Hossein et al. 2019).

At present, the main algorithms used to solve the location problem can be divided into three categories: (1) precise algorithm. For example, branch and bound, cut plane, and other traditional accurate algorithms. Plastrial considers a maximum coverage model using price decisions in a competitive environment and solves the model by exhaustion (Plastria 2009). Although such precise algorithms can obtain global optimal solutions, they are slow (Vazirani 2001). (2) Approximation algorithm. This kind of algorithm can guarantee to find a constant approximation ratio solution in a finite time, but cannot find the optimal solution (Shehadeh and Tucker 2022). (3) Heuristic algorithm. For example, Davari et al. designed a heuristic algorithm combining variable neighborhood search to solve the maximum coverage problem of fuzzy coverage radius (Davari et al. 2013). Galvao and Revelle proposed a Lagrangian heuristic for the maximum coverage problem (Galvao and Revelle 1996). At present, most of the research on the algorithm for solving the location problem focuses on the design of the heuristic algorithm, but the heuristic algorithm usually cannot find the optimal solution to the problem (Guo et al. 2018a; Atta et al. 2018).

The above model only considers the problem of coverage demand and does not consider the problem that "over-coverage" and "pseudo-coverage" lead to excessive concentration and idling of site demand, resulting in additional negative effects. Therefore, Gendreau et al. put forward the "double coverage model", which satisfies that the demand point is covered by two or more sites, and uses multiple sites to serve multiple demand points, so as to realize the linkage between multiple sites and reduce the phenomenon (Gendreau et al. 1997). Liu et al. improved the traditional double coverage model and discussed the practicability of the model. Liu et al. improved the double coverage model by covering demand points at least twice within the minimum coverage criteria, making reasonable allocation planning for demand and sites, and solving it with CPLEX. The results show that demand coverage can be effectively improved and response time reduced by relocating existing facilities without increasing resources (Liu et al. 2016a). Lamontagne et al. in order to find the optimal location of electric vehicle charging stations in multiple time periods, a stochastic discrete selection maximum coverage model was proposed, and a heuristic algorithm was designed to solve it (Lamontagne et al. 2023). Blanco and Gázquez combined the standby coverage model, the problem of discrete and continuous maximum coverage siting is considered from the perspective of the facility (Blanco and Gázquez 2023). Xu et al. considered the elastic demand, determined the optimal deployment of battery electric vehicle charging stations by maximizing coverage, and proposed a customized two-stage method to solve the model (Xu et al. 2020).

The multi-cover location strategy is put forward under the background that the number of sites is limited and each site only serves one demand point, resulting in the unsatisfied demand. Meeting multiple demand points is covered by multiple sites, and multiple sites serve multiple demand points. It can effectively avoid the demand priority problem in P-median and P-center site selection and the problem of uneven site coverage demand in the maximum coverage model. It can ensure that the demand is covered many times under the premise of maximizing coverage and also ensure the timeliness, convenience, and fairness of services.

Although the point demand positioning method has achieved success in many applications, it still faces problems such as fuel station location and charging station location in transportation, such as demand occurring in the journey rather than the fixed location, cost index, and so on. Drivers may prefer to meet the needs of their customers on long trips, rather than making specialized trips between private facilities and customer locations to obtain service. Also, travel distance/time as a point demand location method cost is not an appropriate measure of system cost in traffic location modeling. Therefore, both the point demand representation and the coverage definition are inaccurate in this case. Therefore, a new location model is needed to deal effectively with this type of location problem.

#### 1.2.2.2 Intercept Location

The traffic demand location approach assumes that consumers are looking for nearby recharge sites on their way to their destination. Represent the demand as a flow along the travel route. Hodgson proposes a Flow Capturing Location Model (FCLM) based on traffic, with the purpose of locating a predetermined number of facilities to maximize the use of traffic (Hodgson 1990). FCLM is a generalization of the maximum coverage location model, with the only difference being the definition of demand coverage. Traffic on a given trip is considered covered or captured if there is at least one charging station on the trip. However, the model does not take into account the main constraint that drivers of electric vehicles face, which is the limited driving range. Therefore, Kuby and Lim introduce a Flow-refueling Location Model (FRLM) that takes into account the driving range limits of EVs. If a sufficient number of charging stations are placed on the corresponding trip to allow the EV driver to travel from the origin of the trip to the destination and back without charging, then consideration is given to covering the EV flow (Kuby and Lim 2005). Most of the literature regards the maximization of vehicle flow coverage as an objective function. This can be explained by the limited financial resources available in EV charging infrastructure projects, so selecting the best location for a predetermined number of stations may be the best way to achieve the highest traffic coverage within a given budget.

Other literature adopts a centralized coverage approach that focuses more on cost minimization while ensuring full coverage of EV traffic, i.e. all EV drivers can complete the trip without running out of power costs, including the cost of charging station construction (Guo et al. 2018a), the cost of electric energy storage system (Huang et al. 2015), the cost of charging station operation (Wang and Lin 2009), the cost of battery investment (He et al. 2019a), and the cost of transformer investment (Kweon et al. 2017). Network expansion. Greenhouse gas emissions are also considered, reducing negative effects such as charging and waiting times (Wang et al. 2019a) or travel times (Zhang et al. 2018a) measured by cost. Cost minimization measures can be replaced by profit maximization measures if it is considered that private investors are building toll infrastructure and the profits generated by charging services are related to station location decisions (Deb et al. 2022). Other negative reduction targets relate to power or energy consumption (Zhang et al. 2010), taking into account the driver's reduction of deviations from the shortest path, etc. (Wang et al. 2019b).

#### 1.2.2.3 Research on Hybrid Location Selection Strategy

The site selection strategy is a combination of point demand and interception location strategy, and a site selection model is established according to the research needs and other factors. In most of the assumptions, the charging station capacity is usually unlimited, that is, a large number of drivers can charge their batteries at each charging station at the same time. Wang et al. (2019c), Wang et al. (2013), Lin (2018), and Zheng and Peeta (2017), an extension of the path segment-based siting model for charging stations introduced by Hosseini and MirHassani (2017) in MirHassani and Ebrazi (2013) Charging station capacity is taken as a decision variable, and the impact of these decisions on charging infrastructure costs is considered. In addition to budget constraints and station capacity constraints, a number of other types of constraints are considered in the charging station siting literature. Ngo and others have incorporated equity constraints into their models to ensure equitable use of charging stations between regions with different needs and reduce negative effects (Ngo et al. 2020).

Chung et al. use a form of fair constraint based on traffic to ensure that gas stations in each region capture the same percentage of traffic (Chung et al. 2018). Kuby et al. took into account policy considerations to mandate minimum coverage thresholds in each European country to obtain a more equitable distribution (Kuby et al. 2017). In terms of station capacity limits, some literature uses service capacity limits to ensure that customers are satisfied with station availability and the total time spent waiting and charging. Davidov and Pantowitz require that the total time for EV batteries to be fully charged be below a certain value to maintain power system stability (Davidov and Pantoš 2019). Yang et al. introduced quality-of-service constraints to ensure that the probability of EV taxis being charged at least once a day is above a certain level (Yang et al. 2017). The flow-controlling FRLM studied by Scheiper et al. takes into account the flow constraints that ensure the safe operation of the power system, maximize the benefits, and minimize the negative effects (Scheiper et al. 2019). He et al. introduce an example of side constraints related to site location constraints to consider nodes where charging stations already exist or where local governments do not allow the construction of stations. Drivers may choose to go to other charging stations because of queuing or known road congestion, and a reasonable location of charging stations will make this detour reduce the negative effect (He et al. 2019b). Kim and Kuby introduced the Departure stream refueling Location Model (DFRLM) as an extension of Kuby and Lim's FRLM, with the idea that drivers will accept departures from their shortest path (Kim and Kuby 2012). This assumption may allow more EV traffic to be covered with the same budget, but it needs to be determined which path should be used on each line. Given longer paths, EV drivers have less opportunity to travel. Hosseini et al. proposed a capacity-limited Departure flow refueling position model (CDFRLM) based on the arc cover method of Capar et al. (2013) Their model is more flexible than Kim and Kuby's because it allows multiple OD pairs to be used on the same road. Yildiz et al. also used the arc covering principle to enhance the original path segment-based formula (Yildiz et al. 2016).

Given the inherent complexity of the replenishment behavior as a random problem, some model simplifications aim to reduce the number of constraints and variables in the problem. Vries and Duijzer (2017), Kchaou-Boujelben and Gicquel (2019), and Lee and Han (2017). In a range-deterministic charging station siting model, the amount of battery power after charging is assumed to be constant and fixed at the maximum battery capacity, while the energy consumption is considered proportional to the distance driven. In practice, however, the amount of charge is subject to uncertainty and depends on the time spent at the charging station, the battery's service life, and the applied technology. In addition, unit energy consumption varies according to row path length, traffic conditions, driving style, use of cooling or heating devices, etc. Therefore, the actual driving distance is variable and its uncertainty should be taken into account in the charging station model in order to obtain a more accurate location solution. EV charging infrastructure planning and facility siting more generally involve strategic decisions, and the parameters of the siting problem (such as demand and cost) can change considerably over the planning period. Therefore, there are also many scholars who have conducted research on the random location problem, and Snyder summarized the model of facility siting problem (FLP) under uncertainty. Moreover, in long-term planning, decision-makers have to deal with parameters that change over time and need to adjust their decisions according to these changes (Snyder 2006). In this case, multi-cycle models offer the possibility of making decisions based on the evolutionary dynamics of the parameters of the problem, and Arabani and Farahani investigated (2012). Generally, in the context of EV charging infrastructure planning, multi-period modeling allows the calculation of the required number of charging plans in each time period based on charging demand and available budget, rather than making one-time decisions for the entire planning scope. It also allows for the adjustment of more operational decisions on traffic flow and charging plans according to network conditions in each period and has practical reference value for the evaluation of real-time negative effects within the network and the siting and deployment of charging stations in relevant areas.

## *1.2.3 Analysis of the Current Situation of Research on the Cooperative Game of Urban Transportation Energy Supply Network Sites*

In recent years, game theory has become a hot topic in various aspects of research. In solving practical contradictions, cooperative game and evolutionary game are used most. By analyzing the income of each player, incentive and punishment measures are adopted to encourage the player to adjust his strategy and achieve the final goal. At the same time, some scholars take the government as the game party and build a three-way game model.

### 1.2.3.1 Two-Party Game

Shared energy storage is a new energy storage application mode in the sharing economy environment, which realizes the shared use of energy storage resources through the separation of ownership and use rights of energy storage equipment, which can improve the utilization rate of energy storage and reduce the investment and operation costs of energy storage equipment. Chakraborty uses cooperative game theory to prove the feasibility of cooperative sharing of energy storage between users (Chakraborty et al. 2018). Xu and Chen divided the investment in NEV resource innovation into human resources, technology acquisition, and environmental support, and proposed that inter-industry partnerships can be used to examine the innovation efficiency of NEVs (Xu and Chen 2020). In the context of the sharing economy, the cooperation of member enterprises in the new energy vehicle supply chain has become a trend. Wang believes that in the case of cooperation, the respective revenue of both parties in the supply chain and the total R&D revenue will reach a higher level (Wang and Hu 2017). Subramanian et al. studied the cooperation between logistics companies and cloud computing service providers and discussed the cooperation between the two sides from the perspective of ability and trust (Subramanian and Abdulrahman 2017). Yang et al. studied the cooperation between two logistics service providers and concluded that the cooperation between the two parties can effectively reduce costs and improve corporate profits (Feng et al. 2016). From the above literature, it can be seen that the cooperative game has been applied to daily life, and the content of cooperation between the two sides is relatively easy, and only the interests of both sides need to be considered.

#### 1.2.3.2 Tripartite Game

With one more player in a three-way game than in a two-way game, there are more points to consider. For the first time, Fang and Cho introduced the network cooperative game theory method into the research on enterprise operation decision-making in the field of OM. We study the problem of inventory transfer cooperation of n retailers under the network relationship, prove the stability of the double price allocation method under the network cooperation alliance, and prove that the stability condition of MJW value under the network cooperation alliance is given (Fang and Cho 2014). Wang considers that the collaboration of participants such as renewable energy suppliers, hydrogen energy storage systems, and combined cooling, heating, and power systems can form complementary resources to improve economic benefits and reduce carbon emissions (Wang et al. 2023b). Rosenthal et al. used the idea of cooperative game to study the toll problem of regional highway network, constructed a cooperative game model based on cost sharing, designed the corresponding cost-sharing mechanism, and compared it with the current toll collection mechanism (Rosenthal 2017). In order to improve the energy self-sufficiency rate, Wang conducted research on urban virtual power grids, and proposed cooperation between demand side and factory to achieve mutual benefit (Wang et al. 2021). Luis examines the impact of the distribution of renewable energy generation on consumers, producers, and producers to analyze their individual contributions to the Alliance (Gomes and Vale 2024). Zhang et al. analyzed an alliance between a manufacturer and two retailers and proposed three different cooperation mechanisms to explore the impact of fairness on the alliance (Zhang et al. 2021). Hao et al. divided the entire regional comprehensive energy system based on the benefit co-creation and sharing mechanism, considered the cooperative alliances of different subjects, and established an optimal scheduling model for the cooperative game of multi-regional alliances (Hao et al. 2019). Teng et al. used the framework of cooperative game theory to establish a fair profit distribution scheme for the stakeholders of the integrated delivery project (Teng et al. 2019). Ding et al. used the cooperative game method to study the mechanism of how to distribute project profits among the members of the engineering service chain (Ding et al. 2019). According to the above literature, in terms of cooperation strategy, there are only two ways of cooperation in the game between the two sides, but the three-way game needs to consider a variety of strategies, and whether the strategy is in line with the actual problem.

### 1.2.3.3 Profit Distribution

With the improvement of the economic level, the sharing economy is considered an effective way to improve the efficiency of resource allocation. Through cooperation, costs can be reduced, competitiveness improved, etc. In order to ensure the stability of the alliance, it is necessary to design a reasonable allocation mechanism. Foreign scholars have made corresponding achievements in the study of income distribution. Krajewska et al. believe that freight carriers can reduce costs and improve market competitiveness through alliance and adopt Shapley value (Krajewska et al. 2008). Frisk et al. analyzed Shapley value and other sharing mechanisms for transportation cooperative cost allocation and proposed a new allocation method (Frisk et al. 2010). Dai and Chen studied the carrier cooperation problem with pickup and delivery requirements and time window constraints and proposed three profit distribution mechanisms to ensure the stability (Dai and Chen 2012). Mario and Mikael analyze the number of members of cooperative logistics, cooperative alliance, and build a mixed integer linear programming model of alliance structure and alliance cost. At the same time, by analyzing the cooperative transportation problem in the literature, various cost allocation methods are sorted out, and the use cases of the main methods are introduced (Guajardo and Rönnqvist 2015, 2016). In order to reduce the dissatisfaction of alliance distribution, Hezarkhani et al. adopted nucleolar solution and equal profit-sharing methods to solve the problem (Hezarkhani et al. 2016). Niknamfar and Niaki analyzed the cooperation between a company and multiple operators and established a dual-objective hub location problem model to fairly optimize the profit (Niknamfar and Niaki 2016). In order to protect the sensitive information of distribution customers, Yu et al. proposed a mathematical model of minimum cooperative travel and calculated the accurate Shapley value (Yu et al. 2017). In order to analyze how the cost of different companies is affected by cooperation, Tinoco et al. put forward different cost-sharing agreements and designed corresponding allocation mechanisms (Tinoco et al. 2017). Wang et al. studied the cooperative multi-center vehicle routing problem and designed a multi-stage hybrid solution algorithm and then established the profit allocation scheme by using the improved Shapley value model. At the same time, we also analyze the cooperative multi-center vehicle routing problem with simultaneous delivery and pickup, propose a hybrid heuristic algorithm, and use the cost gap method to check the profit distribution scheme (Wang et al. 2017, 2018a). Wang et al. believe that the cost of cooperative green pickup and delivery problems may be higher than that of non-cooperative ones, and propose a compensation and profit distribution method (Wang et al. 2018b). Lozano et al. proposed that horizontal cooperative transportation between carriers can reduce costs, and adopted four cooperative game schemes to solve the distribution problem (Lozano et al. 2013). Yang et al. consider the "last kilometer" cooperative vehicle routing problem, using the nucleolar method and Shapley value to fairly distribute the cost (Yang et al. 2020).

In the existing literature, Shapley value and its improvement method are the most common ones. In the application analysis, on the basis of joint distribution, constraints such as multi-center, low carbon, and time window are considered.

Domestic scholars further analyze fresh products, rural logistics, and other scenarios, providing a rich theoretical basis for subsequent research.

#### 1.2.3.4 Evolutionary Game

Evolutionary games no longer assume that all players are completely rational, but reach game equilibrium through trial and error, similar to the principle of biological evolution. In the cooperative competition of enterprises, the players are different, and the players are bounded rationality due to different environments and information. Based on the problems related to the unbalanced profit distribution of the participants in the process of comprehensively promoting rural revitalization, such as the non-obvious income increase of villagers and the imperfect cooperation mechanism, Chen et al. argue that the government has introduced a variety of incentives to promote the development of charging piles, but the effects and differences of these policies are still unclear, so they construct an evolutionary game model of complex networks to analyze the effects of different incentive policies (Chen et al. 2023b). Wu et al. studied the value co-creation between charging service operators and private pile owners, and when the two sides reached a cooperation, they obtained the optimal investment scale of the operator (Wu et al. 2022). Wang et al. believe that it is difficult to install private piles due to the interference of property companies, so they proposed an evolutionary game model based on prospect theory to solve the conflict of interests between users and property companies and promote the construction of private piles (Wang et al. 2022a). Based on the evolutionary game model under the framework of small-world complex networks, Fang et al. analyzed the impact of policy incentives and consumer preferences on the construction of EV charging infrastructure and proved the advantages of balanced dynamic subsidies and tax policies in promoting charging infrastructure (Fang et al. 2020). Based on the different subsidies for pure electric vehicles and hybrid vehicles in China, Shao et al. analyzed the evolutionary game between the government and car-sharing companies and discussed the future development direction of electric vehicles (Shao et al. 2022). Considering that operators' differentiated pricing of different charging piles will affect consumers' charging behavior, CAO uses an evolutionary game to explore the interaction process between EV users and charging operators (Cao et al. 2022). Huang et al. consider the interaction between charging stations and electric vehicle homes in a complex network environment. Therefore, an agent-based evolutionary game model is designed to analyze the impact of users' personal behavior on the investment and construction of charging stations (Huang et al. 2022a). Based on the development of the sharing economy, energy storage sharing can effectively save investment costs. In order to ensure the sustainable development of the energy storage system, an evolutionary game model of supporting energy storage and renewable energy is constructed, and the development path of the shared energy storage model is analyzed (Li et al. 2023). Hirte et al. used an evolutionary game approach to analyze the driving mechanisms of taxation and subsidies, as well as social benefits and costs, on the development of electric vehicles (Hite and Tscharakischiew 2013). Through the analysis, it can be seen that evolutionary game is widely used in the field of economics, and the game subject is limited rationality, which cannot obtain the optimal result in an instant.

#### 1.2.3.5 Site Selection Game

The common methods of site selection include gravity center method, analytic hierarchy process, fuzzy comprehensive evaluation method, etc., but there are also scholars who apply the knowledge of game theory to site selection. Li Weihong and Wang Qiang use the three-stage game model to study the pricing and location decision of duopoly manufacturers when they face asymmetric demand and endogenous product cost. In the first stage, the manufacturers choose the location of their factories. In the second stage, based on the known location, the manufacturers negotiate with their suppliers to decide the price of raw materials. In the third stage, on the premise that the location and the price of raw materials are known, the manufacturer determines the market price (Hotelling 1929). Aiming at the problem that the traditional Hotelling line segment model assumes that the production costs of duopoly enterprises are exactly the same, Liu Yanchun and Gao Liqun apply the two-stage game model and put forward the pricing model (Wendell and McKelvey 1981). Considering the nested characteristics of the population and the spatial distribution conditions of the market, Meng Shangxiong studied the location game of service facilities. In the case of the existence of Nash equilibrium, there is no guarantee that the market shares of both sides are equal (Miller et al. 1991). Starting from the problem of reservoir resettlement, Gao Xiaoyong et al. studied how to achieve stable and harmonious immigration and introduced the public participation model into the decision-making of the location selection of migrant resettlement sites. Therefore, they studied the participatory decision-making of the location (Ahm et al. 2004). Wei Hao studied the site selection game problem of two companies of different sizes in one-dimensional finite linear market. The game model of two companies with different sizes is established, and then the competition system in which one large company invests in one store and another small company invests in several chain stores is introduced. Under the premise of balance, the strategies to be taken in the game of site selection between small-chain companies and large-chain companies are discussed (Ghosh and Craig 1984). Liu Hong et al. aiming at the problem that the current active distribution network planning method cannot reflect the decision of investors, put forward the operation strategy of photovoltaic and energy storage system considering peak-valley electricity price. Then, based on the game theory method, a two-layer optimization model is constructed on (Lederer and Thisse 1990). Considering the NIMby effect, Zhang Xianghe and Peng Xuya, from the perspective of industrialization and marketization of waste disposal, use Hotelling linear city model and Bertrand game to study the site selection and pricing strategy of waste disposal sites and analyze the interest coordination among the government, waste disposal sites, and residents, and put forward the benefit compensation mechanism (Serra and ReVelle 1999). Dong Shiyong and Zhang Dingwen discussed the spatial layout of affordable housing in China, analyzed the social needs of low-income residents, the economic demands of developers, and the balance of comprehensive benefits of the government from the perspective of a three-party game, and proposed a three-party site selection game mechanism (Kress and Pesch 2016). Han Chuanfeng et al. set up a non-cooperative dynamic game model (Friesz et al. 2006). Based on the dynamic game model, Wang Jiguang studied the potential disruption risk caused by the openness and expansibility of the supply chain and constructed a two-stage two-person dynamic game model (Konur and Geunes 2012). Zhang Yong and Xiao Yan believe that the site selection of third-line construction enterprises is very important. As the two main bodies of the site selection game, the central department and third-line enterprises respectively adopt different action strategies, and the various actions taken by the employees of third-line enterprises to change the site will have a certain impact (Berglund and Kwon 2014). Cheng Yukun and Mei Lili believe that the non-payment mechanism design of the facility location game problem is an interdisciplinary subject of combinatorial optimization and theoretical computer science, which has important applications in management science, information science, and social economics and has important theoretical significance and practical application value (Cardinal and Hoefer 2010). According to the conditions of different types and quantities of facilities, different personal preferences, different measurement Spaces, and different overall social goals, this paper introduces various types of facility location game models, lists relevant research results, and summarizes the problems to be solved. Zhu Tao believes that under the market economy, retail enterprises have become the real subject of commercial investment decisions, and whether the location is appropriate or not is the key to the success or failure of enterprise management. Hotelling model and Steinberg model are used to carry out game analysis (Rohaninejad et al. 2017). Song Yan and Teng Chenmei studied the problem of site selection in the game between the government and extremist organizations as participants in terrorist attacks. In this game, terrorists attack a city to maximize its damage effect, and the government, as the leader of the Stark game, anticipates the strategic choice of terrorist organizations and makes site selection decisions to minimize the extent (Karatas 2017). From the above literature, it can be seen that there have been some researches on site selection in game theory, most of which are based on non-cooperative game and choose site selection to maximize their own interests, without considering the balance of system layout.

To sum up, the content of the game theory is divided into five parts, in which the cooperative game is the main discussion content, from two-party game to three-party game, discusses the way of profit distribution, and then expands the evolutionary game and site selection game. In the cooperative game, Shapley presented the solution of cooperative game, and then Gillies proposed the concept of kernel and Nash solution, which were all static, and then Owen proposed the dynamic Owen value. Cooperation is everywhere, and more and more attention is paid to cooperation in economic activities, and evolutionary games can help enterprises to analyze the optimal cooperative equilibrium point.

## 1.2.4 *Analysis of Current Research Status of Vehicle Scheduling in Urban Transportation Energy Supply Network*

### 1.2.4.1 Research on Supplementary Energy Guidance

Domestic and foreign scholars' research on vehicle replenishment guidance mainly focuses on route optimization scheduling, taking the shortest replenishment path or the minimum total travel cost as the optimization goal. Few literatures study the problem of replenishment vehicle guidance from the perspective of minimizing the negative effect of replenishment, and establish the matching scheme between replenishment vehicle and replenishment station. For example, Zhou et al. established an intelligent path-planning strategy for electric vehicles with the lowest voltage deviation rate as the optimization goal and proposed a multi-level, two-layer optimization algorithm to solve it (Zhou et al. 2022). Schoenberg and Dressler proposed a charging and routing strategy considering travel time, waiting time, and charging time based on the user perspective, and proposed a multi-standard shortest path search algorithm with a shrinking hierarchy to solve the problem, so as to reduce the user's time cost (Schoenberg and Dressler 2023). With the optimization goal of minimizing the total travel time, Hung proposed an EV dynamic charging scheduling strategy based on the real-time traffic data of the urban road network, which considered the driving distance and charging service efficiency (Hung and Michailidis 2015). Sun and Yagcitekin also considered the impact of travel time and charging cost on drivers' route selection and established an intelligent charging scheduling strategy for electric vehicles (Sun and Zhou 2016; Yagcitekin and Uzunoglu 2016). Said used the queuing theory method to build the EV charging queuing model, considered the minimum queuing time of EV, and established the EV charging scheduling strategy oriented to the operating benefits of charging stations (Said et al. 2013).

At present, the replenishment guidance of electric vehicles by domestic and foreign scholars is mostly based on real-time traffic information and charging station service information. Shi et al. pointed out that combining real-time information of roads and real-time service information of charging stations with EV navigation system is the key to achieve efficient EV scheduling optimization (Shi et al. 2020). Huberh considers the dynamic characteristics of electric vehicle driving energy consumption and charging time and combines real-time road traffic information to build two kinds of optimal path search algorithms for electric vehicles in a dynamic traffic environment (Huber and Bogenberger 2015). Yang et al. proposed a route optimization algorithm considering dynamic driving energy consumption of electric vehicles by sharing real-time traffic information and designed a travel navigation system for electric vehicles on this basis (Yang et al. 2016). Shao et al., based on the change characteristics of real-time road network traffic state and considering the impact of charging time, built a time-varying road network route optimization model for electric vehicles and proposed a dynamic path search algorithm to solve the model (Shao et al. 2017). Considering the changing characteristics of the traffic network, Liu established a real-time pricing strategy for charging stations and routed users to the optimal charging station (Liu et al. 2014). Yuan et al. constructed a charging guidance framework based on real-time information of the traffic network and established an EV route optimization model targeting queuing time and charging cost (Yuan et al. 2018). Considering real-time power grid and road traffic information, Levering and Zeng et al. built an EV charging scheduling optimization model for intelligent interactive systems of electric vehicles, distribution network, and road traffic network, and explored the interaction among electric vehicles, road network, and power grid through the model (Levering et al. 2022; Zeng et al. 2022). Qian et al. proposed to combine the information of electric vehicles, distribution network, and transportation network to establish a two-tier scheduling model, so as to determine the optimal charging service fee for electric vehicles, guide the charging behavior of electric vehicles, and minimize the total cost (Qian et al. 2020).

In many studies, when planning the optimal path between OD pairs, path planning in a traffic network can be abstracted as the problem of finding the shortest distance on a directed graph. At present, the path planning algorithms commonly used in research mainly include Dijkstra algorithm, A* algorithm, online search algorithm, hierarchical search algorithm, and so on. Algorithms commonly used in research include A* algorithm and hierarchical algorithm. A* algorithm is a very classical best-first search algorithm. In order to find the optimal value, the evaluation function is constantly used to evaluate each node in the search process. The smaller the deviation between the estimated distance value and the actual value, the faster the search speed (Hart et al. 1968). The hierarchical design idea of the hierarchical algorithm is to extract the network hierarchy through the characteristics. Its search process is to realize the switch between different levels of network layers along the starting point until the termination rule is established (Rajagopalan et al. 2008). Khosroshahi et al. proposed A method to obtain road condition information based on an on-board network and sensors, proposed a path cost function, and then used A* algorithm to plan the optimal path (Khosroshahi et al. 2011). Jayapal et al. proposed A weighted shortest path algorithm A-Dijkstra based on ANN by using the prediction model based on neural networks to predict the real-time traffic conditions at a given location and plan the optimal traffic path (Jayapal and Roy 2016).

To sum up, most domestic and foreign scholars' researches on replenishment vehicle guidance are based on real-time traffic information and station service information to optimize the driving path of replenishment vehicle, with the shortest path or minimum total travel cost as the optimization goal. Few literatures study the problem of replenishment vehicle guidance from the perspective of minimizing the negative effect of replenishment and establish the matching scheme between replenishment vehicle and replenishment station.

#### 1.2.4.2 Research on Energy Replenishment Network Equilibrium

The traditional transportation allocation is based on the complete rationality of travelers, considering the shortest travel time or the smallest travel cost. Based on this assumption, Wardrop proposed traveler's route selection criteria and introduced User Equilibrium (UE) and System Optimum (SO) (Wardrop 1952). In view of this, Zhong et al. assumed the traveler was familiar with the road network. The influence of traveler's path choice behavior on assignment is studied (Zhong et al. 2011). Based on the traditional model, Jiang et al. considered traveler's behavior and proposed the influence of individual behavior choice on distribution (Jiang et al. 2014a). In the actual travel process, travelers have their own travel habits, which will directly affect the distribution results. Hu et al. specifically considered driver personality, risk preference, and other behavioral factors, proposed a distribution model based on cumulative prospect theory and found that individual behavior has a direct impact on distribution results (Hu et al. 2019). Pan et al. considered travelers' driving risk preference and absenteeism attributes, divided travelers into risk and risk taking, developed a mixed choice model, and explored the impact of individual behavior choice on distribution (Pan et al. 2019).

Due to the time-varying traffic conditions in the real road network, the distribution results also change with time, and travelers cannot fully grasp the road information, so there is a certain error between the prediction of path impedance and the actual impedance. To eliminate the influence of randomness on route selection, Chen et al. apply the Stochastic User Equilibrium (SUE) method by logit to solve stochastic problems in real networks. Compared with Determined User Equilibrium (DUE), the practicability of random user equilibrium is proved (Chen et al. 2014). Rasmussen et al. relaxed the assumptions of the DUE model, established the SUE model, and increased travelers' perceived impedance to the path, so that the results were more consistent with the reality (Rasmussen et al. 2015). In order to solve the stochastic user equilibrium model, Liu et al. proposed linearization of nonlinear problems for continuous road network design problems with SUE constraints and obtained the global optimal solution (Liu and Wang 2015). Watling et al. combined the unused routes in SUE and the perceived errors in SUE to propose two model families, recombining the unused routes with the used routes randomly to achieve an equilibrium (Watling et al. 2015). Huang et al. analyzed drivers' route selection behavior under vehicle demand elasticity, established a cost minimization distribution model, and obtained a method to maximize profits (Huang and Kara 2020).

Jiang et al. established a hybrid user equilibrium model considering the random charging behavior of electric vehicles and the selection of the shortest path on the road network where electric vehicles and fuel vehicles run together (Jiang et al. 2014b). Zhang et al. proposed to consider the impact of carbon emission constraints of road network and EV travel costs on the location of charging stations in a hybrid network (Zhang et al. 2018b). AMX et al. studied the user balance of hybrid pure electric vehicles and traditional fuel vehicles in a traffic network constrained by battery-changing stations and road slopes (Xu et al. 2017). Liu said that when choosing a route, drivers of pure electric vehicles should not only minimize travel costs but also consider the feasibility of the route. A set of user equilibrium conditions is proposed to describe the route selection behavior of electric vehicle drivers considering flow-dependent power consumption (Liu and Song 2018). Jing et al. proposed a user equilibrium model for the mixed traffic network and considered two ways of multinomial logit and multinomial probit to solve the traffic allocation problem (Jing et al. 2017). He et al. analyzed the impact of charging time, traveler's mileage anxiety, and user's risk attitude on charging demand on travel path and in-transit charging behavior, and proposed a network equilibrium model based on travel chain (He et al. 2015a). Under different network Settings and scenarios, Nan et al. formulated and solved the network equilibrium problem of destination, route, and parking combination selection, which is affected by the driving range limit of pure electric vehicles and the composition of alternative travel costs (Nan et al. 2014). Huang proposed that the layout and design of charging stations are the key to influencing the utilization rate of pure electric vehicles. Under the condition of balanced user traffic allocation, a site selection model of fast electric vehicle charging stations with random charging demand was built (Huang and Kockelman 2020). Hong et al. studied the traffic balance problem of electric vehicles and the optimal configuration of charging stations under distance constraints, which is similar to the network design problem with traffic balance (Zheng et al. 2017). Agrawal et al. selected travel paths based on travelers' previous experience and traffic information, considered users' preference for travel speed, and established a dynamic user equilibrium model for mixed traffic flow (Agrawal et al. 2016). Based on the actual survey data, Jou uses the Probit model to analyze the impact of travelers' mastery of traffic information on route selection and departure time and finds that understanding traffic information can change travelers' departure time or path (Jou 2001). Similarly, Zhang et al. analyzed the traveler's path inertia choice behavior caused by limited traffic information and further analyzed the impact of providing traffic information on alleviating inertial user equilibrium (Zhang and Yang 2015).

Most of the above literatures are based on travelers' unilateral perception to obtain the path with the shortest travel time or the lowest travel cost. Due to the limited cognitive ability and information acquisition ability of travelers, they usually seek satisfactory solutions rather than optimal solutions in the decision-making process. Therefore, they should timely understand the situation of the entire road network and take this as supplementary decision-making. It is the basis of reducing the negative effect of replenishment energy. To sum up, most domestic and foreign scholars only consider queuing time and road section travel time as two negative effects to conduct network balance modeling and do not comprehensively model all negative effects.

#### 1.2.4.3 Research on Incentive Mechanism of Replenishment Scheduling

In order to make drivers better respond to scheduling policies and improve their satisfaction with scheduling schemes, it is necessary to establish scheduling incentive mechanism. At present, scholars at home and abroad have studied and applied incentive means in various fields. For example, Park et al. proposed an incentive algorithm considering distribution system constraints and established a random arrival EV response model considering demand and price elasticity, time, and charging pressure to reduce energy waste and economic loss (Park et al. 2022). Aung proposed a new dynamic traffic congestion pricing and EV toll management system by introducing incentives, which rewarded drivers who chose non-congested roads and non-congested charging stations and proved that the system could effectively reduce traffic congestion and total charging time at charging stations (Aung et al. 2021). Wang et al. introduced two incentive schemes, vehicle and station, so that users could change the use and parking of vehicles, and combined with the stations in the system, the most appropriate scheduling route was obtained by sorting, so as to improve the scheduling efficiency of the system (Wang et al. 2019d). Xiong et al. considered user travel information, established personalized incentive measures for users, and guided users to adjust travel routes and times to reduce congestion (Xiong et al. 2020).

Considering the sensitivity of users to the price of electricity, some studies take the charging price as the entry point to study the charging scheduling strategy of electric vehicles, so as to realize the coordination between electric vehicles and charging stations. Considering the influence of electricity price on charging behavior, Erdogan et al. proposed a charging scheduling strategy with minimum energy cost and TOU demand cost (Erdogan et al. 2022). Yang et al. set up a charging scheduling model of TOU based on real-time charging prices considering the differences in electricity prices in different time periods (Yang et al. 2015). Zhang et al. (2022, 2023), considering the influence of price factors on charging selection behavior, established an EV charging route optimization model according to charging prices in different periods. Wang et al. considered the cluster scheduling of electric vehicles under real-time electricity price and optimized the orderly charging and discharging behavior of electric vehicles by using deepened intensity learning (Wang et al. 2022b). Huang et al., considering the influence of charging price on charging behavior, proposed a charging guidance strategy that minimizes the cost of electric vehicles, including charging time, charging energy, charging station, and charging route selection (Huang et al. 2022b). Barco et al. and Sassi et al. designed the optimal travel path and charging scheme for electric vehicles with the goal of minimizing the cost of charging stations (Barco et al. 2017; Sassi et al. 2015). Zhang and Wei introduced game theory into EV charging scheduling and conducted a detailed study on charging strategy (Zhang and Li 2016; Wei et al. 2015).

To sum up, most of the current studies on the guided scheduling incentive of supplementary vehicles are aimed at inducing drivers to go to appropriate supplementary stations by reducing the driver's supplementary energy cost. In the induction of charging vehicles, TOU is used to guide EV off-peak charging, but static TOU may lead to a new charging peak when the electricity price is low. The negative effect of reducing the negative effect is not obvious. In recent years, some scholars have begun to study the dynamic TOU price or other dynamic incentive methods.

#### 1.2.4.4 Research on Vehicle Networking Communication Technology

At present, the research on vehicle networking communication technology is mainly focused on providing efficient and real-time route planning for vehicles in various situations, improving urban traffic conditions, and realizing communication between vehicles, so that vehicles can dynamically correct routes according to current traffic conditions, so as to achieve uniform traffic capacity of all roads. In order to find a fast real-time route in case of sudden road congestion, Wu et al. proposed a dynamic route selection algorithm based on obtaining road condition information (Wu et al. 2016). Guo et al. proposed TTE for travel time estimation, which avoids congested roads by comparing path planning algorithms, estimates real-time travel time, and carries out dynamic path planning through real-time traffic information sharing mechanism (Guo et al. 2018b). El-Wakeel et al. calculated the travel time of each path according to the fuzzy model to recommend the optimal path (ElWakeel et al. 2018). Considering that not all traffic congestion should be avoided, Souza et al. proposed a route planning algorithm Chimera, which avoids roads with large congestion index by classifying congested roads, avoids congestion through classification mechanism and rerouting algorithm, and adjusts all vehicles and possible traffic network overload in the road network in real time (Souza et al. 2016). Considering the overload of traffic network, Akabane et al. proposed APOLO (context-Aware and People-centric vehicular traffic rerouting), which uses historical traffic information to plan routes for vehicles when the network is offline. However, offline vehicles cannot obtain real-time road congestion and deal with it (Akabane et al. 2017). He et al., based on the current position of the vehicle and the vehicle's action prediction, adopted the group sensing system to maximize the coverage of the system (He et al. 2015b). Similarly, Wang et al. proposed a vehicle selection strategy with maximum spatio-temporal coverage based on the vehicle travel route and combined with the spatio-temporal coverage (Wang et al. 2018c). Liu et al. collected spatio-temporal data on vehicles and proposed a new vehicle selection strategy based on system group perception (Liu et al. 2016b). Based on the limited perception budget, Dong et al. proposed two incentive strategies to attract more vehicles to participate in the perception task and obtain more data (Dong et al. 2014). Similarly, Zhou et al. proposed a recruitment strategy that allows vehicles to actively participate, thereby improving perceived coverage and service efficiency (Zhou et al. 2018). In order to improve the accuracy of data, Peng et al. proposed a quality-based crowdfunding incentive strategy and paid each participant according to their effective contribution (Peng et al. 2018). Wu et al. use wireless communication network to realize information transmission between cloud platform, power grid, charging station network, traffic road network, and electric vehicle through the vehicle networking cloud platform system, analyze and calculate the obtained information, and then guide and dispatch electric vehicles to improve the benefits of drivers, charging station and power grid (Wu and Pang 2023). Similarly, Yang et al. contacted various entities through the vehicle networking platform system and then proposed an orderly charging and discharging strategy with comprehensive cost minimization considering the interests of multiple entities (Yang et al. 2023).

To sum up, although the research of vehicle networking communication technology is distributed in various fields, there are few researches on the guidance and scheduling of supplementary vehicles. In order to obtain real-time traffic information and recharge station service information to recommend suitable recharge station and recharge routes for drivers, vehicle networking communication technology is a suitable method.

### *1.2.5 Analysis of Current Situation of Simulation Research on Urban Transportation Energy Supply Network*

Urban transportation energy supply network system is a complex system. In the study of complex systems, Agent-based modeling (ABM) technology is more prominent, which provides effective and practical techniques for hidden order mining, state prediction, and auxiliary decision-making of the system. The huge theoretical and technical advantages of multi-agent computer modeling make it more and more widely used in the fields of economy, society, and control.

The negative effect of urban transportation energy supply network is applicable to the simulation of traffic flow, in which the traffic flow is the energy replenishment traffic flow. Traffic flow simulation is an important tool for designing new facilities or evaluating existing ones. Traffic engineers often use simulation tools to analyze the impact on traffic flow. In the automotive industry, simulation is used to develop and analyze autonomous driving features and their functions. The steps can be summarized as a three-step calibration process to achieve effective model calibration by focusing on macroscopic traffic flow mechanics and microscopic driving behavior (Hofinger et al. 2023). The traffic simulation model can be analyzed from macroscopic and microscopic perspectives, and the volume of traffic flow on the traffic road can be studied from macroscopic perspective. From the microscopic point of view of a single vehicle, using cellular automata and queuing theory, and other models to study vehicle travel behavior. Since the 1990s, computer technology has evolved and described the actions of vehicles on a microscopic level, and the development of microscopic vehicle traffic such as cellular automata models and social force models has provided a solid foundation (Wang et al. 2023c).

The existing research on station traffic service facility system modeling mainly focuses on two aspects: First, based on the specific physical form of the transportation service facility system and the simulation modeling of individual parameters at the microscopic level, such as cellular automata model (Małecki et al. 2022), social force model, efficiency model and magnetic force model, etc., micro-simulation software for pedestrians and vehicles such as Legion, Evacnet, Paxport, SimWalk, and AnyLogic has been developed (Helbing and Molnar 1995). The second is to consider the relationship between the transportation service system and the individual to constitute the service and the service, and abstract it into a kind of queuing network system. According to the design parameters of the facility system and the statistical law of the passenger group, the simulation and analysis model are carried out at the meso level.

AnyLogic is a widely used tool for modeling and simulating decentralized, continuous, and hybrid systems (XJ Technologies 2009). It is an object-oriented commercial software that can effectively describe the use of a variety of more complex languages through hybrid patterns. It supports multi-agent simulation, system dynamics simulation, and discrete event simulation in the fields of control, social systems, asset management, supply chain, warehousing and logistics, and transportation. AnyLogic mainly uses agents to describe various things in the real world, such as gas stations, roads, vehicles, etc. (Muravev et al. 2021). AnyLogic is a Java-based locale where active objects are used to create arbitrary logical models. It has its own properties and behavior states. The corresponding properties can be adjusted through the change of parameters and variables, while the behavior state is determined by the upstream and downstream logical model, writing function objects can scan down to other objects, can interact and transmit messages through ports, and transmit information from the real world to simulate the transmission of information in real time. Designing AnyLogic models is actually to simplify various objects in the real world and define the relationships between these objects. Running the model is the process of displaying the designed model and showing the activity of the object.

Site simulation can be divided into three schemes for vehicle requirements. The first is the normal scheme, which optimizes the layout of the facilities in the operating energy supply station. The second is a prediction scheme, which predicts whether the incoming traffic will exceed the maximum service capacity. The third is the dredging scheme, when too many queued vehicles cause congestion, the station to the queued vehicles has reasonable guidance. The process taken by these three schemes is basically the same, but the traffic flow and parameters are adjusted accordingly. In general, site simulation is divided into four main processes: site data acquisition, simulation modeling, simulation display, and evaluation and optimization, as shown in Fig. 1.3.

The purpose of site simulation is to reflect the replenishment activity of energy supply station in real time. Reasonable simulation parameters can make the simulation effect closer to the actual situation (Jordán et al. 2022). Through the analysis of

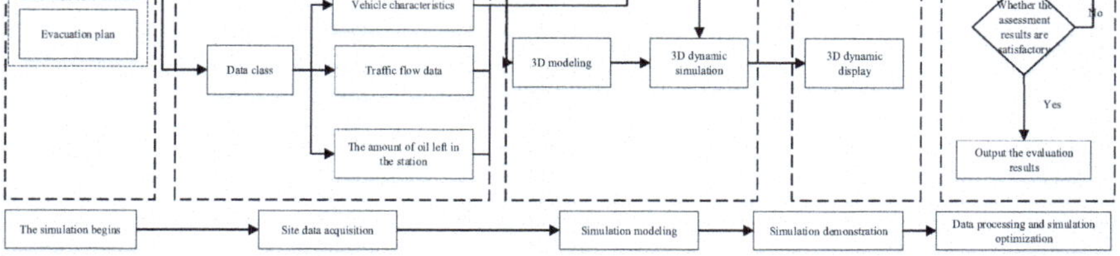


**Fig. 1.3** Technical process of station simulation

AnyLogic function, parameters can be set into four categories: vehicle individual, traffic flow distribution, station guidance and station layout, as shown in Table 1.1.

Hueper applied computer simulation in the field of traffic simulation, using interpolation method based on iterative learning to simulate the missing ramp flow data (Hueper et al. 2009). Compared to numerical analysis and algorithmic models, simulation can be added with more detail and can therefore be used as a means of analysis. Traffic simulation can be divided into continuous and discrete. According to the research object, the traffic simulation model can be divided into three types: microsimulation, middle simulation, and macrosimulation.

Microcosmic is the unit of vehicle and involves many detailed elements. The Federal Highway Administration has proposed seven traffic flow analysis models. In addition to the three models mentioned earlier, there are also numerical analysis models, which in a broad sense is also a form of simulation. It is not difficult to see that traffic simulation plays an important role in traffic flow analysis. The microscopic model is derived from vehicle following theory and lane change theory, focusing on the simulation of individual vehicle behavior in the road network. When a vehicle enters the road network, its corresponding vehicle characteristics are recorded and sent to the vehicle model. Many results have been produced in the study of micro models, such as the simulation of vehicle travel behavior. In micro traffic simulation, the driver's parameters are the most difficult to simulate. Therefore, the debugging of driver parameters is an important verification work to ensure the effectiveness of the model. Micro traffic flow simulation requires a lot of data and usually requires a large data storage device. Therefore, the scale of the simulation is usually small, such as within a road section. If large-scale simulations are required, mesoscopic and macroscopic models are needed.

The research object of Mesoscale simulation is no longer a single vehicle, but a group. Compared with the micro traffic simulation model, the description of traffic

**Table 1.1** Configuration table of station simulation parameters

| Number | Parameter type | Parameter role | Parameter |
|---|---|---|---|
| 1 | Individual vehicle characteristics | Describe the individual travel characteristics of vehicles in the station | Fuel tank capacity, vehicle length, initial speed of the vehicle |
| 2 | The law of traffic distribution | Describes the distribution of passenger traffic within a station | The actual traffic volume of the road network and the probability of vehicle pit stops |
| 3 | Vehicle guidance in the station | Arrange the orderly refueling of the vehicle according to the service status of the tanker | The number of lanes in the station, the queue capacity in the station |
| 4 | Site facility layout | Reflects the layout information of the site | The number of refueling (charging) guns, the service capacity of a single tanker, the remaining amount of fuel |

flow in mesoscopic simulation is not as detailed as that in micro, and the speed change of vehicles is not considered, so the simulation effect in mesoscopic simulation is lower than that in micro. But the meso can deal with some tasks that the micro cannot achieve, such as the interaction between vehicles, vehicle queuing problems, vehicle time delays, etc. At present, the development of traffic simulation is divided into three stages: the first stage is the initial stage, traffic simulation mainly focuses on the optimization of urban traffic lights. The second stage is the rapid development stage, where algorithms such as traffic flow and following theory are gradually applied to traffic simulation, such as DYNAMIT model. The third stage is the mature stage, the rapid development of information networks, the combination of traffic simulation and more advanced computers, which has gradually developed into intelligent transportation systems.

The macroscopic traffic simulation model redefines the three parameters of velocity, density, and flow, which is suitable for analyzing the whole road network from the macroscopic point of view. Macro traffic analyzes the information state of the whole road network, and no longer studies from the perspective of vehicles. Macro traffic mainly studies the equilibrium of the traffic network, which is generally built on the basis of total speed and total demand. The purpose of macrosimulation is to reproduce the traffic congestion on the road, so as to analyze the corresponding congestion characteristics and find the scheme to reduce the congestion. Compared to micro and meso models, macro models describe less detail and therefore do not take up as much storage space.

### *1.2.6 Brief Review of Domestic and International Research*

According to the research content, relevant literature at home and abroad were analyzed from the aspects of negative effect identification and formation mechanism, negative effect evaluation method, layout optimization of energy supplement stations, multi-agent cooperation game of energy supply stations, and vehicle scheduling of energy supplement.

First of all, the current studies on the negative effects of urban TESN mainly focus on the following aspects: (1) The negative effects of traffic congestion, which generally refers to the adverse effects of traffic congestion on the economy and society, such as environmental pollution and time waste. (2) Research on the negative effect of traffic carbon emissions, motor vehicle carbon emissions model mainly includes energy statistics, emission factors, and emission rates, through the measurement, simulation, and physical formula derivation of a combination of methods to determine the overall vehicle carbon emissions calculation method. (3) The research of road network carbon emission model mainly focuses on two situations: road section environment, and intersection environment. The research on the negative effect of urban transportation energy supply mainly focuses on four aspects: site service, safety, environment, and traffic impact. However, the main problem is that the research directions of various scholars are relatively scattered, and few scholars have conducted comprehensive evaluations on the negative effects of urban traffic stations. The problem of site negative effects is composed of many factors, not just a single aspect. Therefore, constructing the comprehensive evaluation of urban transportation energy supply stations is the source of the negative effect, which can play a basic role in the analysis of network negative effect. Secondly, in terms of TESN negative effect evaluation, although domestic and foreign scholars have taken into account the impact of all aspects and the evaluation of operation level in the evaluation and research process of transportation energy supply stations, there are few comprehensive negative effects evaluation of energy supply stations. Comprehensive negative effect evaluation is conducive to the analysis and observation of the overall state of the station and its impact on the surrounding area and can compare the correlation of multiple factors, providing a reference for the overall layout optimization and coverage adjustment of the urban transportation energy supply system.

The research on the negative effect reduction mechanism of urban TESN involves three aspects: layout optimization of energy stations, multi-agent game of energy supply network, and replenishment vehicle scheduling. The research on the reduction mechanism of urban transportation energy supply network mainly focuses on: (1) Site selection studies, such as refueling and charging station selection, which are relatively abundant and mature, but few studies on supply station layout optimization from the perspective of TESN negative effect minimization. (2) The multi-agent game research of energy supply network: the research of government policy in reducing the negative effect of TESN mainly focuses on the optimization of transportation energy structure and the sustainable development of transportation energy. In addition, according to the literature collection, there is no research on the collaborative services of TESN supply stations, and few existing studies study the burden reduction mechanism of urban energy supply network from the perspective of game. (3) Most studies on replenishment vehicle scheduling are based on traffic information and station service information. Travelers rely on unilateral perception to obtain the path with the lowest time or cost in the travel process. However, due to the limited cognition and information acquisition ability of travelers, travelers usually make decisions based on satisfactory solutions rather than optimal solutions, so they can timely understand the entire road network. Take this as the supplementary energy decision, which is the basis for reducing the negative effect of supplementary energy.

The above research status at home and abroad shows that the research on urban transportation energy supply network has been gradually carried out. Scholars have conducted researches on the formation of urban TESN negative effect, optimization of supply station layout, multi-agent game of energy supply network, vehicle energy supply behavior, etc., and obtained a number of valuable research works, which have laid a certain research foundation for this book. However, there is a lack of research on the established TESN. At the same time, most of the existing studies are from the perspective of energy economic efficiency, and there is no quantitative study on the impact of TESN negative effect. With the enhancement of time cost and environmental protection, these undesirable consequences have become important factors affecting TESN service efficiency. In addition, few existing studies have established the TESN burden reduction mechanism from the perspective of minimizing negative effects, which should be paid sufficient attention to. By searching topics and keywords in China National Knowledge Network, EBSCO, Spring, EiServer, Web of Science, and other Chinese and foreign databases, there are few literatures with the same ideas as this book. Therefore, the research in this book has a certain degree of pioneering and innovative in academia.

## References


Agrawal S, Zheng H, Peeta S et al (2016) Routing aspects of electric vehicle drivers and their effects on network performance. Transp Res Part D: Transp Environ 46:246–266

Ahm HK, Cheng SW, Cheong O, et al (2004). Competitive facility location: the Voronoi game. Theor Comput Sci 310:457–467

Akabane AT, Gomes RL, Pazzi RW, et al (2017) Apolo: A mobility pattern analysis approach to improve urban mobility. 2017 IEEE global communications conference, 1–6

Arabani AB, Farahani RZ (2012) Facility location dynamics: An overview of classifications and applications. Comput Ind Eng 62:408–420

Atta S, Mahapatra PRS, Mukhopadyay A (2018) Solving maximal covering location problem using genetic algorithm with local refinement. Soft Comput 22(1):3891–3906

Aung N, Zhang W, Dhelim S et al (2021) Dynamic traffic congestion pricing and electric vehicle charging management system for the internet of vehicles in smart cities. Digit Commun Netw 7(4):492–504

Barco J, Guerra A, Muñoz L, et al (2017) Optimal routing and scheduling of charge for electric vehicles: A case study. Math Probl Eng 2017:1–16

Beckmann M, McGuire CB, Winsten CB (1956) Studies in the economics of transportation. Econ J 67(265):116–118

Berglund P, Kwon C (2014) Solving a location problem of a Stackelberg firm competing with Cournot-Nash firms. Netw Spat Econ 14(1):117–132

Blanco V, Gázquez R (2023) Fairness in maximal covering location problems. Comput Oper Res 157:106287

Cahyono S (2016) Spatial service of petrol filling stations in Surabaya city. Procedia Soc Behav Sci 227(1):124–131

Cao S, Pan L, Mao X (2022) Research on the pricing strategies of electric vehicle charging operators based on an evolutionary game model. Energy Sustain Dev 71:378–388

Capar I, Kuby M, Leon VJ et al (2013) An arc cover–path-cover formulation and strategic analysis of alternative-fuel station locations. Eur J Oper Res 227(1):142–151

Cardinal J, Hoefer M (2010) Non-cooperative facility location and covering games. Theoret Comput Sci 411:1855–1876

Cen X, Lo HK, Li L et al (2018) Modeling electric vehicles adoption for urban commute trips. Transp Res Part B: Methodol 117(1):431–454

Chakraborty P, Baeyens E, Poolla K et al (2018) Sharing storage in a smart grid: A coalitional game approach. IEEE Trans Smart Grid 10(4):4379–4390

Chen J (2020) Research on Beijing urban traffic operation efficiency-based on entropy weight and fuzzy comprehensive evaluation. Int Core J Eng 6(10):204–209

Chen A, Ryu S, Xu XD et al (2014) Computation andapplication of the paired combinatorial logit stochastic userequilibrium problem. Comput Oper Res 2(43):68–77

Chen L, Chen SJ, Chen WK et al (2023a) Efficient presolving methods for solving maximal covering and partial set covering location problems. Eur J Oper Res 311(1):73–87

Chen R, Fan R, Wang D et al (2023b) Effects of multiple incentives on electric vehicle charging infrastructure deployment in China: an evolutionary analysis in complex network. Energy 264:125747
Chin SM, Greene DL, Hopson J et al (1999) Toward national indicators of vehicle travel and traffic congestion based on real-time traffic data. Transp Res Rec 1660(1):132–139
Chung BD, Park S, Kwon C (2018) Equitable distribution of recharging stations for electric vehicles. Socioecon Plann Sci 63:1–11
Church RL, ReVelle CS (1974) The maximum covering location problem. Pap Reg Sci Assoc 32(2):101–118
Dai B, Chen H (2012) Profit allocation mechanisms for carrier collaboration in pickup and delivery service. Comput Ind Eng 62(2):633–643
Davari S, Zarandi MHF, Turksen IB (2013) A greedy variable neighborhood search heuristic for the maximal covering location problem with fuzy coverage radi. Knowl-Based Syst 41(41):68–76
Davidov S, Pantoš M (2019) Optimization model for charging infrastructure planning with electric power system reliability check. Energy 166:886–894
De Filippo G, Marano V, Sioshansi R (2014) Simulation of an electric transportation system at the Ohio State University. Appl Energy 113:1686–1691
Deb S, Tammi K, Gao XZ et al (2022) A robust two-stage planning model for the charging station placement problem considering road traffic uncertainty. IEEE Trans Intell Transp Syst 23(7):6571–6585
Dhakal S (2009) Urban energy use and carbon emissions from cities in China and policy implications. Energy Policy 37(11):4208–4219
Ding H, Chen X, Lin K et al (2019) Collaborative mechanism of project profit allotment in petroleum engineering service chain with customized integration. Int J Prod Econ 214:163–174
Dong S, Zhou JB (2020) A comparative study on drivers' stop/go behavior at signalized intersections based on decision tree classification model. J Adv Transp 2020:1–13
Dong Z, Li XY, Ma H (2014) How to crowdsource tasks truthfully without sacrificing utility: Online incentive mechanisms with budget constraint. Infocom, IEEE.
Du X, Pei Y, Wang Z et al (2018) Highway safety influencing factors in cold regions based on attribute recognition theory. Adv Mech Eng 10(12):1687814018818337
Dupont A, Hayel Y, Jiménez T et al (2023) A loss queueing game for electric vehicle charging performance evaluation. Perform Eval 161:102350
ElWakeel AS, Noureldin A, Hassanein HS, et al (2018.) Idrivesense: Dynamic route planning involving roads quality information. 2018 IEEE global communications conference, 1–6
Erdogan N, Kucuksari S, Murphy J (2022) A multi-objective optimization model for EVSE deployment at workplaces with smart charging strategies and scheduling policies. Energy 254:124161
Ermagun A, Levinson D (2018) Spatiotemporal traffic forecasting: review and proposed directions. Transp Rev 38(6):786–814
Fang X, Cho SH (2014) Stability and endogenous formation of inventory transshipment networks. Oper Res 62:1316–1334
Fang Y, Wei W, Mei S et al (2020) Promoting electric vehicle charging infrastructure considering policy incentives and user preferences: an evolutionary game model in a small-world network. J Clean Prod 258(6):156–165
Feng Y, Mingming Y, Qiong X et al (2016) Collaborative distribution between two logistics service providers. Int Trans Oper Res 23(6):1025–1050
Friesz T, Rigdon M, Mookherjee R (2006) Differential variational inequalities and shipper dynamic oligopolistic network competition. Transp Res Part B 40(6):480–503
Frisk M, Göthe-lundgren M, Jörnsten K et al (2010) Cost allocation in collaborative forest transportation. Eur J Oper Res 205(2):448–458
Galvao RD, Revelle C (1996) A Lagrangean heuristic for the maximal covering location problem. Eur J Oper Res 88(1):114–123

Gendreau M, Laporte G, Semet F (1997) Solving an ambulance location model by tabu search. Locat Sci 5(2):75–88
Ghaffarinasab N, Kara BY, Campbell JF (2022) The stratified p -hub center and p -hub maximal covering problems. Transp Res Part B: Methodol 157:120–148
Ghosh A, Craig CS (1984) A location allocation model for facility planning in a competitive environment. Geogr Anal 16(1):39–51
Gomes L, Vale Z (2024) Costless renewable energy distribution model based on cooperative game theory for energy communities considering its members' active contributions. Sustain Cities Soc 101:105060
Guajardo M, Rönnqvist M (2015) Operations research models for coalition structure in collaborative logistics. Eur J Oper Res 240(1):147–159
Guajardo M, Rönnqvist M (2016) A review on cost allocation methods in collaborative transportation. Int Trans Oper Res 23(3):371–392
Guan X, Zhou H, Li M et al (2021) Multilevel coverage location model of earthquake relief material storage repository considering distribution time sequence characteris tics. Jourmal Traffie Transp Eng 8(2):26–34
Guo F, Yang J, Lu J (2018a) The battery charging station location problem: Impact of users' range anxiety and distance convenience. Transp Res Part E: Logist Transp Rev 114:1–18
Guo C, Li D, Zhang G et al (2018b) Realtime path planning in urban area via vanetassisted traffic information sharing. IEEE Trans Veh Technol 67(7):5635–5649
Hakimi SL (1964) Optimum locations of switching centers and the absolute centers and medians of a graph. Oper Res 12(3):450–459
Hao C, Wang X, Huanwen J et al (2019) Robust coalitional game theoretic optimization for cooperative energy hubs with correlated wind power. IET Renew Power Gener 13(13):2391–2399
Hart PE, Nilsson NJ, Raphael B (1968) A formal basis for the heuristic determination of minimum costpaths. IEEE Trans Syst Sci Cybern 4(2):100–107
Havranek T, Kokes O (2015) Income elasticity of gasoline demand: A meta-analysis. Energy Economics 47:77–86
He F, Yin Y, Zhou J (2015a) Deploying public charging stations for electric vehicles on urban road networks. Transp Res Part C: Emerg Technol 60:227–240
He Y, Song Z, Liu Z (2019a) Fast-charging station deployment for battery electric bus systems considering electricity demand charges. Sustain Cities Soc 48:101530
He Z, Cao J, Liu X (2015b) High quality participant recruitment in vehicle-based crowdsourcing using predictable mobility. 2015 international conference on computer communications. Hong Kong, IEEE, 2542–2550
He Y, Kockelman KM, Perrine KA (2019) Optimal locations of U.S. fast charging stations for long-distance trip completion by battery electric vehicles. J Clean Prod 214:452–461
Helbing D, Molnar P (1995) Social force model for pedestrian dynamics. Phys Rev E 51(5):4282
Hezarkhani B, Slikker M, Van WT (2016) A competitive solution for cooperative truckload delivery. Or Spectrum 38(1):51–80
Hite G, Tscharakischiew S (2013) The optimal subsidy on electric vehicles in german metropolitan areas: a spatial general equilibrium analysis. Energy Economics 40:515–528
Hodgson MJ (1990) A flow-capturing location-allocation model. Geogr Anal 22(3):270–279
Hofinger F, Haberl M, Fellendorf M et al (2023) 3-step calibration process of a microscopic traffic flow simulation platform for mixed traffic scenarios. Transp Res Procedia 72:1778–1785
Hossein MR, Biswajeet P, Maryam AS (2019) Allocation of emergency response centres in response to pluvial flooding—prone demand points using integrated multiple layer perceptron and maximum coverage location problem model. Int J Disaster Risk Reduct, 38.
Hosseini M, MirHassani SA (2017) A heuristic algorithm for optimal location of flow-refueling capacitated stations. Int Trans Oper Res 24(6):1377–1403
Hotelling H (1929) Stability in competition. Econ J 39:41–57
Hu L, Dong J, Lin Z (2019) Modeling charging behavior of battery electric vehicle drivers: a cumulative prospect theory based approach. Transp Res Part C: Emerg Technol 102(2):474–489

Huang YT, Kara MK (2020) Electric vehicle charging station locations: Elastic demand, station congestion, and network equilibrium. Transp Res Part D 78(1):102–129
Huang Y, Kockelman KM (2020) Electric vehicle charging station locations: Elastic demand, station congestion, and network equilibrium. Transp Res Part D: Transp Environ 78:102179
Huang Y, Li S, Qian ZS (2015) Optimal deployment of alternative fueling stations on transportation networks considering deviation path. Netw Spat Econ 15:183–204
Huang X, Lin Y, Lim MK et al (2022a) Electric vehicle charging station diffusion: an agent-based evolutionary game model in complex networks. Energy 257:124700
Huang J, Wang X, Shao C et al (2022b) An improved charging navigation strategy of electric vehicles via optimal time-of-use pricing. Electr Power Syst Res 210:108077
Huber G, Bogenberger K (2015) Long-trip optimization of charging strategies for battery electric vehicles. Transp Res Rec 2497(1):45–53
Hueper J, Dervisoglu G, Muralidharan A et al (2009) Macroscopic Modeling and Simulation of Freeway Traffic Flow. IFAC Proc Vol 42(15):112–116
Hung Y, Michailidis G (2015) Optimal routing for electric vehicle service systems. Eur J Oper Res 247(2):515–524
Jayapal C, Roy SS (2016) Road traffic congestion management using vanet. 2016 international conference on advances in Human Machine Interaction (HMI), 1–7
Jiang N, Xie C, Duthie JC et al (2014a) A network equilibrium analysis on destination, route and parking choices with mixed gasoline and electric vehicular flows. EURO J Transp Logist 3(1):75–98
Jiang N, Xie C, Duthie JC et al (2014b) A network equilibrium analysis on destination, route and parking choices with mixed gasoline and electric vehicular flows. EURO J Transp Logist 3(1):55–92
Jing W, Kim I, Ramezani M et al (2017) Stochastic traffic assignment of mixed electric vehicle and gasoline vehicle flow with path distance constraints. Transp Res Procedia 21:65–78
Johnson J, Chowdhury M, He Y et al (2013) Utilizing real-time information transferring potentials to vehicles to improve the fast-charging process in electric vehicles. Transp Res Part C 26(1):352–366
Jordán J, Palanca J, Martí P et al (2022) Electric vehicle charging stations emplacement using genetic algorithms and agent-based simulation. Expert Syst Appl 197:116739
Jou RC (2001) Modeling the impact of pre-trip information on commuter departure time and route choice. Transp Res Part D: Methodol 35(10):887–902
Jung J, Chow JYJ, Jayakrishnan R et al (2014) Stochastic dynamic itinerary interception refueling location problem with queue delay for electric taxi charging stations. Transportation Research Part c: Emerging Technologies 40:123–142
Karatas M (2017) A multi-objective facility location problem in the presence of variable gradual coverage performance and cooperative cover. Eur J Oper Res 262(3):1040–1051
Kariv O, Hakimi SL (1979) An algorithmic approach to network location problems. SIAM Joumal Appl Math 37(3):513–538
Karkatsoulis P, Siskos P, Paroussos L et al (2017) Simulating deep $CO_2$ emission reduction in transport in a general equilibrium framework: The GEM-$E_3$T model. Transp Res Part D 55(1):343–358
Karlaftis MGK, Vlahogianni EIV, Kepaptsoglou KK (2011) Nonlinear autoregressive conditional duration models for traffic congestion estimation. J Probab Stat 2011:1–13
Kathuria V (2004) Impact of CNG on vehicular pollution in Delhi: a note. Transp Res Part d: Transp Environ 9(5):409–417
Kchaou-Boujelben M, Gicquel C (2019) Efficient solution approaches for locating electric vehicle fast charging stations under driving range uncertainty. Comput Oper Res 109:288–299
Khosroshahi AH, Keshavarzi P, KoozehKanani ZD, et al (2011) Acquiring real time traffic information using vanet and dynamic route guidance. 2011 IEEE 2nd international conference on computing, control and industrial engineering, 9–13

Kim JG, Kuby M (2012) The deviation-flow refueling location model for optimizing a network of refueling stations. Int J Hydrogen Energy 37(6):5406–5420

Knay MB, Gzara F, Alumur SA (2021) Full cover charging station location problem with routing. Transp Res Part B Methodol 144(4):1–22

Konur D, Geunes J (2012) Competitive multi-facility location games with non-identical firms and convex traffic congestion costs. Transp Res Part E: Logist Transp Review 48(1):373–385

Krajewska MA, Kopfer H, Laporte G et al (2008) Horizontal cooperation among freight carriers: request allocation and profit sharing. Journal of the Operational Research Society 59(11):1483–1491

Kress D, Pesch E (2016) Competitive location and pricing on networks with random utilities. Netw Spat Econ 16:837–863

Kristensson A, Johansson C, Westerholm R et al (2004) Real-world traffic emission factors of gases and particles measured in a road tunnel in Stockholm. Sweden. Atmospheric Environment 38(5):657–673

Kuby M, Lim S (2005) The flow-refueling location problem for alternative-fuel vehicles. Socioecon Plann Sci 39(2):125–145

Kuby M, Capar I, Kim JG (2017) Efficient and equitable transnational infrastructure planning for natural gas trucking in the European Union. Eur J Oper Res 257(3):979–991

Kweon SJ, Hwang SW, Ventura JA (2017) Electrification of a city bus network—an optimization model for cost-effective placing of charging infrastructure and battery sizing of fast-charging electric bus systems. Int J Sustain Transp 11(10):707–720

Lai J, Yu L, Song G et al (2013) Development of city-specific driving cycles for transit buses based on VSP distributions: Case of Beijing. J Transp Eng 139(7):749–757

Lamontagne S, Carvalho M, Frejinger E et al (2023) Optimising electric vehicle charging station placement using advanced discrete choice models. INFORMS J Comput 35(5):1195–1213

Lederer PJ, Thisse JF (1990) Competitive location on networks under delivered pricing. Oper Res Lett 9(3):147–153

Lee C, Han J (2017) Benders-and-price approach for electric vehicle charging station location problem under probabilistic travel range. Transp Res Part B: Methodol 106:130–152

Lesage S, Xu H, Novakowski KS (1997) Distinguishing natural hydrocarbons from anthropogenic contamination in ground water. Groundwater 35(1):149–160

Levering N, Boon M, Mandjes M et al (2022) A framework for efficient dynamic routing under stochastically varying conditions. Transp Res Part B: Methodol 160:97–124

Li J, Reniers G, Cozzani V et al (2017) A bibliometric analysis of peer-reviewed publications on domino effects in the process industry. J Loss Prev Process Ind 49:103–110

Li X, Chen L, Sun F et al (2023) Share or not share, the analysis of energy storage interaction of multiple renewable energy stations based on the evolution game. Renewable Energy 208:679–692

Lin CC (2018) The p-center flow-refueling facility location problem. Transp Res Part B: Methodol 118:124–142

Liu Z, Song Z (2018) Network user equilibrium of battery electric vehicles considering flow-dependent electricity consumption. Transp Res Part C Emerg Technol 95:516–544

Liu C, Wu J, Long C (2014) Joint Charging and Routing Optimization for Electric Vehicle Navigation Systems. IFAC Proc Vol 47(3):2106–2111

Liu X, Ma S, Tian J et al (2015) A system dynamics approach to scenario analysis for urban passenger transport energy consumption and $CO_2$ emissions: A case study of Beijing. Energy Policy 85:253–270

Liu Y, Li ZZ, Liu JX et al (2016a) A double standard model for allocating limited emergency medical service vehicle resources ensuring service reliability. Transp Res Part C 69:120–133

Liu Y, Niu J, Liu X (2016b) Comprehensive tempo-spatial data collection in crowd sensing using a heterogeneous sensing vehicle selection method. Pers Ubiquit Comput 20(3):397–411

Liu D, Xu Z, Wang Z et al (2020) Regional evaluation of fire apparatus requirements for petrol stations based on travel times. Process Saf Environ Prot 135:350–363

Liu HX, Wang DZW (2015) Global optimization method fornet work design problem with stochastic user equilibrium. Transp Res Part B: Methodol 72(1):20–39
Lozano S, Moreno P, Adenso-díaz B et al (2013) Cooperative game theory approach to allocating benefits of horizontal cooperation. Eur J Oper Res 229(2):444–452
Lucas A, Silva CA, Neto RC (2012) Life cycle analysis of energy supply infrastructure for conventional and electric vehicles. Energy Policy 41(1):537–547
Ma G, Huang Y (2019) Safety assessment of explosions during gas stations refilling process. J Loss Prev Process Ind 60:133–144
Małecki K, Kamiński M, Wąs J (2022) A multi-cell cellular automata model of traffic flow with emergency vehicles: effect of a corridor of life and drivers' behaviour. J Comput Sci 61:101628
Mangones SC, Jaramillo P, Rojas NY et al (2020) Air pollution emission effects of changes in transport supply: the case of Bogotá, Colombia. Environ Sci Pollut Res 27(29):35971–35978
Miller T, Tobin R, Friesz T (1991) Stackelberg games on a network with Cournot-Nash oligopolistic competitors. J Reg Sci 31(4):435–454
MirHassani SA, Ebrazi R (2013) A flexible reformulation of the refueling station location problem. Transp Sci 47(4):617–628
Mirheli A, Hajibabai L (2023) Charging network design and service pricing for electric vehicles with user-equilibrium decisions. IEEE Trans Intell Transp Syst 24(3):2888–2902
Mohamed M, Farag H, El-Taweel N, Ferguson M (2017) Simulation of electric buses on a full transit network: Operational feasibility and grid impact analysis. Electr Power Syst Res 142:163–175
Mohammadfam I, Zarei E (2015) Safety risk modeling and major accidents analysis of hydrogen and natural gas releases: A comprehensive risk analysis framework. Int J Hydrogen Energy 40(39):13653–13663
Muravev D, Hu H, Rakhmangulov A et al (2021) Multi-agent optimization of the intermodal terminal main parameters by using AnyLogic simulation platform: case study on the Ningbo-Zhoushan Port. Int J Inf Manage 57:102133
Nan J, Chi X, Duthie JC et al (2014) A network equilibrium analysis on destination, route and parking choices with mixed gasoline and electric vehicular flows. Euro J Transp & Logist 3(1):55–92
Narain U, Krupnick A. The impact of Delhi's CNG program on air quality. 2007.
Ngo H, Kumar A, Mishra S (2020) Optimal positioning of dynamic wireless charging infrastructure in a road network for battery electric vehicles. Transp Res Part D: Transp Environ 85:102385
Niknamfar AH, Niaki STA (2016) Fair profit contract for a carrier collaboration framework in a green hub network under soft time-windows: Dual lexicographic max–min approach. Transp Res Part E: Logist Transp Rev 91:129–151
Ottosen TB, Kakosimos KE, Johansson C et al (2015) Analysis of the impact of inhomogeneous emissions in the Operational Street Pollution Model (OSPM). Geosci Model Dev 8(10):3231–3245
Pan L, Yao E, MacKenzie D (2019) Modeling EV charging choice considering risk attitudes and attribute non-attendance. Transp Res Part C: Emerg Technol 102:60–72
Pandian S, Gokhale S, Ghoshal AK (2009) Evaluating effects of traffic and vehicle characteristics on vehicular emissions near traffic intersections. Transp Res Part D: Transp Environ 14(3):180–196
Park SW, Cho KS, Hoefter G et al (2022) Electric vehicle charging management using location-based incentives for reducing renewable energy curtailment considering the distribution system. Appl Energy 305:117680
Pelegrín M, Xu L (2023) Continuous covering on networks: Improved mixed integer programming formulations. Omega 117:102835
Peng D, Wu F, Chen G (2018) Data quality guided incentive mechanism design for crowdsensing. IEEE Trans Mob Comput 17(2):307–319
Plastria LF (2009) Maximal covering location problem with price decision for revenue maximization in a competitive environment. OR Spectr: Quant Approaches Manag 31(3):555–571
Qian T, Shao C, Li X et al (2020) Enhanced Coordinated Operations of Electric Power and Transportation Networks via EV Charging Services. IEEE Trans Smart Grid 11(4):3019–3030

Rajagopalan R, Mehrotra KG, Mohan CK et al (2008) Hierarchical path computation approach for large graphs. IEEE Trans Aerosp Electron Syst 44(2):427–440
Rasmussen TK, Watling DP, Prato CG et al (2015) Stochastic user equilibrium with equilibrated choice sets: Part I-solving there stricted SUE for the logit family. Transp Res Part B: Methodol 77(3):146–165
Rodriguez SA, De la Fuente RA, Aguayo MM (2020) A facility location and equipment emplacement technique model with expected coverage for the location of fire stations in the Concepcion province, Chile. Comput & Ind Eng 147(2):23–35
Rohaninejad M, Navidi H, Nouri BV et al (2017) A new approach to cooperative competition in facility location problems: Mathematical formulations and an approximation algorithm. Comput Oper Res 83:45–53
Rosenthal EC (2017) A cooperative game approach to cost allocation in a rapid-transit network. Transp Res Part B: Methodol 97:64–77
Said D, Cherkaoui S, Khoukhi L (2013) Queuing model for EVs charging at public supply stations. In 2013 9th international wireless communications and mobile computing conference (IWCMC), 65–70
Sassi O, Cherif-Khettaf W, Oulamara A (2015) Multi-start iterated local search for the mixed fleet vehicle routing problem with heterogenous electric vehicles. European conference on evolutionary computation in combinatorial optimization, 138–149
Scheiper B, Schiffer M, Walther G (2019) The flow refueling location problem with load flow control. Omega 83:50–69
Schoenberg S, Dressler F (2023) Reducing Waiting Times at Charging Stations With Adaptive Electric Vehicle Route Planning. IEEE Transactions on Intelligent Vehicles 8(1):95–107
Serra D, ReVelle C (1999) Competitive location and pricing on networks. Geogr Anal 31(2):109–129
Shan X, Hao P, Chen X et al (2018) Vehicle energy/emissions estimation based on vehicle trajectory reconstruction using sparse mobile sensor data. IEEE Trans Intell Transp Syst 20(2):716–726
Shao S, Guan W, Ran B et al (2017) Electric vehicle routing problem with charging time and variable travel time. Math Probl Eng 2:1–13
Shao J, Xue W, Wang J et al (2022) Advancing to full electrification of rideshare vehicles: Applying differentiated subsidy phase-out policies to plug-in hybrid electric vehicles and battery electric vehicles through an evolutionary game analysis. J Clean Prod 377:134265
Shehadeh KS, Tucker EL (2022) Stochastic optimization models for location and inventory prepositioning of disaster relief supplies. Transp Res Part C: Emerg Technol 144:103871
Shi X, Xu Y, Guo Q et al (2020) A distributed EV navigation strategy considering the interaction between power system and traffic network. IEEE Trans Smart Grid 11(4):3545–3557
Snyder LV (2006) Facility location under uncertainty: a review. IIE Trans 38:537–554
Souza A, Yokoyama, R Maia G, et al (2016) Realtime path planning to prevent traffic jam through an intelligent transportation system. IEEE symposium on computers and communication (ISCC), 726–731
Subramanian N, Abdulrahman MD (2017) Logistics and cloud computing service providers' cooperation: a resilience perspective. Prod Plan & Control 28(11):919–928
Sun Z, Zhou X (2016) To save money or to save time: Intelligent routing design for plug-in hybrid electric vehicle. Transp Res Part D: Transp Environ 43:238–250
Sun X, Lin K, Jiao P et al (2020) The dynamical decision model of intersection congestion based on risk identification. Sustainability 12(15):5923
Tauseef SM, Abbasi T, Pompapathi V et al (2018) Case studies of 28 major accidents of fires/explosions in storage tank farms in the backdrop of available codes/standards/models for safely configuring such tank farms. Process Saf Environ Prot 120:331–338
XJ Technologies (2009) Anylogic pedestrian library tutorial. St. Petersburg, Russian Federation
Teng Y, Li X, Wu P et al (2019) Using cooperative game theory to determine profit distribution in IPD projects. Int J Constr Manag 19(1):32–45
Tinoco SVP, Creemers S, Boute RN (2017) Collaborative shipping under different cost-sharing agreements. Eur J Oper Res 263(3):827–837

Toregas C, Swain R, ReVelle C et al (1971) The location of emergency service facilities. Oper Res 19(13):1363–1373
Vafa-Arani H, Jahani S, Dashti H et al (2014) A system dynamics modeling for urban air pollution: A case study of Tehran, Iran. Transp Res Part D: Transp Environ 31:21–36
Vazifeh MM, Zhang H, Santi P et al (2019) Optimizing the deployment of electric vehicle charging stations using pervasive mobility data. Transp Res Part A: Policy Pract 121:75–91
Vazirani V (2001) Approximation algorithm. Springer, Berlin Heidelberg
Vaziri M (2002) Development of highway congestion index with Fuzzy set models. Sage J 1802(1):16–22
Vries H, Duijzer E (2017) Incorporating driving range variability in network design for refueling facilities. Omega 69:102–114
Wallace HW, Jobson BT, Erickson MH, et al (2012) Comparison of wintertime CO to NOx ratios to MOVES and MOBILE6.2 on-road emissions inventories. Atmos Environ 63:289–297
Wang C, Hu Q (2017) Knowledge sharing in supply chain networks: effects of collaborative innovation activities and capability on innovation performance. Technovation 12(2):25–37
Wang YW, Lin CC (2009) Locating road-vehicle refueling stations. Transp Res Part E: Logist Transp Rev 45(5):821–829
Wang G, Xu Z, Wen F et al (2013) Traffic-constrained multi-objective planning of electric-vehicle charging stations. IEEE Trans Power Delivery 28:2363–2372
Wang Y, Ma X, Li Z et al (2017) Profit distribution in collaborative multiple centers vehicle routing problem. J Clean Prod 144:203–219
Wang Y, Zhang J, Assogba K et al (2018a) Collaboration and transportation resource sharing in multiple centers vehicle routing optimization with delivery and pickup. Knowl-Based Syst 160:296–310
Wang J, Yu Y, Tang J (2018b) Compensation and profit distribution for cooperative green pickup and delivery problem. Transp Res Part B: Methodol 113:54–69
Wang C, Li C, Qin C et al (2018c) Maximizing spatial–temporal coverage in mobile crowd-sensing based on public transports with predictable trajectory. Int J Distrib Sens Netw 14(8):1–10
Wang X, Shahidehpour M, Jiang C, Li Z (2019a) Coordinated planning strategy for electric vehicle charging stations and coupled traffic-electric networks. IEEE Trans Power Syst 34(1):268–279
Wang C, He F, Lin X et al (2019b) Designing locations and capacities for charging stations to support intercity travel of electric vehicles: An expanded network approach. Transportation Research Part c: Emerging Technologies 102:210–232
Wang X, Shahidehpour M, Jiang C et al (2019c) Coordinated planning strategy for electric vehicle charging stations and coupled traffic-electric networks. IEEE Trans Power Syst 34(1):268–279
Wang L, Jin Y, Wang L et al (2019d) Incentive-based approach to control demands for operation of one-way carsharing system. Transp Res Rec J Transp Res Board 2673(4):427–438
Wang Y, Gao W, Qian F et al (2021) Evaluation of economic benefits of virtual power plant between demand and plant sides based on cooperative game theory. Energy Convers Manage 238:114180
Wang K, Wang H, Yang J, Feng J, Li Y, Zhang S, Okoye MO (2022b) Electric vehicle clusters scheduling strategy considering real-time electricity prices based on deep reinforcement learning. Energy Rep 8:695–703
Wang Y, Zhao Y, Gan S et al (2023a) Optimization of charging stations integrated with dynamic transportation systems in metropolises. Transp Res Part D: Transp Environ 119:103726
Wang Y, Qin Y, Ma Z et al (2023b) Operation optimization of integrated energy systems based on cooperative game with hydrogen energy storage systems. Int J Hydrogen Energy 48(95):37335–37354
Wang J, Lv W, Jiang Y, Huang G (2023c) A cellular automata approach for modelling pedestrian-vehicle mixed traffic flow in urban city. Appl Math Model 115:1–33
Wang Y, Fan R, Du K, Wang Y, et al (2022a) Private charger installation game and its incentive mechanism considering prospect theory. Transp Res Part D: Transp Environ 113:103508.
Wardrop JD (1952) Some theoretical aspects of road traffic research. Institution of Civil Engineers 2(1):325–378

Watabe A, Leaver J, Ishida H et al (2019) Impact of low emissions vehicles on reducing greenhouse gas emissions in Japan. Energy Policy 130(1):227–242

Watling DP, Rasmussen TK, Prato CG et al (2015) Stochastic user equilibrium with equilibrated choice sets: Part I-Model formulations under alternative distributions and restrictions. Transp Res Part B Methodol 77(1):166–181

Wei W, Liu F, Mei S (2015) Charging strategies of EV aggregator under renewable generation and congestion: a normalized Nash equilibrium approach. IEEE Trans Smart Grid 7(3):1630–1641

Wendell RE, McKelvey RD (1981) New perspectives in competitive location theory. Eur J Oper Res 6(2):174–182

Weninger D, Wolsey LA (2023) Benders-type branch-and-cut algorithms for capacitated facility location with single-sourcing. Eur J Oper Res 310(1):84–99

Wu S, Pang A (2023) Optimal scheduling strategy for orderly charging and discharging of electric vehicles based on spatio-temporal characteristics. J Clean Prod 392:136318

Wu W, Deng L, Tao Y et al (2022) Research on evolutionary game of value co-creation behavior of shared private charging piles of electric vehicles. Discret Dyn Nat Soc 2022:1–13

Wu S, Li D, Zhang G, et al (2016) Densitybased dynamic revision path planning in urban area via vanet. International conference on machine learning and intelligent communications, 129–138.

Xiao D, An S, Cai H et al (2020) An optimization model for electric vehicle charging infrastructure planning considering queuing behavior with finite queue length. J Energy Storage 29:101317

Xiong C, Shahabi M, Zhao J et al (2020) An integrated and personalized traveler information and incentive scheme for energy efficient mobility systems. Transp Res Part C: Emerg Technol 113:57–73

Xu X, Chen H (2020) Exploring the innovation efficiency of new energy vehicle enterprises in China. Clean Technol Environ Policy 22(8):1671–1685

Xu M, Meng Q, Liu K (2017) Network user equilibrium problems for the mixed battery electric vehicles and gasoline vehicles subject to battery swapping stations and road grade constraints. Transp Res Part B: Methodol 99:138–166

Xu M, Meng Q, Mannering F (2020) Optimal deployment of charging stations considering path deviation and nonlinear elastic demand. Transp Res Part B: Methodol 135:120–142

Yagcitekin B, Uzunoglu M (2016) A double-layer smart charging strategy of electric vehicles taking routing and charge scheduling into account. Appl Energy 167:407–419

Yang S, Cheng W, Hsu Y et al (2013) Charge scheduling of electric vehicles in highways. Math Comput Model 57(1):2873–2882

Yang H, Yang S, Xu Y et al (2015) Electric vehicle route optimization considering time-of-use electricity price by learnable partheno-genetic algorithm. IEEE Trans Smart Grid 6(2):657–666

Yang J, Chou L, Chang Y (2016) Electric-vehicle navigation system based on power consumption. IEEE Trans Veh Technol 65(8):5930–5943

Yang J, Dong J, Liang H (2017) A data-driven optimization-based approach for siting and sizing of electric taxi charging stations. Transp Res Part C: Emerg Technol 77:462–547

Yang F, Dai Y, Ma Z-J (2020) A cooperative rich vehicle routing problem in the last-mile logistics industry in rural areas. Transp Res Part E: Logist Transp Rev 141:102024

Yang A, Wang H, Li B et al (2023) Capacity optimization of hybrid energy storage system for microgrid based on electric vehicles' orderly charging/discharging strategy. J Clean Prod 411:137346

Yi Z, Liu XC, Wei R (2022) Electric vehicle demand estimation and charging station allocation using urban informatics. Transp Res Part D: Transp Environ 106:103264

Yildiz B, Arslan O, Karasan OE (2016) A branch and price approach for routing and refueling station location model. Eur J Oper Res 248(3):815–826

Yu Y, Lou Q, Tang J et al (2017) An exact decomposition method to save trips in cooperative pickup and delivery based on scheduled trips and profit distribution. Comput Oper Res 87:245–257

Yuan K, Song Y, Shao Y et al (2018) A charging strategy with the price stimulus considering the queue of charging station and EV fast charging demand. Energy Procedia 145:400–405

Zarei E, Azadeh A, Khakzad N et al (2017) Dynamic safety assessment of natural gas stations using Bayesian network. J Hazard Mater 321(5):830–840
Zeng Z, Wang S, Qu X (2022) On the role of battery degradation in en-route charge scheduling for an electric bus system. Transp Res Part E: Logist Transp Rev 161:102727
Zhang L, Li Y (2016) A game-theoretic approach to optimal scheduling of parking-lot electric vehicle charging. IEEE Trans Veh Technol 65(6):4068–4078
Zhang JL, Yang H (2015) Modeling route choice inertia in network equilibrium with heterogeneous prevailing choice sets. Transp Res Part C 57(1):42–54
Zhang X, Rey D, Waller ST (2010) Multitype recharge facility location for electric vehicles. Int J Hydrogen Energy 35(10):4652–4667
Zhang X, Rey D, Waller ST (2018a) Multitype recharge facility location for electric vehicles. Computer-Aided Civil and Infrastructure Engineering 33(11):943–965
Zhang X, Rey D, Waller ST (2018b) Multitype recharge facility location for electric vehicles. Comput-Aided Civ Infrastruct Eng 1:1–23
Zhang R, Ma W, Si H et al (2021) Cooperative game analysis of coordination mechanisms under fairness concerns of a green retailer. J Retail Consum Serv 59:102361
Zhang ZY, Lu LJ, Wang XH (2023) Researh on dynamic time-sharing tariff orderly charging strategy based on NSGA2 in PV-Storage-charging stations. Electr Power Syst Res 225:109784
Zheng H, Peeta S (2017) Routing and charging locations for electric vehicles for intercity trips. Transp Plan Technol 40:393–419
Zheng J, Zhao Y, Zhao L et al (2014) Queuing-based approach for optimal dispenser allocation to hydrogen refueling stations. Int J Hydrogen Energy 39(15):8055–8062
Zheng H, He X, Li Y et al (2017) Traffic equilibrium and charging facility locations for electric vehicles. Netw Spat Econ 17(2):1–23
Zhong RX et al (2011) Dynamic user equilibrium with side constraints for a traffic network: theoretical development and numerical solution algorithm. Transp Res Part B: Methodol 45(7):1035–1061
Zhong J, Yang N, Zhang X et al (2022) A fast-charging navigation strategy for electric vehicles considering user time utility differences. Sustain Energy, Grids Netw 30:100646
Zhou Z, Liao H, Gu B et al (2018) Robust mobile crowd sensing: when deep learning meets edge computing. IEEE Network 32(4):54–60
Zhou Z, Liu Z, Su H et al (2022) Intelligent Path Planning Strategy for Electric Vehicles Combined With Urban Electrified Transportation Network and Power Grid. IEEE Syst J 16(2):2437–2447

# Chapter 2
# Analysis of the Current Situation of Urban Transportation Energy Supply Network

## 2.1 Analysis of the Current Situation of Urban Transportation Energy Supply

### *2.1.1 Energy Supply Structure of Urban Transportation*

As of 2021, the number of gas stations in China is 119,000, as shown in Fig. 2.1. It is mainly divided into three categories: state-owned enterprises, private enterprises, and foreign enterprises. State-owned enterprises' gas stations are represented by "three barrels of oil". Sinopec has 30,713 gas stations, CNPC has 22,619 gas stations, CNOOC has more than 800 gas stations, and CNPC has more than 600 gas stations, and "three barrels of oil" accounts for about 55.6% of the number of gas stations in the country.

Private enterprises account for about 50%, and foreign enterprises account for about 5%. From the perspective of refined oil sales, Sinopec and CNPC contributed more than 70% of the refined oil sales, and the remaining less than 30% came from CNOOC, CNPC, and other social gas stations.

In the domestic refined oil market, the original market pattern of CNPC and Sinopec concentrated wholesale refined oil is changing. Among them, gas stations as the main channel of refined oil retail, currently two barrels of oil occupy about 50% of the market share. In addition, about 80% of gas stations are distributed in national roads, provincial roads, county roads, urban areas, and other high-traffic routes. As shown in Table 2.1 and Fig. 2.2, more than 80% of gas stations in China are distributed on high-traffic routes such as national and provincial roads, county and township roads, and urban areas. Among them, provincial and national roads accounted for 33.7%, county and township roads accounted for 23.8%, and urban areas accounted for 24%.

From the national distribution point of view, natural gas filling stations are divided into CNG and LNG, and there are also some joint stations, which can be pressurized

 
X. Ge, *Urban Transportation Energy Supply Network*, Industrial Ecology and Environmental Management 5, https://doi.org/10.1007/978-3-031-64619-5_2

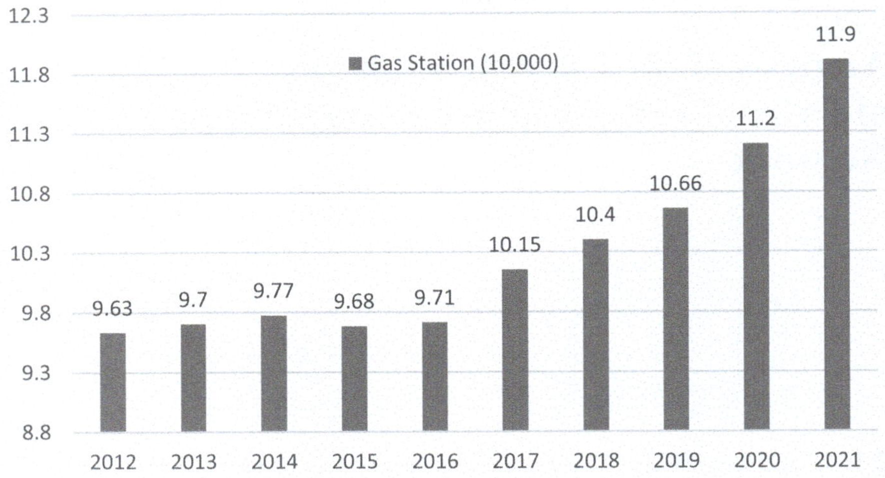


**Fig. 2.1** Change trend of the number of gas stations in China

**Table 2.1** Distribution ratio of domestic gas stations

| Grade of road | National and provincial road | County road | City center | Village | Freeway | Other |
|---|---|---|---|---|---|---|
| Proportion | 33.7% | 23.8% | 24.5% | 12% | 4.8% | 1.2% |

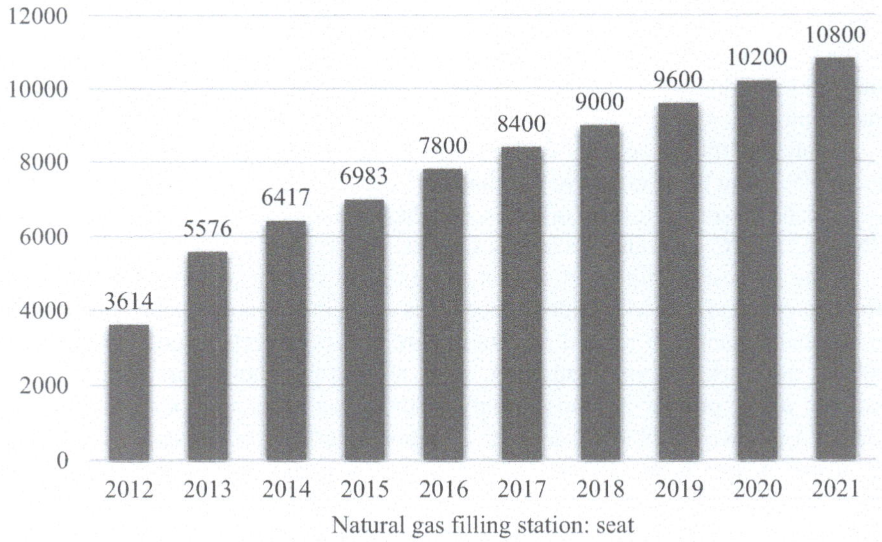


**Fig. 2.2** The number of natural gas filling stations in China from 2012 to 2021

natural gas and liquefied natural gas, known as L-compressed natural gas filling stations. The number of filling stations in the world has grown steadily, from 11,100 in 2007 to 33,338 by the end of 2020. The reason for the obvious decline in the growth rate of China's LNG filling stations in 2015 is that a series of problems such as the economic slowdown, the downturn in the logistics market, and the poor operation of related industries have made the demand for vehicles decline rapidly, and the overall market is flagging. Affected by this, the attitude of mainstream LNG filling station enterprises has changed, the intensity of market expansion has declined, and the number of new filling stations has continued to shrink.

From the perspective of national distribution, China's LNG filling stations are mostly concentrated in the coastal areas of North and East China, overlapping with downstream consumption blocks. The top three provinces and regions are Shandong (564), Jiangsu (334), and Xinjiang (283). Judging from the partial distribution of LNG refueling stations, China's LNG refueling stations are clustered in the provinces with large LNG consumption and the provinces with large LNG output such as Inner Mongolia, Shaanxi, and Shanxi, among which Shandong is developing more rapidly. As shown in Fig. 2.2, the number of natural gas vehicle filling stations in China in 2020 will be about 10,200, an increase of about 600, an increase of 6.3%. By the end of 2021, the number of natural gas filling stations in China will be about 10,800.

As shown in Fig. 2.3, in 1988, China began to introduce the full set of CNG filling stations, refit automobile parts and high-pressure gas cylinders, and build CNG filling stations for automobiles. By the end of 2020, the number of CNG filling stations in China is about 6083.

As shown in Fig. 2.4, China's LNG refueling station equipment manufacturing started late, with the increasing demand and consumption level of natural gas, led to

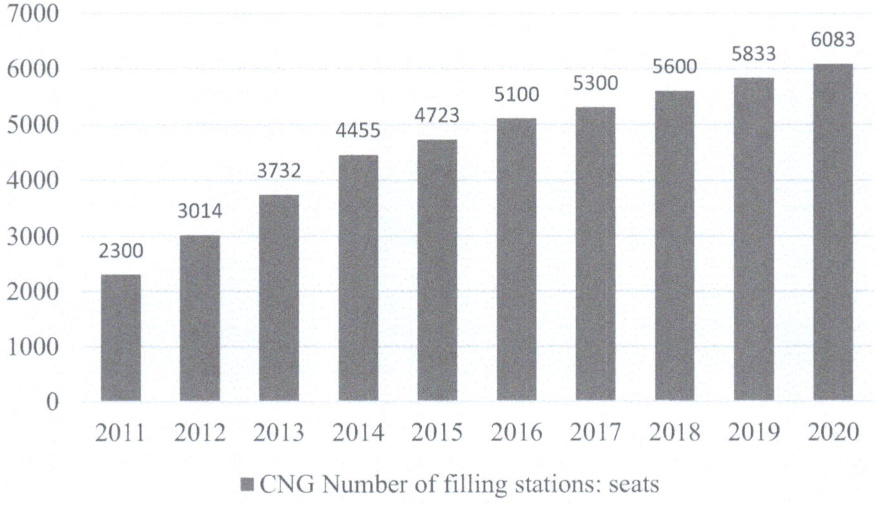


**Fig. 2.3** The number of CNG filling stations in China from 2011 to 2020

the rapid development of LNG refueling station equipment industry, by the end of 2020, China's LNG refueling station holdings of about 4800.

As shown in Fig. 2.5, by the end of 2021, the number of natural gas vehicles in China is about 7.02 million, a decrease of 50,000 vehicles, a decrease of 0.71%, the reason for the decline is that many places call off the "oil to gas" and blindly implement the vehicle "gas to electricity", but still ranked first in the world.

As shown in Fig. 2.6, China's natural gas (LNG) automobile industry began to develop in 2003. With the continuous development of technology and the support

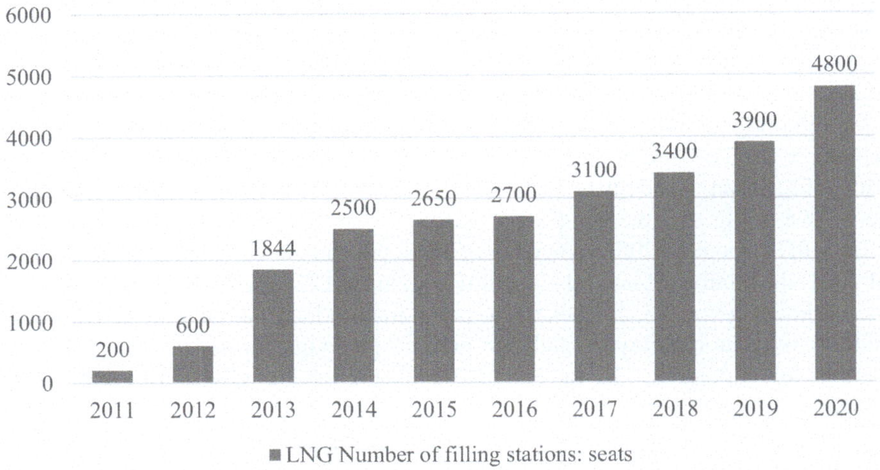


**Fig. 2.4** The number of LNG filling stations in China from 2011 to 2020

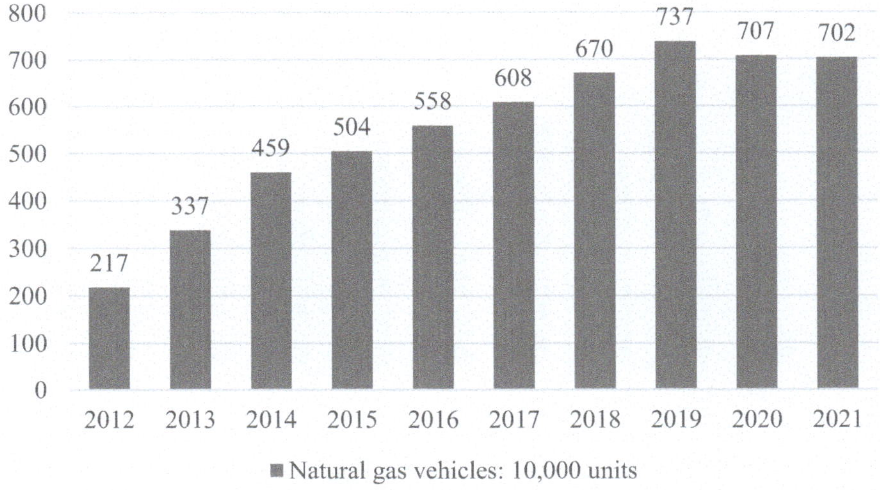


**Fig. 2.5** The number of natural gas vehicles in China from 2012 to 2021

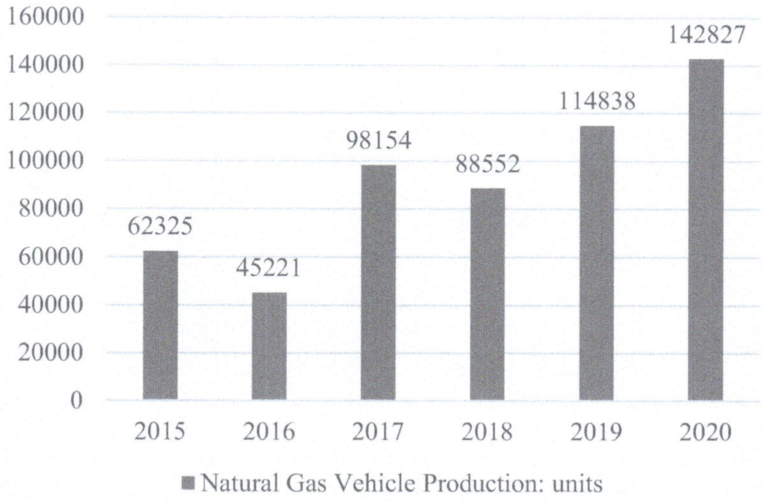


**Fig. 2.6** Production of natural gas vehicles in China, 2015–2020

of policies, the output of natural gas automobiles has grown rapidly. By 2020, the output of natural gas automobiles in China will be 142,827.

As shown in Fig. 2.7, under the vigorous promotion of policies, China's natural gas automobile industry has developed rapidly in recent years, and the sales volume of natural gas passenger cars and commercial vehicles in China in 2015 is not much different. Since 2017, the sales of natural gas commercial vehicles have been increasing, and by 2020, the sales of natural gas commercial vehicles have reached 140,210, but the natural gas passenger car market has gradually shrunk, and the sales of natural gas passenger vehicles in 2020 have plummeted to only 274.

As shown in Fig. 2.8, by the end of 2021, the number of new energy vehicles in China has reached 7.84 million, accounting for 2.6% of the total number of vehicles. Compared with the end of 2020, the number has increased by 2.92 million, an increase of 59.25%. Among them, there are 6.4 million pure electric vehicles, accounting for 81.63% of the total number of new energy vehicles. In 2021, up to 2.95 million new energy vehicles were newly registered, accounting for 11.25%, compared with 2020, the growth rate is rapid, an increase of 151.61% in volume. The number of newly registered new energy vehicles from 650,000 in 2017 to 2.95 million in 2021 shows an exponential growth trend of new energy vehicles.

As shown in Fig. 2.9, while the number of charging piles is increasing, the number of charging stations is also increasing. According to data released by the China Electric Vehicle Charging Infrastructure Promotion Alliance, the number of charging stations in China has increased from 1,069 in 2015 to 68,000 in 2021.

As shown in Fig. 2.10, according to the cumulative sales volume of new energy vehicles and the possession of charging piles, the ratio between new energy vehicles and charging piles in China can be obtained. According to the data released by the

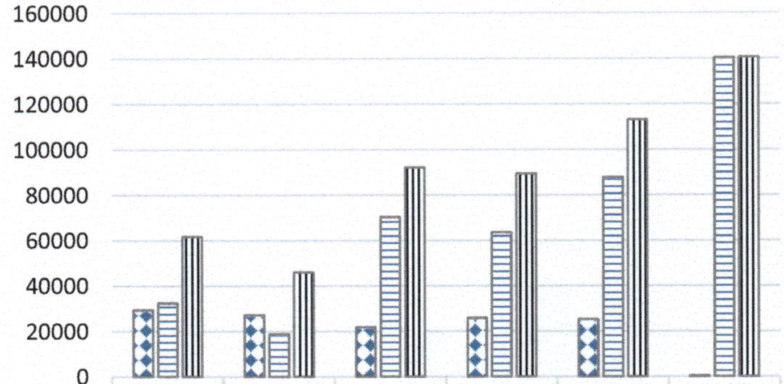


| | 2015 | 2016 | 2017 | 2018 | 2019 | 2020 |
|---|---|---|---|---|---|---|
| Natural Gas Passenger Vehicles: units | 29349 | 27165 | 21821 | 25890 | 25287 | 274 |
| Natural Gas Commercial Vehicles: units | 32338 | 18770 | 70330 | 63564 | 87869 | 140210 |
| Natural Gas Vehicle Sales: units | 61687 | 45935 | 92151 | 89454 | 113156 | 140484 |

**Fig. 2.7** Sales volume of natural gas vehicles in China, 2015–2020

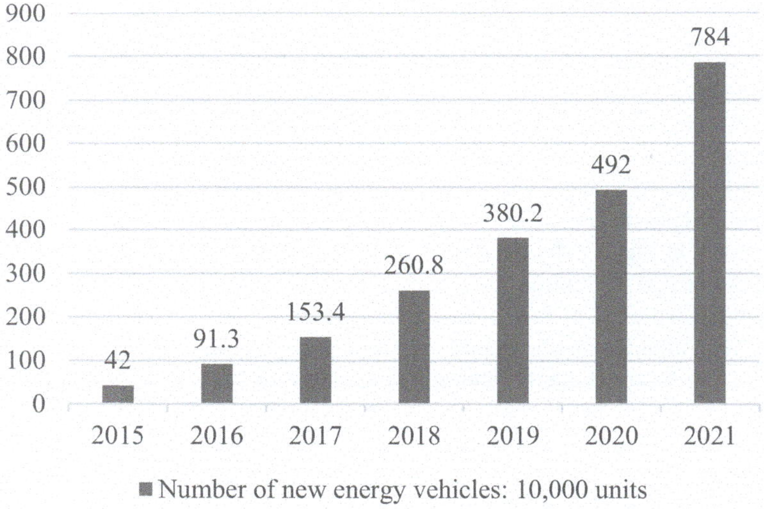


**Fig. 2.8** Trend of new energy vehicle ownership in China from 2015 to 2021

China Electric Vehicle Charging Infrastructure Promotion Alliance, the number of charging piles in China in 2021 is 2.617 million units, an increase of 55.7%, and the ratio of car piles in China in the same year has been increased from 6.36:1 in 2015 to about 3:1, a significant improvement.

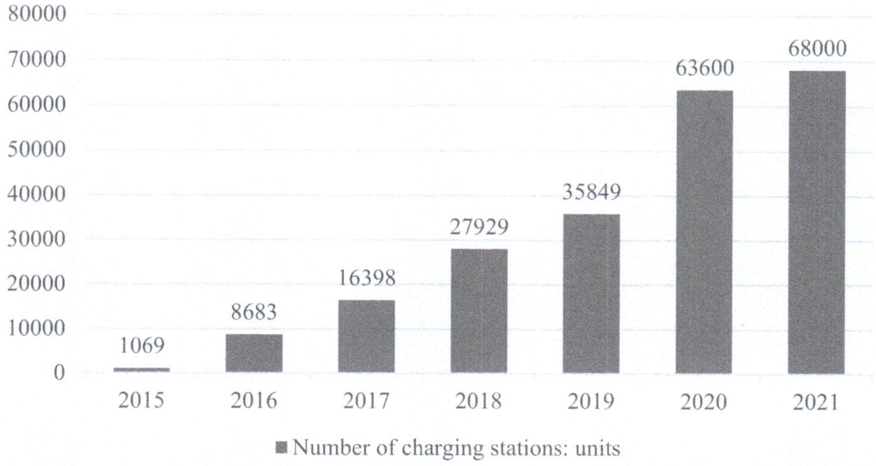


**Fig. 2.9** Number of charging stations for new energy vehicles from 2015 to 2021

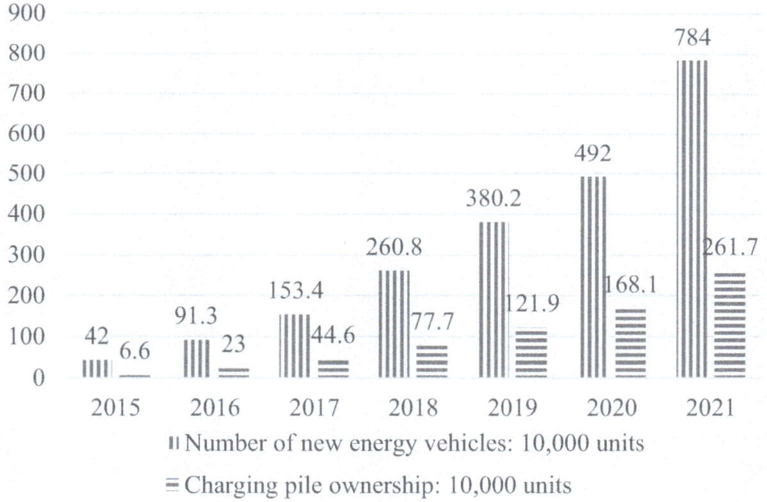


**Fig. 2.10** The ratio between new energy vehicles and charging piles

### 2.1.2 *Current Situation of Domestic Transportation Energy Supply*

At present, in addition to the traditional energy of gasoline and diesel, which still occupy the dominant position in China's transportation energy. There are also non-petroleum liquefied natural gas, compressed natural gas, alcohol fuels (methanol and

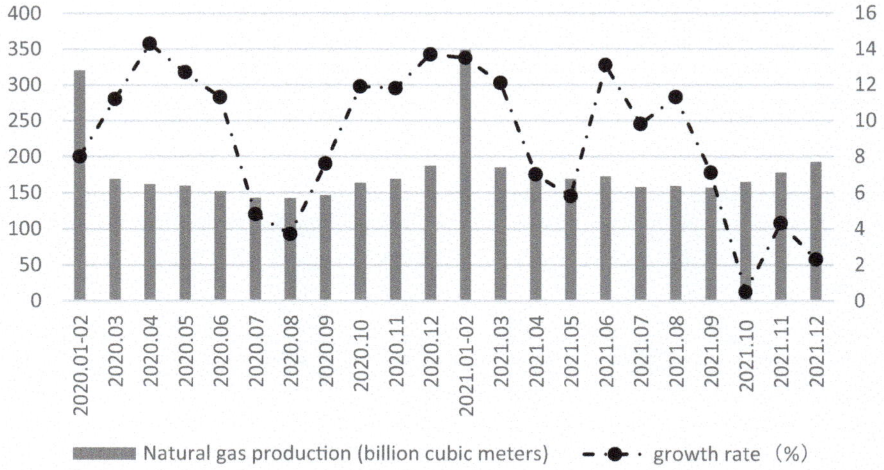


**Fig. 2.11** National crude oil production and growth by month in 2020–2021

ethanol), ether fuels (DME), biodiesel, hydrogen, and other clean alternative fuels, as well as electricity.

As shown in Fig. 2.11, in 2021, domestic crude oil production was 19.98 million tons, an increase of 2.4% year-on-year, an increase of 4.0% over 2019, and a two-year average increase of 2.0%. In the first quarter, crude oil output was 49.176 million tons, up 3.3% year-on-year. In the second quarter, crude oil output was 50.103 million tons, up 9.7% year-on-year. In the third quarter, crude oil production was 50.51 million tons, up 8% year-on-year.

As shown in Fig. 2.12, China's crude oil resources are small and its dependence on foreign imports is serious. In 2020, China's crude oil imports will be 542.39 million tons, an increase of 7.62% compared with the same period last year. In 2021, China's domestic crude oil imports will be 513 million tons, down 5.4% compared with the same period last year. Since 2020, affected by the decline in international oil prices, China's crude oil imports have maintained rapid growth. The first three quarters imported 420 million tons of crude oil, up 12.7% year-on-year, 2.8 percentage points faster than the first half of the year.

As shown in Fig. 2.13, China's GDP growth rate has been maintained at a high level in recent years, and oil consumption has been increasing. According to data from the National Statistical Yearbook and the General Administration of Customs of China, China's crude oil consumption has grown gently from 2012 to 2020. By 2020, China's annual crude oil consumption will be 736 million tons, an increase of 5.56%. In the first half of 2021, China's apparent consumption of crude oil was 462 million tons, and the annual apparent consumption of crude oil was as high as 111.52 million tons, an increase of 51.2% over the same period last year.

As people pay more and more attention to environmental quality, the development of new transportation energy has been stimulated to a certain extent. As shown in

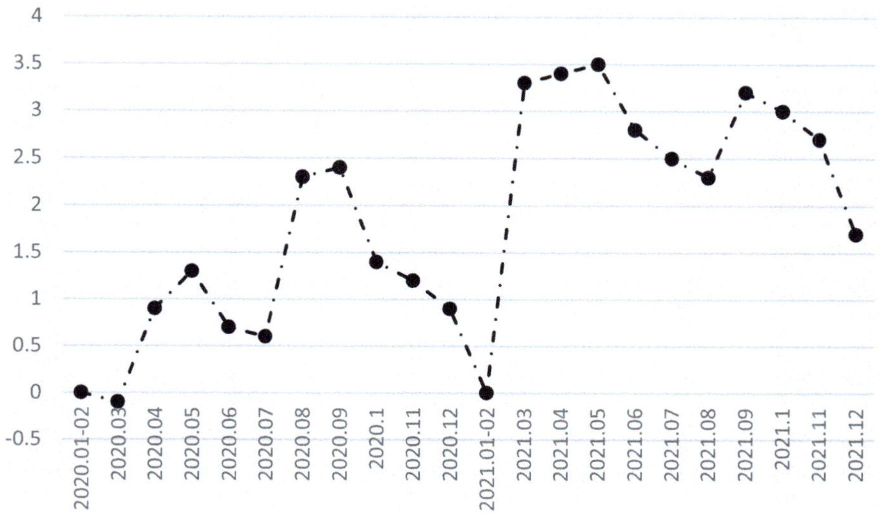


**Fig. 2.12** Crude oil production and growth rate 2020–2021

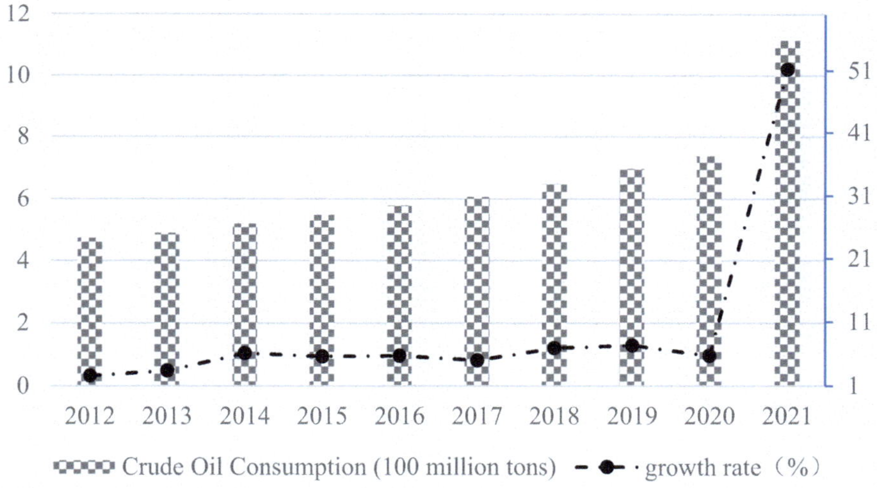


**Fig. 2.13** China's crude oil consumption and growth rate from 2012 to 2021

Fig. 2.14, in 2021, natural gas production was 205.3 billion cubic meters, an increase of 8.2% year-on-year and 18.8% over 2019, with an average increase of 9.0% over the two years. In the same year, China imported 121.36 million tons of natural gas, of which the import volume of gaseous natural gas was 42.43 million tons, down 36.8%. LNG imports were 78.93 million tons, up 17.6% year-on-year. In the first quarter of 2021, natural gas production was 48.27 billion cubic meters, an increase of 10.4%, of which unconventional natural gas made an increasingly significant contribution, accounting for 33.6% of natural gas, an increase of 1.6 percentage points over the same period last year.

With the soaring demand for new energy electric vehicles, the demand for battery power has risen sharply. As shown in Fig. 2.15, the annual power battery shipment in 2021 was 22 gigawatt-hours, an increase of 175% year-on-year in 2020, and the installed capacity of new energy vehicle batteries in China showed explosive growth in 2021. From January to November, the cumulative load of vehicles was 12.83 billion watt-hours, up 153.1% year-on-year. In December, the installed capacity of new energy vehicle batteries was 2.62 billion watt-hours, an increase of 102.4% year-on-year and 25.9% month-on-month.

From the perspective of different vehicle types, the electricity demand of passenger cars is still to occupy the majority. As shown in Fig. 2.16, the battery demand for passenger cars in 2021 reached 135.19 million KWH, an increase of 2.8 times over 2020, and the battery consumption for passenger cars was 10.74 million KWH, down 10 percentage points from the same period last year. Whether it is the output of new energy vehicles or the electricity demand of new energy vehicles, for the power battery of new energy vehicles, the passenger car market contributes the main power battery installed capacity, followed by buses, and then special vehicles.

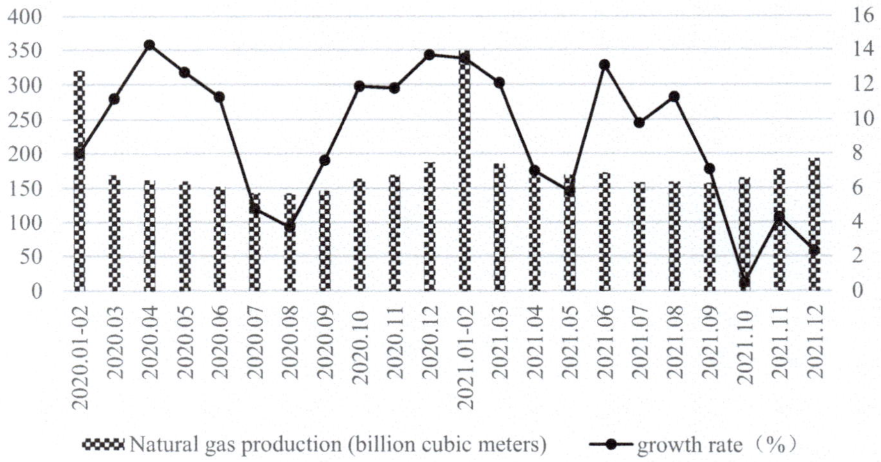


**Fig. 2.14** Natural gas production in 2020–2021

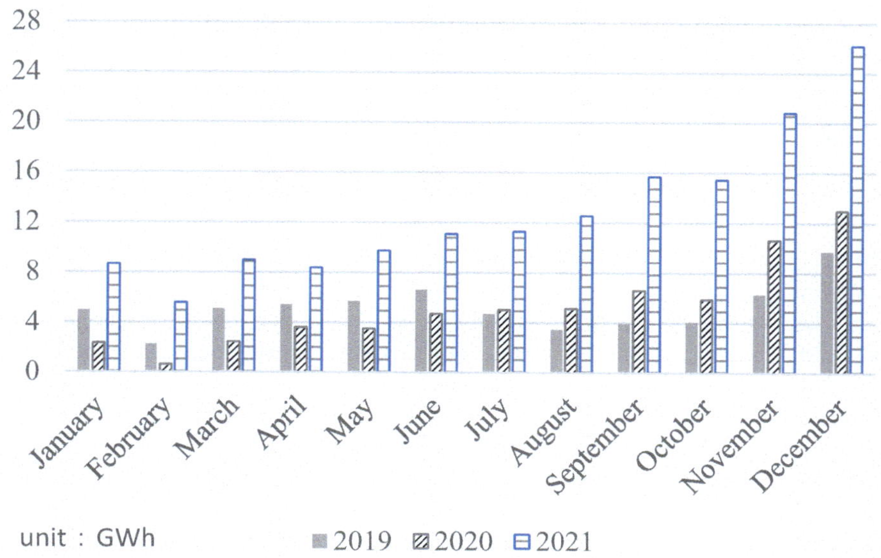


**Fig. 2.15** Installed capacity of power batteries 2019–2021 (gigawatt-hour)

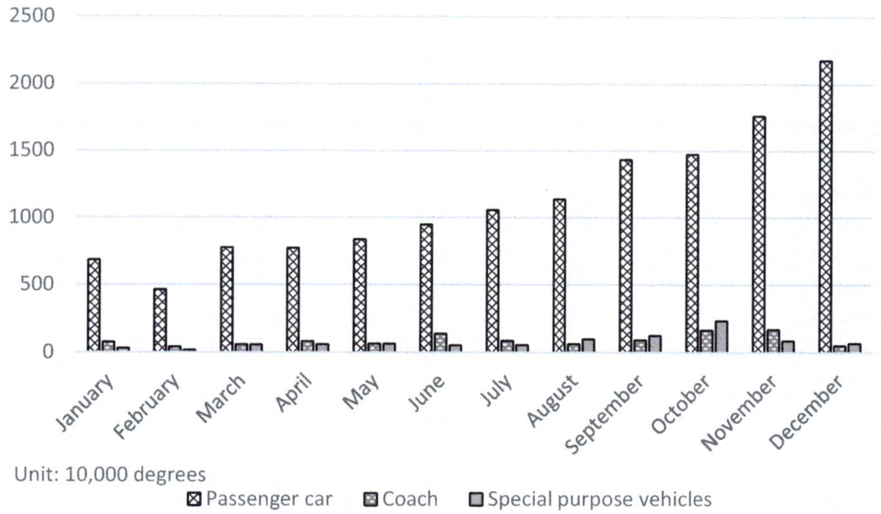


**Fig. 2.16** Battery demand from January to December 2021 Battery demand by vehicle type (10,000 KWH)

### *2.1.3 Analysis of Urban Transportation Energy Supply Network Structure*

#### 2.1.3.1 Road Network Structure of Typical Cities

The "Ring Plus Radiation" Road Network Structure in Chengdu

The road network of Chengdu is obviously concentric circle structure, and the central city relies on the "16 high and 25 fast" high-speed roads to build an integrated and networked road network system. According to the interconnection idea of "building large passageways outside, building large networks inside and building large hubs together", Chengdu has built a comprehensive three-dimensional transportation system with internal traffic going to each other in the same city and fast and smooth external traffic. The transportation network of external passageways has been built, including 17 radial external passageways, which effectively integrates the national "eight horizontal and eight vertical" high-speed railway network and the layout of 23 external high-speed lanes, forming a "three rings and seventeen shots" high-speed road network.

The Grid Network Structure of Xi'an City

Xi'an takes the checkerboard road network of the old city as the core, and on the basis of maintaining the characteristics of the square road network, it extends to the west, south, and east. Road network to "nine-palace pattern, checkerboard network, axis prominent, one city multi-heart" layout development. Xi'an City has built a nine-palace urban road network relying on the "meter-shaped" traffic skeleton. The traffic network of Xi'an shows the "checkerboard" type traffic network structure inside the ancient city wall. Outside the ancient city walls, the road network structure shows a grid and radial form. At present, the main skeleton of the urban road network is a "two-axis, three-ring, eight-radiation" road skeleton. The road network in the central city of Xi'an is densely distributed, forming a network covering the whole situation. The overall road network has a high probability of being presented in the form of a network, and the four types of roads are densely distributed, covering all corners of Xi'an, forming a network of traffic with high efficiency and strong connectivity. Especially in the city center, the road network is densely distributed, and the closer the road network is to the central city, the denser the road network distribution pattern in space.

Chongqing "Narrow" Road Network Structure

The road network structure of Chongqing is a "narrow and long" road network, which is limited by special terrain. Each road relies on the topography of the city to develop, twists and turns, and cannot form a specific geometric figure, so it is "forced" to appear as a north–south or east–west narrow urban structure. During the "13th Five-Year Plan" period in Chongqing, the central urban area has basically formed a "five horizontal, six vertical, one ring, three connections" rapid road network structure, adding "3 Bridges and 4 tunnels", and building a total of "30 Bridges and 19 tunnels". The high-speed road network in the central urban area has reached a new high and basically formed a "five horizontal, six vertical, one ring, three connections" rapid road network structure. The "14th Five-Year Plan for Urban Road Construction in Chongqing (2021–2025)" proposes that by 2025, the rapid road network structure built in the central urban area will be upgraded from "five horizontal, six vertical, one ring and three connections" to "six horizontal, seven vertical, one ring and six connections".

#### 2.1.3.2 Energy Supply Layout of Several Typical Cities

Beijing Transportation and Energy Supply Layout

The number of petrol stations in Beijing leads the country. As of July 2021, Beijing is mainly dominated by Sinopec and CNPC brand gas stations, with more than 690 Sinopec gas stations and 304 CNPC gas stations. Among foreign gas stations, there are 27 Shell gas stations and 5 Total gas stations. Although the development status of Beijing gas station is in the leading position in the country, there are still some shortcomings. The main problems include:

1. The number of gas stations basically meets the market demand but the distribution is uneven, the density of gas stations in some areas is too large, and some new urban areas and radioactive trunk roads have no gas stations.
2. The construction speed of gas stations in Beijing is slow, and the investment cost is high (land acquisition cost and demolition cost). The speed of urban development is accelerating, the change of planning leads to the demolition of gas stations, and the approval of new projects is difficult. The diversified construction of gas station service is affected by the contradiction between urban land use and other reasons.
3. Some gas stations have problems such as incomplete approval procedures and contradictions with urban planning and layout, which require the linkage of relevant government departments to take effective measures to speed up the solution of historical problems left over from some gas stations and improve the relevant procedures for gas stations with conditions and in line with the planning.

Distribution of Transportation and Energy Supply in Shanghai

By the end of 2020, Shanghai has a total of 844 refueling (gas) stations (no water refueling [gas] stations, pure refueling stations). The total scale of the city's gas stations is 977, including 720 current stations are retained, 124 are relocated, and

257 new stations are planned to ensure service density and make up for service blind areas. The planned site scale increased by 133 compared with the current situation, mainly concentrated in areas outside the central city, including 39 new city areas, an increase of 35%. The number of highway gas stations is 47. Among them, 44 existing sites are retained, and 3 new sites are planned. The city plans to build a total of 47 oil and hydrogen stations. The scale of planned sites is mainly increased in the five new cities, of which Pudong New Area, Songjiang District, Qingpu District, Jiading District increased 47, 25, 20, and 15, a total of 107, accounting for 80%. Baoshan District and Minhang District added 7 and 9, respectively. The station scale of the central city and Chongming district basically maintains the status quo.

In the past construction of gas stations in Shanghai, due to the lack of corresponding planning and management, the layout is not reasonable. Moreover, in the planning and construction of such facilities, emphasis is placed on its traffic and risk, and insufficient consideration is given to its commercial service attributes, resulting in a mismatch between demand and supply, which on the one hand causes difficulty in refueling, and on the other hand, some sites have depressed sales and waste of resources. From the perspective of Shanghai gas station layout planning practice, it is very important to plan the annual total station demand forecast. Secondly, on the basis of considering the principles of safety and balance, gas station layout planning should also consider the differentiation of refueling demand in different regions and the current land resources.

Distribution of Transportation and Energy Supply in Chengdu

The current distribution of supplementary energy stations in Chengdu has the following problems:

1. Uneven distribution of gas stations. Due to the policy orientation and market demand, the distribution of gas stations has a high density in some areas of urban areas, while remote towns, some new districts, and industrial parks are short of gas stations. For example, the downtown area of Chengdu accounts for 5.5% of the city's administrative area, but there are 184 gas stations, accounting for 25.1% of the total number of gas stations in the city, of which the gas stations within the third ring road account for more than 70% of the total number of gas stations in the central area.
2. Uneven operating efficiency. At present, the retail of refined oil products is mainly operated by CNPC, Sinopec, social gas stations, and some agricultural machinery gas stations. The operating efficiency of gas stations is uneven. Within the Third Ring Road in the central city, the average daily sales volume of most gas stations is more than 10 tons, while the average daily refueling volume of some gas stations in townships is less than 1 ton. The reason for the great difference is that their respective geographical locations and operating levels are different, resulting in great differences in benefits.

3. Some gas station facilities are backward. Compared with other areas, Chengdu has great differences in the hardware facilities of gas station construction. Some gas station refueling facilities are aging, firefighting facilities are not fully equipped, and the process does not meet the environmental protection requirements, and there are certain security risks.
4. A small number of gas stations have substandard planning and design and illegal construction. Some gas stations in Chengdu are not constructed strictly according to the current capital construction procedures. However, in the context of rapid urban development, in order to solve the current situation of short supply of gas stations, there are still illegal construction phenomena such as blindly seeking new and fast, not following procedures, unauthorized construction or construction without approval, unauthorized relocation, unauthorized reconstruction and expansion. This will inevitably cause some gas stations to have problems such as unreasonable location layout, substandard scale, and non-standard design.

Lanzhou Transportation Energy Supply Layout

In Lanzhou city, the distribution of gas stations with good potential value in urban areas and national and provincial highways shows a "group phenomenon". For example, within 5 km from G212 city to Xiguoru, there are 6 gas stations in Longhui, Hanjiahe, Hanjiahe, Jincheng, Shengrong, and Zhenxing, which are higher than the standard requirements. The core area of the city is in shortage of land and the investment cost is too high. The gas stations in Chengguan District are mainly distributed in the east side of Tianshui Road, there are no gas stations in the north of Nanshan Road on the west side of Tianshui Road, the gas stations in Anning District are mainly concentrated in the middle, and there are no gas stations near Shazhichuan Outlet in the west, and the uneven distribution of stations is extremely obvious. The development area is not synchronized with the construction of gas station land, such as Yanchi, Kyushu area. The demand for gas station construction in districts, counties, towns, rural areas, and remote areas is high, but the single-station sales volume of gas stations in remote areas is low, the return on investment of enterprises is limited, and the construction of gas stations is difficult.

*Distribution of Transportation and Energy Supply in Chongqing*

1. Chongqing Gas Station Network

   Through the field investigation of eight districts in Chongqing city, the statistics show that there are 334 gas stations in Chongqing city. Among them, 34 in Banan District, 16 in Dadukou District, 33 in Jiangbei District, 74 in Jiulongpo District, 46 in Nanan District, 40 in Shapingba District, 77 in Yubei District, and 14 in Yuzhong District.
2. Chongqing Charging Station Network

   Through the field investigation of eight districts in Chongqing city, the statistics show that there are 851 charging stations in Chongqing city. Among them,

**Table 2.2** Types and quantities of supply stations in Chongqing main urban area

| Type of supply station | Gas station | Gas station | Charging station |
|---|---|---|---|
| Quantity (PCS) | 334 | 85 | 851 |
| Percent | 26 | 7 | 67 |

98 were in Banan District, 22 in Dadukou District, 102 in Jiangbei District, 76 in Jiulongpo District, 118 in Nanan District, 252 in Shapingba District, 84 in Yubei District, and 99 in Yuzhong District.

## *2.1.4 Classification and Network Distribution of Transportation Energy Supply Stations*

Taking Chongqing transportation energy supply station network as the object, this paper studies the classification and distribution of Chongqing transportation energy supply station network.

### 2.1.4.1 Classification and Quantity of Supply Stations

Energy supply station is the base of vehicle refueling, its main function is to fuel vehicles efficiently, accurately, and safely. For the investment and financing personnel of the supply station, it should have the function of creating economic wealth. For automobile users, supply stations should be widely distributed in the city to provide convenient and safe filling services. For macro managers, the supply station, as a service enterprise, should generate tax economic benefits and minimize the impact on traffic, environment, and security, which determines the multi-purpose and complexity of the planning layout and evaluation of urban energy supply station.

According to the requirements of economic development and environmental protection in Chongqing main urban area, the main types of supply stations in Chongqing main urban area are gas stations, filling stations, and charging stations. This book collects the data of all supply stations within the Chongqing Expressway. As shown in Table 2.2, there are 851 charging stations, 334 gas stations, and 85 filling stations.

### 2.1.4.2 Supply Station Layout

According to the layout principles of urban energy supply stations, factors that should be taken into account in site selection, and the management mechanism of the national government, as shown in Fig. 2.17, the relevant classification standards in this book collect the longitude and latitude coordinates of all gas stations, gas stations, charging stations, and other supply stations within the inner ring highway of Chongqing, and analyze the layout characteristics and existing problems of Chongqing supply stations.

This book collects the longitude and latitude coordinates of 1302 energy supply stations including charging stations, filling stations, and gas stations through Baidu maps and actual investigations. This book mainly collects the information of supply stations in 8 urban areas, including Nanan District, Dadukou District, Shapingba District, Banan District, Jiangbei District, Yubei District, Yuzhong District, and Jiulongpo District. The supply station data collected for each municipality are shown in Table 2.3.

#### Insufficient Attention Has Been Paid to the Layout of Supply Stations

At present, the commercial activities of the supply station are in an extremely embarrassing position in the commercial activities of the city. From the point of view of the government, the function of the supply station only needs to meet the needs of automobile refueling and contains specialized functional grass-roots commercial activity outlets, which can be fully delivered to the corresponding construction units for independent construction, and the government does not need to support relevant resources. From the perspective of urban planners, a supply station is a basic commercial activity facility. Most of them regard gas station as a commercial activity concept and a basic resource to obtain economic benefits from urban planning. From the perspective of petrochina and Sinopec enterprises, since the planned economic system has been open for a long time, Sinopec sales enterprises have been in the business management thinking, paying attention to light retail and heavy wholesale, and lacking the understanding of the construction and development of the energy network of supply stations, thus forming the "proportion imbalance" of the operation structure and "top-heavy" in the marketing strategy.

#### There Is No Uniform Layout Standard in the Industry

Looking at the indicators of China's current urban master planning, supply stations are not considered to be included in the planning and allocation of public supporting facilities, and the local government does not have clear regulations on the location and layout of supply stations to restrict and guide, which leads to the disordered location selection of supply stations. As a result, many problems such as low entry threshold of gas stations, large differences in the comprehensive quality of operators, and uneven service levels have presented more difficult challenges to the scientific and rational construction of supply stations.

Due to the connection between the construction of gas stations and the lack of planning for urban development and re-layout, and the failure to adapt the scale control to the actual local needs, there are empty facilities for gas stations and insufficient scale supply of some supply stations, resulting in the phenomenon of car users

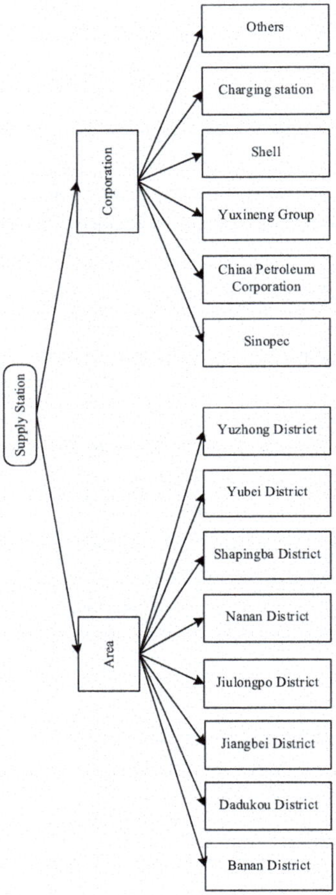


**Fig. 2.17** Data classification standard

Table 2.3 Quantity of supply stations in Chongqing main urban area

| | Nanan | Dadukou | Shapinba | Banan | Jiangbei | Yubei | Yuzhong | Jiulongpo | Total |
|---|---|---|---|---|---|---|---|---|---|
| Gas station | 46 | 16 | 40 | 34 | 33 | 77 | 14 | 74 | 334 |
| Gas station | 11 | 5 | 9 | 9 | 5 | 26 | 6 | 14 | 85 |
| Refueling station | 5 | 3 | 2 | 1 | 0 | 13 | 3 | 5 | 32 |
| Charging station | 98 | 22 | 102 | 76 | 118 | 252 | 84 | 99 | 851 |
| Total (number) | 160 | 46 | 153 | 120 | 156 | 368 | 107 | 192 | 1302 |

waiting in line for a long time because of the lack of connection between the overall planning plans of the construction of supply stations.

### The Layout Method Is Not Scientific

For a long time, the management of supply station and the lack of norms have been studied. The development of social economy and supply station precedes the development of norms. Although the government and related departments have also introduced some supply station management systems, there are some problems such as too rough management framework and insufficient practical feasibility. In this case, most supply stations only rely on the daily experience of operators and relevant planners to carry out site selection planning or competition, and this subjective randomness directly leads to unreasonable gas station planning. For example, the organic connection between the scale of gas stations and the scale of users is extremely thin, and the organic connection between the radiation range of gas stations and the service is also extremely thin. The scale estimation errors and traffic organization errors affect the operational efficiency and economic benefits of supply stations.

### The Overall Layout of the Supply Station Is Not Reasonable

Looking at the overall situation in China, about 10% of the number of supply stations have significant operating losses, and about 15% of supply stations have a long payback period. The reason is that the supply station is concentrated in several areas with large traffic flow, which is the key to the long payback period or operating loss of the supply station. It can be seen from the exported water notes that the supply stations are almost all on the main line, and only a few branches are occupied. For example, if Sinopec builds a new supply station somewhere, petrochina will quickly build another nearby. Driven by economic interests, a small number of units or individuals independently build new buildings around the city, making the network of some branches in the core city and some surrounding areas too concentrated, and the network layout is especially unreasonable. There are fewer gas stations in the central area of the city, and the service intensity is high, but the service level is low. In the morning and evening peak hours, the queue of replenishment vehicle fuel is difficult to effectively meet the needs of users, affecting the road traffic near the gas station.

### 2.1.5 Typical Layout of the Replenishment Site

Through online Baidu map real scene search and offline field research, the collected structural survey and statistics of all the supply stations, including the number of external lanes of the supply station, the number of entrance and exit lanes of the station, and the layout structure of the internal refueling pile. Understanding the number of external lanes of the supply station, that is, whether the supply station is on the main road or the branch road, is helpful to understand the flow of the external network of the station and analyze the cause and mechanism of queuing at the supply station. The number of entrance and exit lanes in the station and the layout of internal refueling piles directly affect the service efficiency of the supply station. Reasonable planning of the number of entrance and exit lanes in the station is helpful to dredging dispersed vehicles entering the station for replenishing energy, and an appropriate internal structure layout can also improve the service rate of the station to a certain extent. Several supply stations in main urban areas with large traffic flow are analyzed, and several typical distribution and structure forms of supply stations are extracted, as shown in Table 2.4. It can be seen from the table that supply stations in Chongqing main urban area mainly have $2 \times 2$, $2 \times 3$, and $3 \times 3$ structural layout forms (Figs. 2.18, 2.19, and 2.20). The internal structure is $2 \times 2$, indicating that there are two rows of two columns of refueling columns. Gas stations with different layout structure forms will form different forms of queues, resulting in different degrees of congestion.

### 2.1.6 Development Trend of Transportation Energy Supply Station

#### 2.1.6.1 Gas Station Development Trend

Changes in the Pattern of China's Gas Station Industry

In the "Special Management Measures for Foreign Investment Access (Negative List) (2018 edition)", the restrictions on more than 30 foreign-funded chain gas stations that need to be controlled by the Chinese side are clearly eliminated, and the sales

**Table 2.4** Typical supply station structure in Chongqing main urban area

| Internal structure | Number of outside lanes | Typical site name | Owning region |
|---|---|---|---|
| 2 × 2 | 2 | China Petroleum Lanxin gas station | Yubei District |
| | | CNPC Yibowan Water Gas Station (Yinxin Road Store) | |
| | | Sinopec gas Station (Renhe Station) 1 | |
| | | Petrochina Hui Xing gas station | |
| | | Sinopec (Lijia Gas Station) | |
| | | Petrochina (Qinglong Road Gas Station) | Nanan District |
| | | Sinopec Gas Station (Haitangxi Station) | |
| | | CNPC Shalangya Gas Station (Dashi Store) | |
| | | Petrochina (Hongsheng Road Branch) | |
| | | Petrochina | |
| | | Shell Nine Middle Road gas station | Jiulongpo |
| | | Sinopec Gas Station (Yinyan Road Branch) | |
| | | Petrochina Huatian gas station | |
| | | Sinopec Gas Station (Huangjueping Station) | |
| | | East Oil (Longteng Gas Station) | |
| | | Sinopec Gas Station (Huade Station) | |
| | | Shell Wing Loong Road petrol station | |
| | | Sinopec Gas Station (Chenjiaping Station) | |
| | | CNPC Gas Station (Mulan Road) | Jiangbei District |
| | | Petrochina (Xugong Gas Station) | |
| | | Xuanyu Petroleum (Bafu Gas Station) | |
| | | China Petroleum (Mafengliang Gas Station) | |

(continued)

**Table 2.4** (continued)

| Internal structure | Number of outside lanes | Typical site name | Owning region |
|---|---|---|---|
| 2 × 2 | 3 | Petrochina (Xingsheng Avenue Branch) | Yubei District |
| | | China Petroleum gas station | |
| | | China Petroleum (through shop) | |
| | 4 | Petrochina (Daxing Village Gas Station) | Jiangbei District |
| | | Petrochina Co., LTD. (Shima Gas Station, Chongqing Sales Branch) | |
| | | Petrochina Baiziqiao gas station | Nanan District |
| | | Petrochina (5 km gas Station) | |
| | | CNPC 7 km Gas Station (South Chongqing Avenue) | |
| | | Petrochina (Jiulong Park Gas Station) | Jiulongpo |
| | | Petrochina (Dalong Gas Station) | |
| | | Chenjiaping gas station | |
| | | Chongqing Kaiyuan Chenjiaping refueling station | |
| | | Shell Kinkai Avenue gas station | Yubei District |
| 2 × 3 or 3 × 2 | 2 | Sinopec (Dashiba Gas Station) | Yubei District |
| | | Sinopec (Jintong Road Branch) | |
| | | Petrochina (Xinnan Road Branch) | Jiangbei District |
| | | Sinopec refueling Station (Tanzishi New Street) | Nanan District |
| | | Petrochina (Panlong Gas Station) | |
| | 3 | Sinopec refueling station | Yubei District |
| | | Sinopec gas Station (Renhe Station) | |
| | | Sinopec Gas Station (Xiazi Dawn Station) | Jiangbei District |
| | | Petrochina (Black Stone Gas Station) | |
| | | Tea garden gas station | Nanan District |

(continued)

**Table 2.4** (continued)

| Internal structure | Number of outside lanes | Typical site name | Owning region |
|---|---|---|---|
| | | Sinopec Gas Station (Water Grinding Station) | Jiulongpo |
| | | Chongqing Jiaoyun Shiqiao Pu gas station | |
| | 4 | CNPC Jiahua West gas station | Jiangbei District |
| | | Sinopec agricultural Park gas station | Yubei District |
| | | Sinopec (Tenglong refueling Station) | Nanan District |
| | | Petrochina (Tenglong Gas Station) | |
| 2 × 4 or 4 × 2 | 1 | Hui Xing refueling station | Yubei District |
| | | Longba CNG filling station | |
| | 3 | China Shi Hua Long Yu refueling station | Yubei District |
| 3 × 3 | 3 | Sinopec refueling Station (Renhe Station) | Yubei District |
| | | Petrochina (Shimen Gas Station) | Jiangbei District |

**Fig. 2.18** 2 × 2 structure layout of gas station

**Fig. 2.19** 2 × 3 structure layout of gas station

**Fig. 2.20** 3 × 3 structure layout of gas station

scope of the downstream link where the oil is located is fully opened, and at the same time, driven by the huge domestic retail market share and benefits and profits, Foreign gas stations are rapidly advancing in China's scale layout, the main body of retail competition will increase day by day. In 2019, the "Decision of The State Council on canceling and delegating a number of administrative license Matters" pointed out that the "approval of the wholesale and storage operation qualification of petroleum products" was included in the catalogue of administrative license matters decided to cancel, and the "approval of the retail operation qualification of petroleum products" was included in the catalogue of administrative license matters decided to delegate

**Table 2.5** Regional proportion of petrochina and Sinopec

| District | China National Petroleum Corporation (%) | Sinopec (%) | Other (%) |
|---|---|---|---|
| Northeast region | 32 | 5 | 62 |
| Northwest China | 45 | 15 | 40 |
| Southwest China | 27 | 22 | 50 |
| North China | 25 | 33 | 42 |
| Central China | 14 | 32 | 54 |
| South China | 22 | 43 | 35 |
| East China | 13 | 33 | 55 |

the approval level. Cancel the application for the wholesale storage and operation qualification of petroleum products, and the municipal government should do a good job in the application for the retail operation qualification of petroleum products, and do a good job in the supervision and management of the transaction of refined oil products before and after, and do a good job in the implementation of safety protection policies. Lower the investment threshold, increase efforts to improve the layout of county and township gas stations, and increase the consumption of refined oil products. At the same time, with the increasing growth of refined oil wholesale business, China's oil market will enter a period of rapid reform in 2021. The General Office of the Ministry of Commerce of the People's Republic of China issued the Guidelines on the Management of the Petroleum Products Circulation Industry to implement The State Council's reform strategy of "discharge management service", remove the application process for the wholesale storage of petroleum products and do a good job in the application process for the retail operation of petroleum products, so as to promote the rapid development of the marketization process of petroleum products.

Petrochina and Sinopec Are Divided into the North and South

The difference in geographical layout makes the two occupy market share advantages in different regions. As shown in Table 2.5, petrochina has an absolute market share advantage in Northeast and Northwest regions and a large market share in Southwest regions. Sinopec has a strong market share advantage in East, South, and Central China, and also has a strong market advantage in North China. In other words, petrochina and Sinopec each have a major market share in their respective regional markets.

#### 2.1.6.2 Integrated Development of Online and Offline

In addition to the policy changes in the refined oil retail industry and the further expansion of terminal layout, the retail industry is also facing transformation in terms of business model operation. Internet+ may open a new chapter in the retail industry, promoting the construction of a new business model for gas stations based on artificial intelligence and big data analysis to realize the integration of online and offline development. The mobile refueling platform thus becomes a link between gas stations and car owners, and improves non-oil business income through online guidance, so as to realize benefit sharing and profit improvement. In addition, with the change of consumption structure, it will create a comprehensive integrated energy supply and service provider such as refueling, charging, gas, non-oil business, automobile beauty maintenance services, and provide convenient settlement and convenient consumption.

#### 2.1.6.3 Gas Station Development Trend

China's natural gas import shows a steady growth trend, and the dependence on foreign countries is further eased. The national pipe network company was restructured and integrated and officially operated, and the reform dividend began to be released. After all, LNG filling stations are still a "youth industry", and in the long run, they will become the main channel for developing LNG downstream terminal demand. At present, in fact, the mismatch between the car and the station is gradually easing.

According to the provisions of the National Development and Reform Commission and the National Energy Administration on the issuance of the "14th Five-Year Plan for Modern Energy System": Speed up the improvement of the complete system of energy production, supply, storage, and marketing, strengthen the ability to balance the allocation of energy resources, implement the construction of energy infrastructure facilities such as power grids and oil and gas pipelines, establish a sound coal reserve system, strengthen the production of oil and gas reserves, focus on the construction of underground gas storage, LNG (liquefied natural gas) receiving stations and other gas storage facilities, and improve the flexibility of energy supply capacity. By 2025, China's oil and gas pipeline network will reach 210,000 km. China's integrated gas storage capacity is between 55 billion cubic meters and 60 billion cubic meters, and the proportion of natural gas consumption is about 13%. It is expected that by 2035, the high-quality development of energy has made decisive progress, and a modern energy system has been basically completed. The energy security system has been greatly improved, green production and consumption patterns have been formed on a large scale, and the proportion of non-fossil energy consumption has increased by leaps and bounds on the basis of reaching 25% by 2030, and several underground gas storage clusters of 10 billion square meters have been built in Northeast, North, northwest, and Southwest China.

High Degree of Automation

With the development of science and technology in China, under the strong promotion of Internet+, highly automated management, creating unattended, making good use of Internet+ smart energy, and realizing the integration of transport and trade operation are some of the development trends of LN point supply.

Distributed Energy

The emergence of large-scale natural gas distributed energy projects has entered the substantial development stage. Compared with the centralized energy supply mode, natural gas-distributed energy has the advantages of environmental protection, energy saving, and safety. Experts in related industries believe that natural gas distributed energy is a good supplement to the existing energy supply system, and the mutual coexistence of centralized and distributed energy supply is the development trend of the industry, and the "14th Five-Year Plan" also proposes to develop distributed energy.

#### 2.1.6.4 Charging Station Development Trend

China Will Become the World's Largest Market for New Energy Vehicles

Since the "13th Five-Year Plan" period, China has gradually stood in the center of the world's new energy vehicle market stage and is in the core position of the world's new energy vehicle market. According to data, by 2020, China's total sales of new energy vehicles have reached 4 million and has surpassed the United States to become the protagonist of the world's new energy vehicle market stage. Moreover, China's automobile market is not only huge but also has a large variety of automobile products, and China has the world's largest supply of energy. China's new energy vehicle market will become a "red Sea" for a new generation of competition. China's new energy vehicle market is not only a market dominated by personal consumption but also a diversified market, its growth focus not only depends on personal consumption, and in the field of urban taxis, urban logistics vehicles, and leasing, new energy vehicle market prospects are broader.

The Configuration of Charging Facilities Is Gradually Improving

In 2020, the construction of charging infrastructure has been attached to great importance by the country, and part of the charging infrastructure of the new infrastructure includes the "charging pile construction". As of mid-2021, China's public charging piles have reached 924,000 units, including 374,000 DC charging piles and 550 thousand AC charging piles, respectively, with a ratio of about 5:1, which is far from meeting the demand for charging piles for pure electric vehicles. The exponential increase in the number of pure electric vehicles has effectively promoted the development of supplementary charging stations. According to the "14th Five-Year Plan", in the next five years, the market size of China's new energy vehicle charging station will reach 126 billion yuan, and the annual profit of the charging station industry can reach 1.8 billion yuan to 2.7 billion yuan. In the future, after the annual construction scale of charging facilities for new energy vehicles reaches 100 billion yuan, the annual profit of the charging station industry can also reach 7 billion yuan to 10 billion yuan, and the boom in the field of charging facilities will continue to extend.

## 2.2 Analysis of the Current Situation of Urban Transportation Energy Demand

### *2.2.1 Current Situation of Urban Population and Motor Vehicle Ownership*

China has the largest population in the world. According to the Statistical Bulletin of National Economic and Social Development, as shown in Figs. 2.21 and 2.22, the total population of China in 2021 will be 1,412.6 million, an increase of 480,000 over 2020. By the end of 2021, China's population density has reached 148.38 people per square kilometer. As the population continues to grow, energy consumption is also increasing, with total energy consumption reaching 524 million tons of standard coal in 2021, of which crude oil consumption increased by 2.4% over 2020. Population growth will lead to an increase in energy demand, especially in transportation, which will affect the number and distribution of gas stations.

Population urbanization will promote the growth of energy consumption, but also change the structure of energy demand. In 1990, China's urbanization level was 26.41%, and by 2021, the urbanization level has reached 64.72%. Urbanization and population growth increase the size of cities and thus increase energy consumption per unit of output. The larger the city, the more energy is wasted due to commuting and congestion. This poses a challenge for improving the traffic structure and alleviating traffic congestion in the future.

With the development of economy and the improvement of people's living standards, the number of cars continues to increase. As shown in Fig. 2.23, the number of civil vehicles in the country will reach 294.19 million in 2021, an increase of 20.78 million over the previous year. The continuous increase of vehicles makes the transportation network expand continuously, and at the same time, it brings no small challenge to the energy supply security.

The number of cars in each region is different, and the higher the density of vehicles, the higher the requirements of the traffic network. In 2020, the car ownership of the four municipalities directly under the central government of China is 6.032 million in Beijing, 3.291 million in Tianjin, 4.403 million in Shanghai, and 5.044

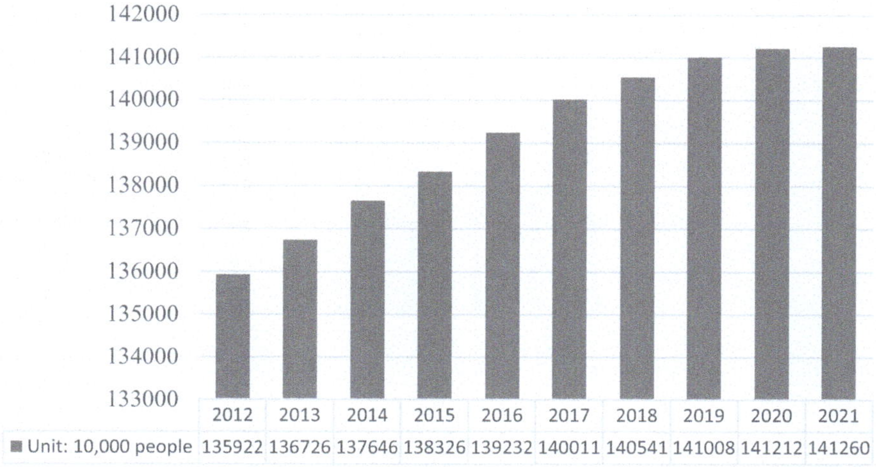


**Fig. 2.21** Population of China, 2012–2021 (unit: 10,000)

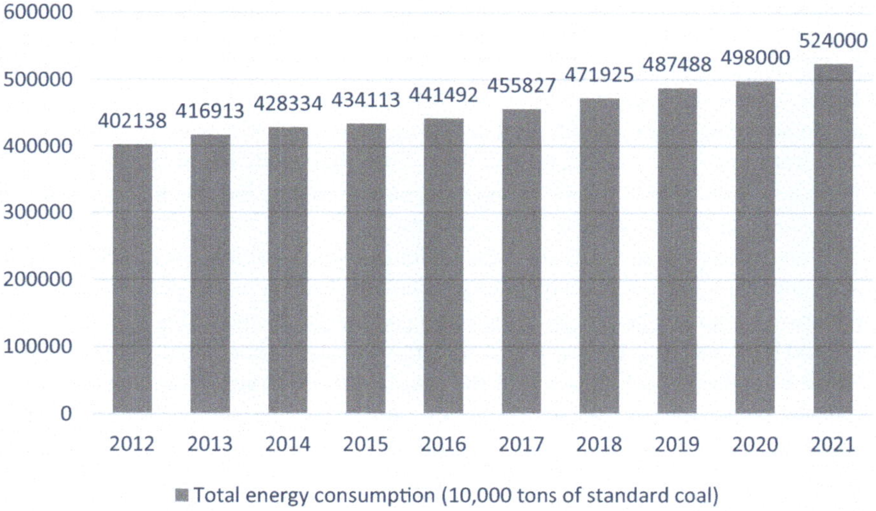


**Fig. 2.22** China's energy consumption 2012–2021 (unit: 10,000 tons of standard coal)

million in Chongqing. As the land area of Shanghai is only 0.63 thousand square kilometers, the road network structure tends to be stable, and the increase in car ownership will inevitably increase the probability of road congestion. From the third quarter of 2020, Chongqing, with an average speed of 25.27 km/h, became the second most congested city in the country's road network during peak periods. With the rapid growth of motor vehicles in big cities, it is easy to cause problems such as slow growth of road resources and a decline in average speed.

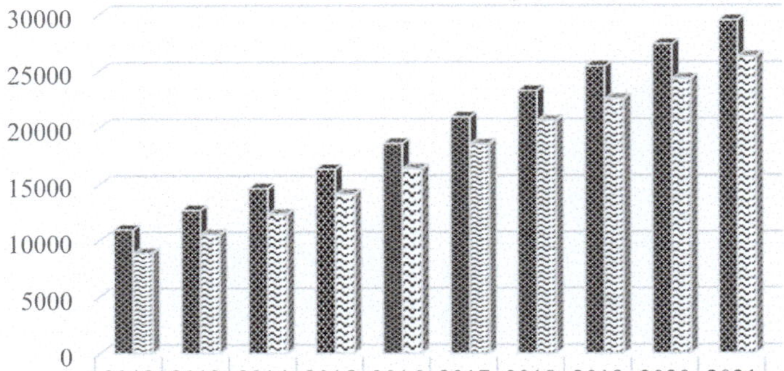


| | 2012 | 2013 | 2014 | 2015 | 2016 | 2017 | 2018 | 2019 | 2020 | 2021 |
|---|---|---|---|---|---|---|---|---|---|---|
| Civilian (10,000 units) | 10933 | 12670 | 14598 | 16284 | 18575 | 20907 | 23231 | 25376 | 27341 | 29419 |
| Private (10,000 vehicles) | 8838.6 | 10502 | 12339 | 14099 | 16330 | 18515 | 20575 | 22509 | 24291 | 26246 |

**Fig. 2.23** National vehicle ownership from 2012 to 2021 (unit: 10,000)

As can be seen from Fig. 2.24, the vehicle density in Shanghai is the highest, reaching 759.17 vehicles per square kilometer, and the negative effects of exhaust pollution and traffic congestion are more obvious. According to the known data, the network layout structure of different road networks has different bearing capacity of vehicles. Shanghai's road network belongs to the hybrid type, which can effectively play the advantages of different road networks, reduce vehicle density, and promote public transport is the current problem that Shanghai needs to solve. Chongqing's vehicle density is only 61.29 vehicles per square kilometer, but due to the impact of terrain, some areas are prone to traffic congestion.

Different driver groups have different replenishment behaviors for vehicles, which has an impact on the location layout of gas stations. From a macro point of view, the number of commuter groups and truck groups is large, so the traffic flow of urban areas and high-speed sections is relatively large, gas stations are mostly concentrated in these two places, and the scale of the station will be larger. From a micro point of view, drivers' personal behaviors will also affect the supply of gas stations. Some drivers attach more importance to the brand of gas stations. For example, if drivers like to replenish energy at CNPC, they would rather drive several thousand meters more even if there are other brands of gas stations nearby. Replenishment time and refueling volume will also have an impact on gas station sales.

### *2.2.2 Analysis of Urban Vehicle Travel Purpose*

The definition of resident travel refers to the displacement process in which residents use certain or multiple modes of transportation and spend a certain amount of time from the starting point through a certain path to the endpoint in order to achieve a

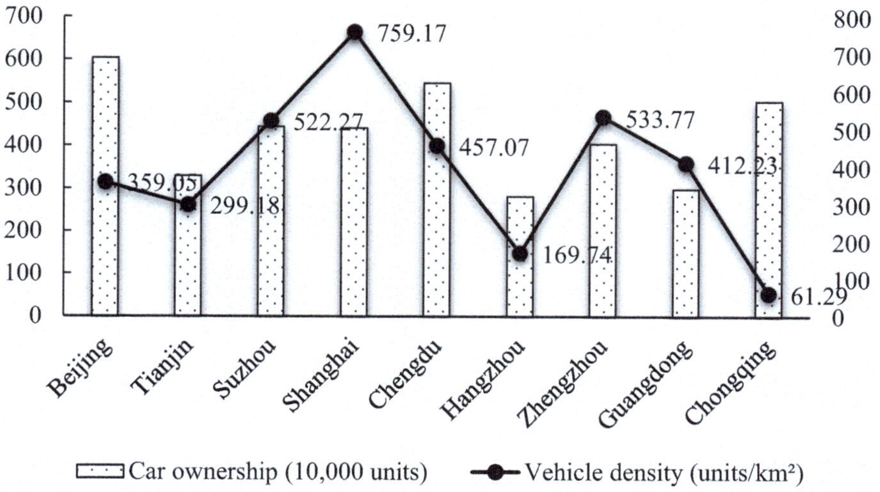


**Fig. 2.24** Car ownership and vehicle density in some provinces and cities in 2020. *Note* Vehicle ownership unit: 10,000 vehicle density unit: vehicle/square kilometer

certain purpose (such as going to work, shopping, traveling, leisure and entertainment, etc.). Travel purpose, which is a pre-planned goal for travelers, is a key attribute of transportation travel behavior and plays a key role in understanding urban residents' travel behavior. The travel purposes of residents are generally divided into the following types: work, school, leisure and entertainment, tourism, medical treatment, visits, shopping, and return trips. With the gradual increase in disposable income of national residents, residents' requirements and needs for life and entertainment are getting higher and higher. Coupled with the continuous improvement and enrichment of various entertainment and catering facilities in residential areas and surrounding areas, more residents choose to eat out, shop, and carry out leisure and entertainment activities, and the proportion of flexible travel demand has also increased significantly. As shown in Fig. 2.25, residents' daily trips for work, shopping, and leisure accounted for a relatively high proportion, totaling 89.19%, especially those for work, which accounted for 48.99%. Holiday travel purposes, because most of their travel purposes are original and different from daily commuting travel purposes, are significantly different from daily traffic in terms of scale, time, and space. Most of the travelers' purposes are leisure, entertainment, and tourism.

### 2.2.3 *Analysis of Urban Vehicle Replenishment Demand*

With the rapid and stable growth of motor vehicle sales and the gradual improvement of vehicle ownership, different motor vehicle travelers have slightly different

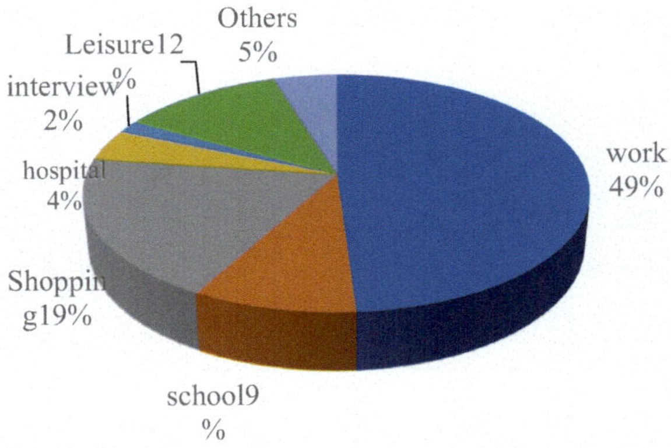


**Fig. 2.25** Daily travel of residents

demands for supplementary energy. The traveler's replenishment means that the traveler determines whether to make up the energy according to the remaining amount of vehicle energy to achieve the purpose of travel. For example, electric vehicle travelers will judge whether they need to be charged according to the remaining power of the electric vehicle and how long it will be charged to meet the purpose of travel. Charging habits are two aspects: the number of daily charges and the minimum State of Charge (SOC) threshold. The My Electric Avenue project in the United Kingdom conducted statistics on the proportion of daily charging times of electric vehicles. Statistics show that the relationship between the number of daily charging and the travel date is not large, 70% of electric vehicle users charge once a day, 20% of users charge twice a day, and the total number of charging times in more days is less than 10%. At the same time, it also shows that a large part of the SOC of electric vehicles is maintained at more than 15%, and more than 60% of electric vehicle users start charging SOC at 20–70%.

According to the data statistics on the charging amount, in January 2021, the total charging amount of Chongqing was 3252.101 million KWH, of which Yubei District accounted for 6.432 million KWH, ranking first. In terms of the distribution of charging time, the charging peak of the whole day occurs from 11:00 to 15:00, and the second wave of the small charging peak is ushered in after 21:00 (Fig. 2.26). During the peak charging period, there is an extremely unbalanced replenishment demand phenomenon, which will aggravate the load of the power grid and the load of public charging piles.

There are also significant differences in the demand for replenishment of different types of vehicles. The average monthly charging frequency of operating electric vehicles (rental passenger cars, public buses, rental passenger cars, logistics special vehicles) is significantly higher than that of non-operating electric vehicles (Fig. 2.27). This generally reflects the high-power demand characteristics of new energy vehicles, and also reflects the high-frequency use characteristics of new energy vehicles, especially operating electric vehicles.

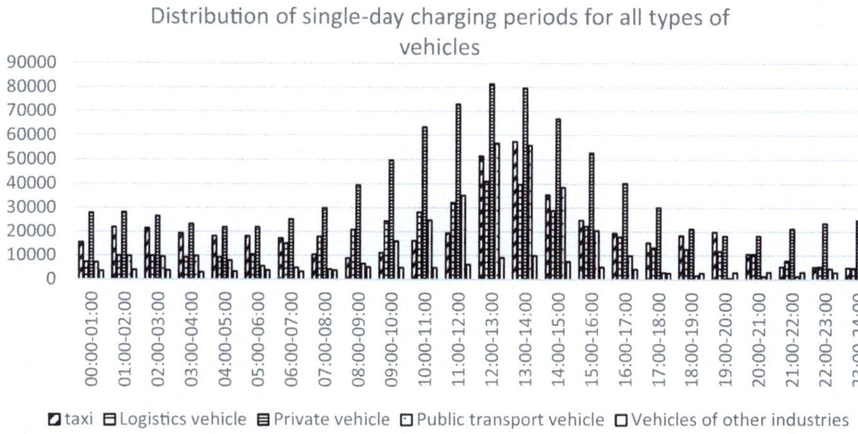


**Fig. 2.26** Single-day charging time distribution of each vehicle type

### *2.2.4 Analysis of the Change Trend of Urban Transportation Energy Demand*

Traditional vehicles use fuel as energy power and use fuel engines to drive vehicles. Electric vehicles are powered by on-board power supply (battery) and drive vehicles by motor. Hybrid electric vehicle refers to the use of fuel or electricity as energy power, using two driving modes to drive the vehicle.

Changes in demand for conventional fuel vehicles. According to the data of the Passenger Association, the market share of traditional fuel vehicles declined from 95.53% in 2018 to 87.27% in 2021 (Fig. 2.28). In terms of sales growth, the incremental sales volume of traditional fuel vehicles declined by 8% in 2019, fell by 8% in 2020, and fell by 6% in 2021. It can be clearly seen that the demand market for traditional fuel vehicles has been in a continuous and rapid decline for several years.

New energy vehicles are becoming more and more popular among automobile consumers, and their sales continue to grow. In 2021, the sales volume of new energy vehicles was 3,334,100, and its penetration rate in China's passenger car market increased from 2.4% in 2017 to 16.0% in 2021 (Fig. 2.29). The number of new energy vehicles in 2020 is only 4.92 million, by the first quarter of 2022 as high as 8.915 million, it can be seen that the demand for new energy vehicles in China is growing rapidly.

However, to calculate the 140 million vehicles in China's car ownership in 2020, it is equivalent to saving 322.9 million tons of oil, replacing 0.311 million tons of oil, saving and replacing a total of 0.639 million tons of oil, which is equivalent to reducing motor vehicle oil demand by 22.7%. Therefore, the rapid growth of new energy vehicles has significantly reduced the oil demand of traditional fuel vehicles (Fig. 2.30), which is also an important starting point for achieving "carbon neutrality" and "carbon peak".

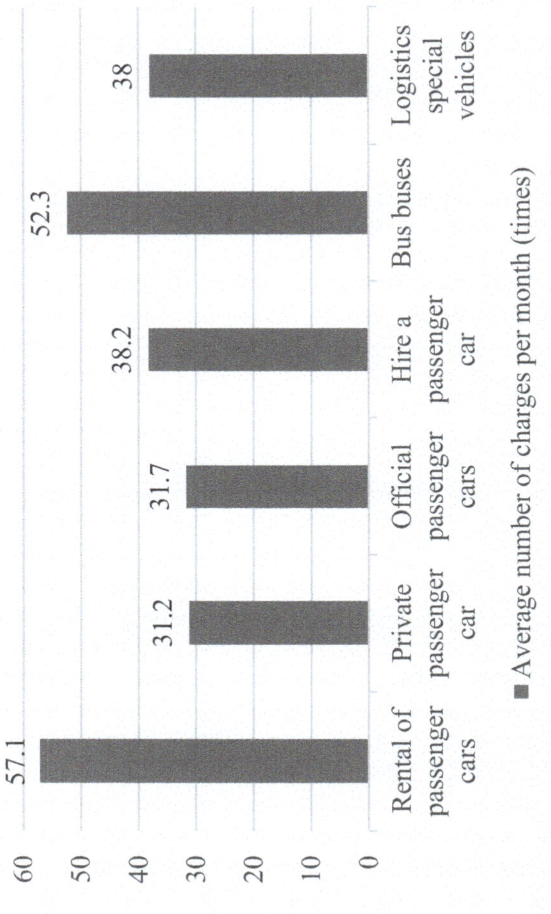


**Fig. 2.27** Average monthly charging times of different types of vehicles

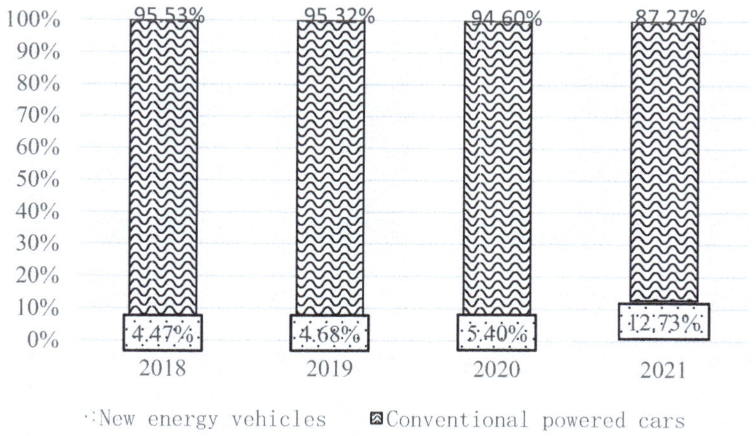


**Fig. 2.28** Trend change of sales proportion of new energy vehicles from 2018 to 2021

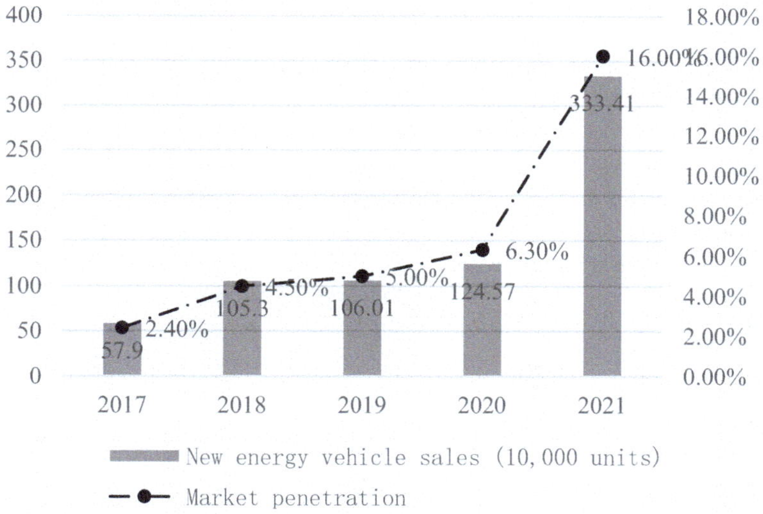


**Fig. 2.29** Change trend of sales volume and penetration rate of new energy vehicles from 2018 to 2021

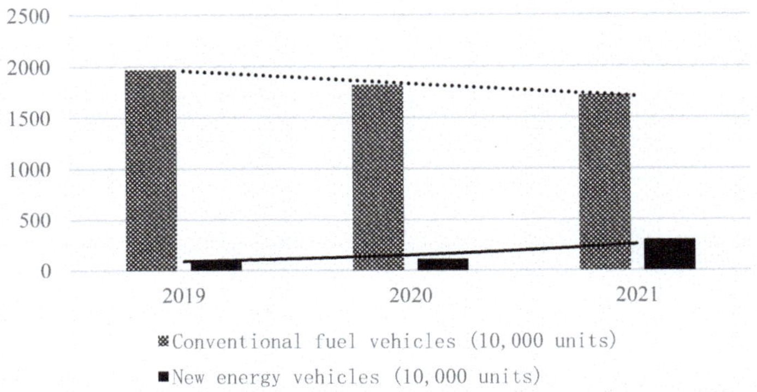


**Fig. 2.30** Retail trends of traditional fuel vehicles and new energy vehicles from 2019 to 2021

## 2.3 Analysis of Current Situation of Urban Transportation Energy Supply and Demand

### *2.3.1 Macro-Analysis of Urban Transportation Energy Supply and Demand Balance*

The overall situation of China's transportation energy supply shows a steady growth trend. The refined oil market in China develops rapidly, and the crude oil consumption increases gently. The oil industry has entered a period of accelerated reform, and crude oil imports have gradually declined. The number of domestic gas stations is increasing year by year. The growth rate of China's natural gas industry has slowed down, but with the growth of natural gas production, natural gas consumption still shows an upward trend, which stimulates the development of new energy vehicles to a certain extent. The domestic filling stations are mainly distributed in the coastal areas of East China and North China and the provinces with large natural gas consumption. China's new energy vehicles, especially pure electric vehicles, have shown a high development trend in the past five years, so the construction of charging infrastructure has received the attention of the state, the number of charging piles has been continuously improved, and the number of charging stations has also been growing. China has entered an all-round well-off society, the rapid development of economy, and the improvement of people's quality of life have made the number of vehicles continue to increase, resulting in an increase in energy demand. The supplementary energy demand of different types of motor vehicle travel is slightly different, and the future trend of motor vehicle types is becoming more and more obvious, and the development of new energy vehicles is the only way for China to move from an automobile country to an automobile power. China has put forward a large number of subsidy policies to promote the development of new energy vehicles, so the future demand for electricity and consumption will increase sharply, the consumption of fuel and natural gas will be relatively reduced. However, in the current car ownership, fuel vehicles account for 90%, and fuel consumption will still dominate in the future.

Relying on non-renewable energy to promote China's economic development has obviously failed to meet the needs of China's current national conditions. On the stage of future energy, the fossil energy era of oil and coal will gradually withdraw, and the "mixed energy era" of new energy and fossil energy complementary will be a new era that China will face in the process of future economic development. In the future, China's energy demand type is bound to realize the reform from fuel consumption to electric energy consumption. On the one hand, the burning of coal and oil is the main factor causing air pollution, especially the burning of loose coal is more harmful. On the other hand, although natural gas is cleaner than coal, it also releases some nitrogen oxides during its use, which is one of the important reasons for the formation of haze. Therefore, improving the use of electric energy in the terminal energy is indeed of great benefit to the prevention and control of air pollution. Although the overall situation of China's energy supply shows a steady upward trend, but our population base is large, the consumption of energy is also increasing. And the development momentum of new energy vehicles is fierce, the future penetration rate will continue to rise, in order to better support the promotion of new energy vehicles, accelerate the construction of charging piles, power stations, and other supporting infrastructure is imminent. As shown in Fig. 2.31, from the perspective of supporting demand, the charging demand for electric vehicles has increased significantly, and the current number of charging equipment in China is far from meeting the design requirements. Although the number of charging facilities in China is increasing year by year, there is still a big gap compared with the annual increase in the number of electric vehicles. The reality of this mismatch between supply and demand will cause problems for the travel of electric vehicle users.

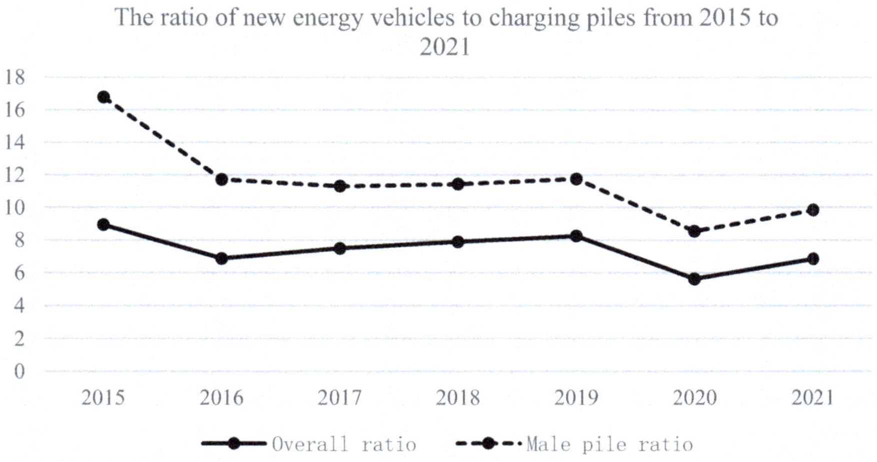


**Fig. 2.31** Ratio of new energy vehicles and charging piles from 2015 to 2021

### 2.3.2 Meso-Analysis of Urban Transportation Energy Supply and Demand Equilibrium

Road is the vein of a city, one of the most important infrastructures of a city, and also the basis and carrier of traffic operation. Each city has its own unique geographical environment and historical accumulation and gradually forms its own unique road network structure combined with its own location advantages. For example, Chengdu's "ring plus radiation" road network structure, Xi'an's "grid" road network structure, Chongqing's "narrow" road network structure, and so on. Urban road network structure also has an important impact on the construction of energy supply network, the location and layout of energy supply stations. The energy supply station is the base of filling fuel for vehicles, which needs to be arranged according to the structural characteristics of the city's road network. It should be widely distributed in the city along the urban vein, providing convenient and safe services for urban residents and minimizing the impact on traffic, environment, and safety. The network structure and layout of the supply station should be combined with the characteristics of the urban road network, and the demand characteristics and demand habits of residents should also be considered. With the gradual increase of disposable income of Chinese residents, the continuous improvement of citizens' pursuit of quality of life and the continuous improvement of various entertainment and catering facilities around them, the proportion of residents' demand for flexible travel has increased significantly. Energy replenishment behavior of residents is very regular, and vehicle energy replenishment behavior will be concentrated in one or several time periods, resulting in energy replenishment peak hours and idle hours. In particular, the charging peak hours generated by EV users' charging behaviors are highly coexisting with residents' daily electricity consumption peak hours, which poses new challenges to the road network and power grid system around energy replenishment supply stations.

By analyzing the distribution of supply station in Chongqing and combining relevant national policies, enterprise strategies, and Chongqing's economic development, it is found that there are unreasonable phenomena in the distribution of supply station in China. The layout method of supply station is not scientific, and the overall layout planning is not reasonable. Therefore, the irregular and unreasonable supply site layout cannot effectively match the daily energy demand of residents, which leads to the contradiction between supply and demand. Then through the analysis of residents' travel purpose and the characteristics of residents' demand for energy replenishment, it can be seen that there are obvious peak hours and idle hours of residents' energy replenishment behavior. During the peak period of energy replenishment, vehicle energy replenishment difficulty or long queue may be caused, which will affect local traffic around the supply station and cause traffic congestion, increase residents' energy replenishment time, and reduce residents' satisfaction with energy replenishment. Idle waste of facility resources in the replenishment supply station during the idle period will reduce the utilization efficiency of supply station. Therefore, it is very important to induce the vehicle to make up energy reasonably during the peak period, so as to alleviate the pressure of the road network and the power grid during the peak period. How to improve the utilization rate of the supply station and reduce the cost of the operator in the idle period should also be considered.

### *2.3.3 Negative Impact of Contradiction Between Supply and Demand of Urban Transportation and Energy*

Urban transportation energy supply network is a transportation equipment energy supply network composed of gas stations, filling stations, charging piles, and other facilities, which is the lifeline to ensure the development of urban transportation, and also an important infrastructure to support the flow of urban people and goods. The overall situation of China's transportation energy supply shows a steady growth trend, but due to the large population base in our country, the number of motor vehicles is increasing year by year, so the energy consumption continues to increase, resulting in the mismatch between transportation energy supply and demand, and the imperfect construction of energy infrastructure has intensified the contradiction between transportation energy supply and demand. The imbalance between supply and demand makes it easy for vehicles to queue up in the process of replenishing energy, which increases the cost of replenishing energy. Unreasonable site layout can easily lead to vicious competition between supply sites. The regularity of residents' energy replenishment behavior will lead to the surge of power grid and network pressure during peak hours, and the demand for energy replenishment of some vehicles cannot be met. The utilization efficiency of idle time replenishment supply station is low, which further increases the operating cost of the supplier. The large-scale replenishment behavior in the road network not only increases the travel time and replenishment cost of the driver but also easily causes congestion of local charging stations, and even affects the traffic, resulting in traffic jams. The queuing phenomenon caused by overload stations will cause traffic congestion on the main roads around the stations, and the congestion delay will aggravate energy consumption and cause serious pollution to the environment.

## 2.4 Future Transportation Demand Development Prospects

In October 2021, the Central Committee of the Communist Party of China and The State Council, in the Opinions on Fully, Accurately and Comprehensively Implementing the New Development Concepts and Doing a Good Job of Carbon Peaking and Carbon Neutrality and the Action Plan for Carbon Peaking before 2030, made arrangements for accelerating the construction of moderately advanced green transportation infrastructure and making overall use of comprehensive transport channel lines, land, and other resources. Orderly progress will be made in the construction of charging piles, supporting power grids, hydrogen refueling stations, and other infrastructure. In the context of "dual carbon", the exploration and practice of traditional fossil energy transformation in the field of transportation is an important part of green transportation. The "New Energy Automobile Industry Development Plan (2021–2035)" points out that the ecology of the new energy automobile industry has entered the "network ecology" involving multiple fields such as automobiles, energy, transportation, and information and communication. According to the report on the Integrated Development of Transportation and Energy 2022, it is expected that by 2030, the annual power generation of new energy in the transportation sector will reach more than 13 billion KWH, the total annual oil consumption will be saved by more than 25 million tons, and the total carbon emissions will be reduced by more than 20 million tons. It is expected that by 2060, the annual power generation of new energy in the transportation sector will reach more than 65 billion KWH, the total annual oil consumption will be saved by more than 120 million tons, and the total carbon emissions will be reduced by more than 100 million tons. In January 2022, the Implementation Opinions on Further Improving the Service Support Capacity of Electric Vehicle Charging Infrastructure issued by the National Development and Reform Commission and other departments clearly proposed to accelerate the construction and installation of charging facilities in residential communities, improve the charging and replacement support capacity in urban and rural areas, and strengthen the research and development and application of new technologies such as vehicle network interaction. In summary, since the "13th Five-Year Plan", China has introduced a number of policies to promote the construction of clean, low carbon, safe, and efficient transportation energy system, the future of China's transportation travel, will accelerate the transformation to the field of electrification, especially in the field of electric vehicles, to achieve vehicle, pile, network, storage intelligence coordination, to safer, greener, smarter travel goals.

In response to the global goal of "net zero emissions by 2050" set by the International Energy Agency, China has actively promoted the transformation of the transportation energy structure. Since 2018, China's traditional fuel vehicle demand has shown a continuous and rapid decline in successive years. The development momentum of new energy vehicles is fierce, among which electric vehicles occupy the main body. The introduction of relevant national and local level policies has effectively promoted the development of new energy vehicles, resulting in a sharp increase in China's electricity demand and consumption, and to a certain extent, alleviating the consumption pressure of non-renewable traditional energy sources such as fuel oil and natural gas. However, due to the large population base in China, with the increase of motor vehicle ownership year by year, the surge of energy consumption and the imperfect construction of energy replenishment infrastructure leads to the imbalance of transportation energy supply and demand. The unscientific layout of the replenishment site, unreasonable induction strategy and incentive mechanism of the replenishment site lead to the low utilization of the replenishment infrastructure. In short, the "mixed energy era" in which new energy and fossil energy complement each other is a new era that China will face in the course of future economic development. In the future, China's energy demand type is bound to realize the reform from fuel consumption to electric energy consumption.

However, China is also facing many problems and challenges in the transition of transportation energy demand. From the perspective of transportation and energy collaboration, the problems of high-power charging, power change, and vehicle network interaction standardization have not yet been cracked, and the potential of new energy vehicles to consume large-scale renewable energy needs to be tapped. From the perspective of the city, the problem of infrastructure placement in the urban space with scarce land resources has become a pain point restricting the development of transportation and energy infrastructure. From the perspective of the urban external transportation system, the spatial distribution of energy supply facilities is uneven, and it is still concentrated in the core urban agglomerations such as the Yangtze River Delta, Beijing-Tianjin-Hebei, and Pearl River Delta, which is difficult to meet the diversified demand of users during the holiday peak period. Finally, charging issues, range issues, and safety issues are still the main factors influencing consumers to buy electric vehicles. In the face of these challenges, it is necessary to carry out cutting-edge technology research in key areas such as the collaborative coupling mechanism of automobile and energy, power balance and network optimization control, and vehicle information interconnection in the future. Accelerate the construction of a policy standard system for orderly charging and vehicle network interaction, promote the application of new technologies and the exploration of business models, thereby driving the coordinated development of new energy vehicles and renewable transportation, and promote the adjustment of the transportation energy structure.

## 2.5 Summary of This Chapter

This chapter starts with the supply and demand of urban transportation energy supply network. Firstly, by collecting the relevant data on domestic urban transportation energy supply network, it sorts out and analyzes the current situation of domestic urban transportation energy supply and the structure of energy supply network. Based on the data of the current situation of the energy supply network in Chongqing, the classification, quantity, and distribution of energy supply stations are summarized, and the development trend of transportation energy supply stations is put forward. Secondly, by collecting the relevant data of domestic population and motor vehicle ownership, this paper analyzes the change in travelers' energy demand during urban travel, obtains the change trend of urban energy demand, and summarizes and analyzes the change trend of urban energy demand. Then, based on the collected data of urban energy supply and demand, the current balance of urban energy supply and demand in China is analyzed from the macro and meso perspectives, and a series of negative impacts caused by the contradiction between urban transportation energy supply and demand are summarized. Finally, the paper analyzes the current situation of transportation energy transformation in China based on relevant policies and gives some relevant suggestions on the problems and challenges faced by the transformation of transportation energy structure in China.

# Chapter 3
# Definition and Evaluation of Negative Effects of Urban Transportation Energy Supply Network

## 3.1 A Conceptual Definition of the Negative Effect of Urban Transportation Energy Supply Network

### *3.1.1 Definition of Negative Effects of Urban Transportation Energy Supply Network*

The effective operation of transportation energy supply is the basic basis for the normal running of vehicles, and also further guarantees the normal development of economy and society. However, due to the large-scale randomness of vehicle replenishment demand, improper site selection, insufficient energy supply, and other reasons, vehicle replenishment behavior will produce negative effects. The negative effect here refers to the adverse factors brought by urban energy supply terminal sites to people and the environment, which are reluctant to produce and need to avoid as much as possible. Specifically, due to the randomness of space and time in the urban traffic network, when the vehicle has the demand for supplementary energy, its driving destination, speed, and lane selection will change, which will affect the normal vehicles and traffic flow on the lane, resulting in behavioral disturbance. In the peak period of energy replenishment, vehicles at the supply station will queue up and wait, which will cause traffic congestion on the main roads around the station, which will aggravate energy consumption, thereby increasing carbon emission and causing serious pollution to the environment. At the same time, it will cause a waste of time and negative emotions for the individual driver, which will in turn interfere with the newly arrived replenishment vehicles and traffic flow, thus aggravating the negative effect. In addition, the waste of resources faced by idle sites also brings losses to urban economy and development.

 
X. Ge, *Urban Transportation Energy Supply Network*, Industrial Ecology and Environmental Management 5, https://doi.org/10.1007/978-3-031-64619-5_3

## *3.1.2 Manifestations of Negative Effects of Urban Transportation Energy Supply Network*

The negative effect of urban Traffic Energy Supply Network (TESN) refers to the adverse factors and negative effects of energy supply stations on people and environment brought by motor vehicles in the urban energy supply network.

Due to the randomness of energy replenishment demand and other reasons, problems such as the mismatch between supply and demand of energy replenishment behavior in time and space, traffic congestion, low efficiency of energy replenishment, and so on have brought a series of negative effects of energy replenishment, mainly in the form of traffic congestion and queuing at energy replenishment stations. The resulting negative effects are: exhaust emissions, time loss, fuel consumption, noise pollution, anxious driving, safety risks, etc.

According to whether negative effects can be quantified, negative effects are divided into two categories:

### 3.1.2.1 Quantifiable Negative Effects

#### Additional Carbon Emissions

From the point of view of social pollution, when the vehicle is queuing up for replenishment at the replenishment station, the engine will run at idle speed for a certain period of time, resulting in additional exhaust emissions, especially carbon emissions. In addition, a large number of vehicles enter and exit the energy replenishment station, causing disturbance to the road traffic flow where the energy replenishment station is located. When a vehicle completes the energy replenishment behavior and joins the main road from the on-ramp, the normal vehicles in the rear will slow down according to the safe vehicle distance, and then gradually accelerate to a stable speed. The disturbance may even propagate sequentially over multiple vehicles in the rear. Due to the disturbance of replenishment vehicles leaving the station, the normal vehicles on the road produce unnecessary deceleration and acceleration behavior, which makes the carbon emission of vehicles near the station increase. When there is a large number of vehicles on the road, this disturbance will even reduce the road capacity, increase the travel time of travelers, accumulate more carbon emissions on the road, increase the level of exhaust pollution of vehicles on the road, and cause physical harm to travelers and residents near the recharge station.

#### Additional Energy Consumption

From the perspective of road traffic, the extra fuel consumption of vehicles caused by supplementary energy stations is due to the disturbance of normal traffic flow caused by supplementary energy vehicles, resulting in traffic congestion, slowing down of traffic speed and other phenomena, which not only increases the travel time of travelers, but also increases the energy consumption of vehicles. In addition, from the perspective of travelers' travel path selection, considering the necessary replenishment demand, travelers' choice of travel path is greatly restricted. In order to meet the replenishment demand, some travelers choose relatively remote travel routes, which increases the energy consumption of travelers.

#### Transit Time Delay

Once the supplementary energy station is set up, the supplementary energy traveler will be limited in the choice of travel path for the consideration of supplementary energy. In addition, due to the establishment of the supplementary energy station, the traffic flow of the main road is bound to be affected when the supplementary energy vehicles enter and exit the supplementary energy station, and the bottleneck node of the traffic flow is formed. In some rush hours, such as commuting, the impact is more serious, such as traffic congestion and slow traffic flow. The traffic speed of the whole road is difficult to reach the ideal state of road design, which eventually leads to the slowdown of road traffic speed and the increase of traffic time.

#### Supplement Energy Time Growth

The randomness of vehicle replenishment demand and the difference of road grade lead to excess replenishment activities in some specific periods at the main road replenishment stations in commercial activity areas and urban commuting routes. Because the station replenishment service capacity is determined by the station replenishment facilities, site layout, and other factors, the replenishment service capacity has an upper limit, and the excessive vehicle replenishment demand forms the phenomenon of station replenishment queuing, increasing the waiting time of vehicle replenishment. For travelers in need of replenishment, these time increases are difficult to accurately estimate before the trip, and additional travel time costs for travelers.

#### Capacity Reduction

Due to the attraction of supplementary energy stations to supplementary energy vehicles, the density of road vehicles increases, road traffic jams, and the speed of vehicles is difficult to reach the standard speed of road design, resulting in the reduction of road capacity.

Supplement Hazardous Materials Risk at Energy Sites

Due to its special function, there are a large number of flammable and explosive dangerous goods in the site. Although the transmission and safety protection of oil, gas and high-voltage electric energy have been greatly improved, there is still a probability of danger. Once the risk occurs, it will cause great losses. The risk is assessed according to whether the facility layout of the replenishment site is reasonable and whether the security precautions are in place. The loss caused by the risk when it occurs varies with the surrounding environment.

Traffic Safety Risk

Due to the attraction of the replenishment station to the travel path of the replenishment vehicle, and the disturbance of the replenishment vehicle to the normal traffic flow, the crowded travel of vehicles leads to the increase of the probability of traffic safety accidents, resulting in the loss of personnel and property. Traffic risk can be reflected in the loss caused by traffic accidents in a period of time.

#### 3.1.2.2 Unquantifiable Negative Effects

Noise Pollution

The noise generated by motor vehicles entering and leaving the site in the urban energy supply station and the honking noise generated by the road attached to the station due to the traffic disturbance caused by the energy recharge vehicle have caused health hazards to the surrounding traffic travelers and nearby residents. Due to the long-term nature of the hazards, and the severity of the hazards varies according to the different physical conditions of the affected people, it is difficult to implement a real specific quantification.

Anxious Driving

In the peak traffic period, the traffic disturbance caused by the replenishment behavior of the vehicle at the replenishment station, the queuing of the replenishment vehicle, etc., causes the traveler (or the modifier) to pay more time in the road section, and this part of the extra time is not predictable or accurately evaluated by the traveler before the trip. Due to the unexpected delay caused by the travel time, drivers have anxious driving, which has a negative impact on the psychological emotions of drivers, and thus increases the probability of irrational driving behavior. Risk of harm to individuals and other traffic participants.

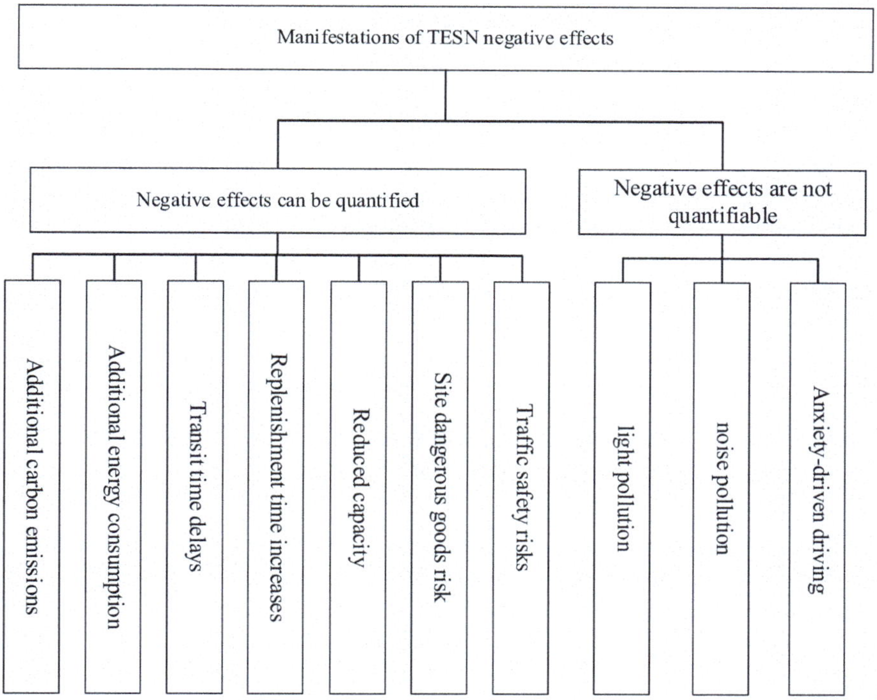


**Fig. 3.1** Manifestation of TESN negative effect

Light Pollution

In cities, multiple reflective devices on motor vehicles (such as rear-view mirrors, special paint for the body, etc.) reflect natural light, especially in sunny weather conditions. The attraction of supplementary energy stations to vehicles with supplementary energy demand, and the disturbance of supplementary energy vehicles to road traffic, lead to the increase of vehicle density near the supplementary energy stations, cause light pollution, and affect the normal activities of surrounding residents and even pedestrians. In addition, the continuous operation of the replenishment station at night also has a certain impact on the rest of the surrounding residents.

The negative effect of urban energy supply network is shown in Fig. 3.1.

## 3.2 Assessment of Negative Effects of Urban Transportation Energy Supply Sites

Urban transportation energy supply network system is a complex system, which involves many evaluation indexes such as road network infrastructure structure and vehicle replenishment behavior. Based on domestic and foreign research, the evaluation methods can be basically summarized into three categories, namely, subjective analysis, objective analysis, and subjective and objective combination analysis. The difference between the two mainly lies in the measurement and calculation of evaluation indicators and the determination of weights.

### *3.2.1 Selection of Negative Effect Evaluation Index of Urban Transportation Energy Supply Station*

The urban transportation energy supply system is one of the key nodes to maintain the operation of the urban transportation system, which integrates the energy demand vehicle, the urban energy supply station, and the relevant government departments. But the occurrence of vehicle replenishment as an extra behavior of motor vehicles will inevitably have a negative impact on the external environment. The construction of negative effect evaluation index of energy supply station is an important premise to produce objective and reasonable evaluation results. Based on the relevant impact studies and field investigations of various scholars on different aspects of urban transportation energy supply stations, this paper comprehensively considers the influencing factors of the negative effect evaluation of the stations, and constructs evaluation indicators according to the principles of systematization, data availability, operability and scientific convenience of quantitative methods. Aiming at the negative effect of urban transportation energy supply station, a comprehensive evaluation index is established from the two aspects of traffic efficiency impact and station service impact, as shown in Fig. 3.2.

### *3.2.2 Data Collection and Calculation of Urban Transportation Energy Evaluation Index*

After clarifying the evaluation index system of the negative effect of urban transportation energy supply, the numerical calculation of each index should be carried out according to the large amount of data collected by field research. The evaluation road section is the road section with transportation energy supply station, and the relevant calculation method is as follows:

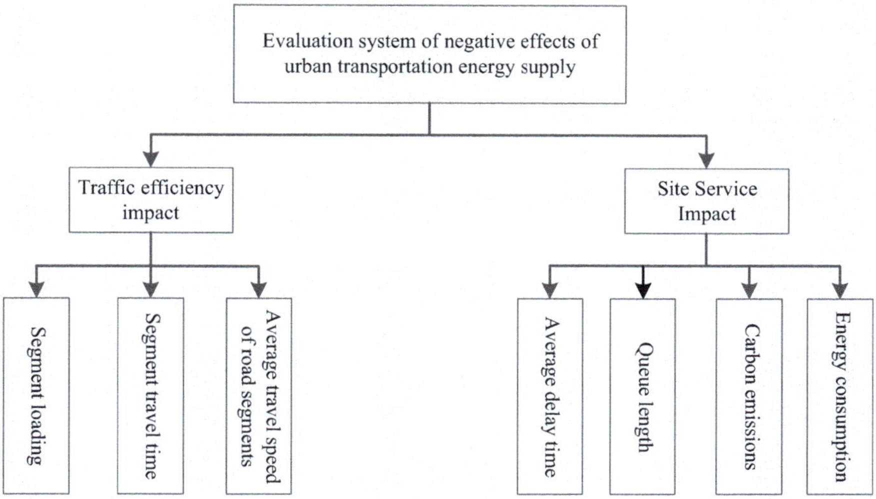


**Fig. 3.2** Index system of negative effect of urban transportation energy supply

#### 3.2.2.1 Impact on Traffic Efficiency

The influence of traffic efficiency represents the influence of the setting of transportation energy supply station on the traffic flow and traffic operation efficiency of the main road, which is mainly evaluated from the aspects of section load degree, section travel time, and section average travel speed.

The load degree of a road section refers to the ratio between the number of vehicles with supplementary energy demand and the maximum traffic volume of the road section, which reflects the traffic load of vehicles with supplementary energy demand on the road section. The higher the load degree, the greater the impact of vehicles with supplementary energy demand on the congestion risk of the road section:

$$\frac{v}{c} = \frac{Q_{ed}}{C_{es}} \tag{3.1}$$

where, $v/c$ is to evaluate the load degree of the road section; $Q_{ed}$ is the number of motor vehicles with supplementary energy demand per unit time (standard vehicles/hour); $C_{es}$ is the capacity per unit time (standard vehicle/hour) of the road section with energy supply station.

Section travel time refers to the time required to evaluate the actual passage of vehicles on the section:

$$T_{es} = t_{ex} - t_{en} \tag{3.2}$$

where, $T_{es}$ is the driving time of motor vehicles (hours) of the evaluated road section; $t_{ex}$ refers to the time (hours) for motor vehicles to leave the road section when the demand for supplementary energy is met; $t_{en}$ is the time (hours) required for vehicles to enter the road section.

Average road speed refers to the average road speed of all vehicles on the road:

$$\overline{V_{es}} = \frac{L_{es}Q_{es}}{\sum_{i=1}^{N} t_{es.i}} \tag{3.3}$$

where, $\overline{V_{es}}$ is the average speed of motor vehicles on the road section (km/h); $L_{es}$ is the length of the section (km); $Q_{es}$ is the traffic flow on the section in unit time (standard vehicles/hour); $t_{es.i}$ is the travel time (hours) of the $i$ motor vehicle passing through the road section in unit time.

#### 3.2.2.2 Site Services Are Affected

The impact of replenishing vehicles in energy supply stations is mainly evaluated in terms of average delay time, queue length, carbon emission, and energy consumption.

The average delay time refers to the difference between the actual time spent in the energy supply station replenishment process and the time spent in the free flow state, which reflects the queuing situation in the energy replenishment process. The average delay time is the average delay time of all replenished vehicles per unit time:

$$D_{ed} = \frac{1}{Q_{ed}} \sum_{i=1}^{n} \left( \frac{L_{ed}}{V_{ed.i}} - \frac{L_{ed}}{V_f} \right) \tag{3.4}$$

where, $D_{ed}$ is the average delay time (hours) of replenishing vehicles; $L_{ed}$ refers to the distance traveled by the vehicle at the energy supply station (km); $V_{ed.i}$ is the driving speed of the i replenishment vehicle in the energy supply station (km/h); $V_f$ is the speed of free flow vehicles in the energy supply station (km/h); $Q_{ed}$ is the volume of traffic in the energy supply station per unit time (standard vehicles per hour).

Queue length refers to the length of the road occupied by the supply station when the queue is generated by the vehicles in the energy supply station:

$$L_{eq} = (Q_{ed} - Q_{ed'})\overline{l_{ed}} \tag{3.5}$$

where, $L_{eq}$ is the queue length (kilometers) of vehicles in the energy supply station for replenishment demand per unit time; $Q_{ed}$ is the number of vehicles that complete the replenishment per unit time (standard vehicles/hour); $l_{ed}$ is the average body length of the supplementary vehicle (km).

Carbon emission refers to the evaluation of $CO_2$ emissions generated by the replenishment process of the vehicle in the energy supply station per unit time (Bowyer et al. 1984):

$$\mathrm{CE}_{ed} = \sum_{j} \mathrm{EC}_{ed}\mathrm{RF}_{ed.j}/1000 \tag{3.6}$$

where, $\mathrm{CE}_{ed}$ is the $CO_2$ emission (kg) of the replenishment vehicle in the energy supply station per unit time; $\mathrm{RF}_{ed.j}$ is the $CO_2$ emission factor for Class $j$ motor vehicles (kg/L).

Energy consumption refers to the evaluation of the energy consumption per unit time of the replenishment vehicle in the supply station during the replenishment process (Lucas et al. 2012):

$$\mathrm{EC}_{ed} = \gamma \sum_{j} \left(P_j L_{ed} \mathrm{FC}_j\right) \times 100 \tag{3.7}$$

where, $\mathrm{EC}_{ed}$ is the energy consumption per unit time of the replenishment vehicle in the energy supply station (kg); $\gamma$ is the conversion coefficient of standard coal; $P_j$ is the contribution rate of Class $j$ motor vehicles (%); $\mathrm{FC}_j$ is the energy consumption per unit of a Class $j$ motor vehicle (kg (100 km)).

### 3.2.3 *Negative Effect Evaluation Model of Urban Energy Supply Station*

#### 3.2.3.1 Standardization of Evaluation Index Data

Since the units and magnitude of each evaluation index data are different, it is necessary to standardize the evaluation index data before the negative effect evaluation of energy supply stations, so that they are at the same magnitude. Here, the range method is adopted. In addition, according to the different positive and negative characteristics of each evaluation index, different standardized formulas should be used. The positive indicator indicates that the larger the value of the indicator, the better, while the negative indicator is the opposite. The following formula is used to deal with it:

$$X_{po} = \frac{x_{ij} - x_{\min}}{x_{\max} - x_{\min}} \tag{3.8}$$

$$X_{ne} = \frac{x_{\max} - x_{ij}}{x_{\max} - x_{\min}} \tag{3.9}$$

where, $X_{po}$ and $X_{ne}$ are positive and negative indicators, respectively; $x_{\max}$ and $x_{\min}$ are the maximum and minimum values of the index sample, respectively.

#### 3.2.3.2 Indicator Weight Determination

There are many methods to determine the index weight, which are mainly divided into subjective weighting method and objective weighting method. Subjective empowerment methods are mainly graded by experts based on experience or individuals are interviewed in the form of questionnaires, which may weaken objectivity when empowering. For the negative effect evaluation of urban transportation energy supply station, the index value can be calculated by observation when the evaluation index system is established, so it is more appropriate to choose the objective weighting method than the subjective weighting method. Here, entropy weight method is more used in the objective weighting method. Calculating index entropy can objectively mine the amount of information and determine the index weight. The main steps are as follows:

Construct Evaluation Index Value Matrix

This book evaluates the negative effects of urban transportation energy supply stations based on real-time data collected from a large number of field investigations, and the value matrix is $n$ evaluation indicators for $m$ evaluation period.

$$R_{m\times n} = \begin{bmatrix} x_{11} & x_{12} & \cdots & x_{1n} \\ x_{21} & x_{22} & \cdots & x_{2n} \\ \vdots & \vdots & & \vdots \\ x_{m1} & x_{m2} & \cdots & x_{mn} \end{bmatrix}_{m\times n} \tag{3.10}$$

Determine the Entropy of Evaluation Indicators

According to $j$ period of evaluation, the entropy value $E_j$ of evaluation index can be determined. The entropy value represents the amount of information contained in the index and is the basis for weight determination (Wang et al. 2015).

$$x_{ij}\prime = \frac{1 + x_{ij}}{\sum_{i=1}^{m} \left(1 + x_{ij}\right)} \tag{3.11}$$

$$E_j = -(\ln m)^{-1} \sum_{i=1}^{m} x_{ij}\prime \cdot \ln x_{ij}\prime \tag{3.12}$$

where, if the calculation is equal to 0, in order to make sense of the formula, a minimum value is taken instead of 0 calculation.

**Table 3.1** Weight of each evaluation index

| Evaluation indicators | Symbol | Weight | Grade |
|---|---|---|---|
| Segment loading | $W_1$ | 0.121 | 3 |
| Segment travel time | $W_2$ | 0.124 | 1 |
| Average travel speed of road segments | $W_3$ | 0.050 | 9 |
| Average delay time | $W_4$ | 0.122 | 2 |
| Queue length | $W_5$ | 0.118 | 5 |
| Carbon emissions | $W_6$ | 0.120 | 4 |
| Energy consumption | $W_7$ | 0.051 | 8 |
| Noise pollution | $W_8$ | 0.045 | 11 |
| Pollutant impact | $W_9$ | 0.059 | 7 |
| Air quality | $W_{10}$ | 0.048 | 10 |
| Tank reserves | $W_{11}$ | 0.044 | 12 |
| Firefighting capabilities | $W_{12}$ | 0.097 | 6 |

Determine the Entropy Weight of Evaluation Indicators

After calculating the entropy value of the $j$-th evaluation index, the entropy weight $W_j$ of the $j$-th evaluation index can be determined, and the entropy weight represents the weight of the evaluation index.

$$W_j = \frac{1 - E_j}{n - \sum_{j=1}^{n} E_j} \quad (3.13)$$

Among them, the molecule is the $j$-th evaluation difference coefficient of the first index.

According to the above steps, the weights of indicators for the negative effect evaluation of urban transportation energy supply stations are obtained, as shown in Table 3.1. Road section travel time has the greatest negative effect on urban transportation energy supply station, followed by road section load and average delay time, and tank reserves have the least effect, which is also consistent with field research. According to the size and grade of the gas station, the basic level of the gas station at different levels is consistent. The low traffic efficiency of the intersection will cause delays and stops for the vehicles entering and leaving the station. The entry and exit of import and export vehicles will also have a great impact on the traffic efficiency of the main road at the intersection, resulting in queuing phenomenon, and increase the carbon emission and energy consumption of vehicles.

#### 3.2.3.3 Evaluation Model Based on GA-BP

BP neural network is a model that simulates biological nervous system and deals with nonlinear problems through continuous learning of sample data. According to the large amount of data and objectivity of the evaluation index of urban transportation energy supply station in the morning and evening peak hours, the evaluation model based on neural network algorithm is more appropriate. Although BP algorithm has many advantages, there are also some problems: (1) Determining the initial value problem: the initial threshold and weight of the neural network are often generated according to experience or randomly, which will directly affect the network; (2) Local optimal solution problem: Because the principle of the algorithm is gradient descent, the algorithm may inevitably appear local optimal solution problems for some problems. By using the global search ability of GA and reducing the influence of BP initial value on the network, the problem can be solved globally.

##### Determine the Negative Effect Evaluation Coefficient of Urban Transportation Energy Supply Station

The sample data here is the evaluation coefficient of the negative effect of urban transportation energy supply station. After analyzing many influencing factors, establishing evaluation index system and determining evaluation index weight coefficient, the evaluation coefficient of the negative effect of urban transportation energy supply station can be obtained.

$$A_j = \sum_{j=1}^{n} W_j X_j \tag{3.14}$$

##### Construct GA-BP Network

Three layers of typical BP neural networks are selected, namely an input layer, a hidden layer, and an output layer. The number of nodes in the output layer depends on the output value, and the transfer function uses the logsig function to output any number [0,1]. Setting of hidden layer according to the rule of hidden layer node number summarized by previous studies and reference formula, the results of hidden layer 4 to 12 layers were selected, as shown in Table 3.2. After comprehensive consideration, the number of hidden layer nodes was determined to be 8, and the purelin function was adopted. trainbfg algorithm was selected under the same error 0.00001 target. The genetic algorithm part is the process of selecting the best initial weights and thresholds. The length of chromosomes is determined by the number of connection weights and thresholds, in which there are $12 \times 8 + 8 = 104$ weights and thresholds from the input layer to the hidden layer, and $8 \times 1 + 1 = 9$ weights and thresholds from the hidden layer to the output layer, so the length of chromosomes

**Table 3.2** Comparison of the number of nodes in different hidden layers

| The number of hidden layer nodes | Mean squared error (MSE) | Average relative error |
|---|---|---|
| 4 | 0.0010447 | 8.66% |
| 5 | 0.00083492 | 7.18% |
| 6 | 0.0008527 | 9.07% |
| 7 | 0.00093837 | 8.88% |
| 8 | 0.00038434 | 5.52% |
| 9 | 0.0044637 | 16.82% |
| 10 | 0.0028892 | 16.95% |
| 11 | 0.0052009 | 19.40% |
| 12 | 0.0055022 | 23.43% |

**Table 3.3** GA-BP related parameter settings

| Name | Numeric value | Name | Numeric value |
|---|---|---|---|
| Input layer | 12 | Population size | 100 |
| Output layer | 1 | Number of iterations | 150 |
| Hidden layers | 8 | Minimum error | 0.00001 |
| Crossover rate | 0.85 | Number of trainings | 2000 |
| Rate of variation | 0.001 | Learning rate | 0.1 |

is 113. The weights and thresholds of BP are converted into populations composed of multiple individuals by real number coding method. The relevant parameters of GA-BP are shown in Table 3.3. The individual with the highest fitness obtained by genetic algorithm is the initial weight and threshold, which also makes the precision of network training the highest.

### Divide the Negative Effect Evaluation Level of Urban Transportation Energy Supply Station

In order to express the negative effect of urban transportation energy supply clearly, the negative effect of urban transportation energy supply is divided into five levels according to the evaluation coefficient. For different transportation energy supply stations, their negative effect levels can be intuitively analyzed according to their evaluation coefficients, as shown in Table 3.4. Level 1 indicates that the supplementary energy flow overflows the energy supply station and seriously disturbs the traffic flow on the main road. Intensive supplementary energy vehicles run and stop, resulting in large energy consumption and carbon emissions. Level 2 indicates that the replenishment vehicles queue up inside the energy supply station and reach the maximum capacity of the energy supply station, which will also produce various

**Table 3.4** Evaluation criteria for negative effects of energy supply in urban transportation

| Grade | Coefficients interval | The degree of negative effects |
|---|---|---|
| 1 | (0.8,1.0] | Very serious |
| 2 | (0.6,0.8] | More serious |
| 3 | (0.4,0.6] | Generally severe |
| 4 | (0.2,0.4] | Less serious |
| 5 | (0.0,0.2] | Not serious |

pollution and energy consumption; Level 3 indicates that the replenishment vehicles have a slight queuing phenomenon at the energy supply station, but the waiting time is short and no vehicles stop on the branch road of the station; Level 4 indicates that the replenishment vehicle basically realizes the replenishment and travel, the traffic flow in the station is relatively smooth, and the pollutants and carbon emissions are maintained within a reasonable range; Level 5 indicates that there is no continuous replenishment vehicle in the energy supply station, the replenishment behavior is scattered or no replenishment behavior occurs, and the pollution and energy consumption are small because the number of replenishment vehicles is small.

## 3.3 Assessment of Negative Effects of Urban Transportation Energy Supply Network

### *3.3.1 Selection of Negative Effect Indicators of Urban Transportation Energy Supply Network*

The negative effect of urban transportation energy supply station mainly refers to the negative impact of the station on the surrounding environment during the operation process, including the vehicle recharge behavior and the station operation and construction itself. Vehicles on urban roads are divided into electric vehicles and fuel vehicles, so transportation energy supply sites can be divided into two categories: rapid refueling stations and charging stations. Rapid refueling stations can quickly fill fossil fuels with a typical refueling time of less than five minutes. In contrast, charging stations take longer to recharge, taking around two hours even for fast charging. This process is bound to cause time delays when there are vehicles queuing, carbon dioxide emissions, energy consumption and surrounding noise, site safety, and other negative effects. Because there are many influencing factors involved in the negative effect, the evaluation model can be used to analyze the periodic change of the negative effect of the site in the time period.

Based on the relevant researches and field investigations of various scholars on different aspects of urban transportation energy supply stations, and comprehensively considering the factors affecting the negative effect evaluation of vehicle replenishment behavior, a total of 12 influencing factors were statistically divided from four aspects: traffic efficiency impact, station service impact, environmental impact, and safety risk impact. After the normalization of the unified dimension, the weight relationship of 12 influencing factors can be obtained by using the entropy weight method, as shown in Table 3.1.

Due to the mutual influence and correlation of each index factor, according to the principles of comprehensiveness and measurement convenience, some factors were combined and discarded, and finally three indicators were selected as representative indicators to measure the negative effect of urban transportation energy supply, namely, road section travel time, average delay time, and carbon emission.

Section travel time refers to the time required for vehicles to actually pass the section before and after the section with the energy supply station node:

$$T_i = \frac{l_a}{v_a} - \frac{l_{a+1}}{v_{a+1}} \tag{3.15}$$

where, $T_i$ is the vehicle traveling time; $l_a$ is the length of the section before the station; $v_a$ is the average driving speed of the section before the station. Then the road section travel time cost $C_t^s$ is:

$$C_t^s = p_v T_i = p_v\left(\frac{l_a}{v_a} - \frac{l_{a+1}}{v_{a+1}}\right) \tag{3.16}$$

where $p_v$ represents the unit time cost of the vehicle.

The average delay time is the average of the delay time of all replenishing vehicles in the unit time, including queuing time and replenishing time. The vehicle replenishing delay time can be obtained according to the queuing model $M/M/S$ (Low 1974):

$$\rho = \frac{x_{j.d}}{\kappa_j \mu} \tag{3.17}$$

$$P_0 = \left[\sum_{k=0}^{\kappa_j - 1} \frac{1}{k_j!}\left(\frac{x_{j.d}}{\mu}\right)^k + \frac{1}{\kappa_j!}\frac{1}{1-\rho}\left(\frac{x_{j.d}}{\mu}\right)^{\kappa_j}\right]^{-1} \tag{3.18}$$

$$L_q = \frac{P_0 \rho (\kappa_j \rho)^{\kappa_j}}{\kappa_j!(1-\rho)^2} \tag{3.19}$$

$$W_{\kappa_j} = \frac{L_q}{x_{j.d}} + \frac{1}{\mu} \tag{3.20}$$

where, $\rho$ represents the system service intensity; $x_{j.d}$ is the replenishment flow per unit time arriving at the replenishment site $j$; $\kappa_j$ is the number of independent servers at site $j$; $\mu$ is the service rate per unit time; $P_0$ indicates the server idle rate. $L_q$

represents the average vehicle waiting time; $W_{\kappa_j}$ represents the average vehicle delay time.

Then the total delay time cost $C_d^s$ can be expressed as:

$$C_d^s = p_v W_{\kappa_j} \tag{3.21}$$

Carbon emissions generated by vehicles driving in the station or frequently starting idle due to queuing are also called cold emissions. Emission cost $C_c^s$ can be expressed as (Macrina 2019):

$$C_c^s = f_c A_i N_i D_{ed} \gamma_{i.d} \tag{3.22}$$

where, $f_c$ is the unit cost of carbon equivalent emissions; $A_i$ is fuel consumption per hour at idle; $\gamma_i$ is the carbon emission conversion coefficient.

According to the weight relationship between the three indicators and the normalization processing, the negative effect coefficient of site $j$ is calculated:

$$NE_j = W_t C_t^s + W_d C_d^s + W_c C_c^s \tag{3.23}$$

By measuring the index data of different stations in each time period, the negative effect coefficient of all stations in the region over time can be obtained as the original data sample, and a cycle can be divided into multiple time periods according to the change relationship of negative effects. The neural network algorithm can be used to evaluate the change trend of negative effects of stations in the region.

### 3.3.2 Assessment of the Negative Effect of Vehicle Routing for Urban Traffic Energy Supplement

The negative effects of vehicle replenishment behavior mainly include vehicle energy consumption, delay time, and driver's psychological anxiety. The additional mileage is incorporated into the total mileage of the driver and the energy consumption is combined to calculate the driving cost of the driver to quantify the negative effect of the driver's path. Therefore, the urban road is considered as a large traffic network $G(N, A, S)$, where $N$ is the set of network nodes, $A$ is the set of road sections in the network, and $S$ is the energy supply station set in the network nodes. Set any entrance to any exit of the network as an OD pair, and divide the replenishment stations supported by the vehicles with replenishment requirements according to the remaining mileage and driving direction of the vehicles after entering the area. For all OD $w \in W$ daily travel vehicles in the network, the traffic flow on route $p \in P$ is mainly composed of three types of vehicles, namely vehicles with no supplementary energy demand, electric vehicles requiring supplementary energy, and fuel vehicles requiring supplementary energy. The path passed by the vehicle between the start and end pairs is $P^w$, and the vehicle returns to its OD path from generating energy demand and completing energy replenishment to its OD path is $p\prime^w$. When the vehicle has no alternative stations on its OD path, path deviation will occur, resulting in additional mileage $\omega_0$, and the vehicle will be updated with a new driving path $\overline{P^w}$, where $p^{'w} \in \overline{P^w}$ and path length $L(\overline{p^w}) \geq L(p^w)$. Therefore, the additional mileage generated by the path deviation is $\omega_0 = L(\overline{p^w}) - L(p^w)$, where $L(\overline{p^w}) = L(p^w)$, $\omega_0 = 0$, indicates that the vehicle is supported by the station on the original OD path of the vehicle after the replenishment demand is generated.

Assuming that the vehicle can recharge energy at most once during the travel process and the demand for replenishment energy of different types of vehicles is related to its flow rate, the ratio between electric vehicle and fuel vehicle is set as r, where the charging ratio of electric vehicle is related to the remaining mileage of the vehicle $\xi_0$ and the psychological anxiety degree of the driver $\psi$ is set as $\alpha_e^w$, the formula is as follows:

$$\alpha_e^w = \frac{2L_e^w}{\xi_0 \psi} \tag{3.24}$$

$$L_e^w = \frac{\sum\limits_{p \in P^w} f_{p.e}^w L(\overline{p^w})}{\sum\limits_{p \in P^w} f_{p.e}^w} \tag{3.25}$$

where $L_e^w$ is the weighted average travel distance of the electric vehicle between the OD pairs, and the charging distance of the daily travelling vehicle during the round trip should be less than the remaining vehicle mileage.

According to the negative effect changes between different time periods $t \in T$ of energy supply stations on path $p \in P$, each time period will be iteratively allocated on the basis of the previous time period. If the vehicles that need to be replenished in the previous period $\Delta t$ fail to arrive at the station for replenishment after the end of the time cycle, they will be allocated together with the vehicles that need to be allocated for replenishment in the next time cycle $\Delta t + 1$, , and the road flow of each time period can be obtained, and the negative effect of the vehicle traveling path can be calculated. Mileage energy consumption needs to be calculated separately according to different vehicle types. According to the Bureau of Public Road (BPR) function, route $p \in P$ consists of section $a \in A$, and the length $l_a$ of each section is often different in the actual road network, and the capacity varies according to different road grades (Bureau of Public Roads, 1964):

$$t_{a.t}(x) = t_{a.t}^0 \left[ 1 + 0.15 \left( \frac{x_{a.t}}{c_{a.t}} \right)^4 \right] \tag{3.26}$$

$$C_{a.t}(x) = C_{a.t}^0 \left[ 1 + 0.15 \left( \frac{x_{a.t}}{c_{a.t}} \right)^4 \right] \tag{3.27}$$

where $t_{a.t}(x)$ is the travel time of time period $t \in T$; $t_{a.t}^0$ is the free flow time of section $a \in A$ in time period $t \in T$; $c_{a.t}$ is the capacity of $t \in T$ sections in the time period; $C_{a.t}^0$ is the free flow cost of section $a \in A$ in time period $t \in T$.

### *3.3.3 Negative Effect Evaluation Model of Urban Transportation Energy Supply Network*

Neural network is the core part of the forward neural network, but there are some defects, such as slow learning convergence, unguaranteed convergence to the global minimum point, and uncertain network structure. Genetic algorithm optimization neural network BP mainly optimizes the weights and thresholds, and then BP is trained and evaluated. A three-layer classical network is used here: an input layer, a hidden layer, and an output layer. The transfer function uses tansig function, the hidden layer node is selected as 5, and the gradient descent method is used for training. The part of the transmission algorithm is the process of selecting the optimal initial weight and threshold. The length of chromosome is determined by the number of link weights and thresholds, and the weights and thresholds of BP are converted into multi-individual populations by real coding method. This method is suitable for network evaluation with complex internal mechanism, and can help to evaluate the negative effect of urban transportation energy supply network.

Prior to the assessment, it is necessary to collect raw data of urban transportation energy supply sites in the weekday peak area, and also observe the overall trend of traffic flow for negative cost calculation. According to the trend characteristics of traffic flow of traffic energy stations, it can be roughly inferred that the trend of negative effect should be consistent with the traffic flow, and the periodic traffic flow will show obvious time series periodic characteristics. The final evaluation results can be obtained by GA-BP algorithm, as shown in Table 3.5.

**Table 3.5** Comparison of the number of nodes in different hidden layers

| The number of hidden layer nodes | 3 | 4 | 5 | 6 | 7 |
|---|---|---|---|---|---|
| Absolute average error | 0.318 | 0.279 | 0.272 | 0.298 | 0.318 |

### 3.3.4 Construction of Negative Effect Model of Urban Transportation Energy Supply Network During Holidays

#### 3.3.4.1 Model Hypothesis and Symbol Description

Based on the time series, the model allocates the replenishment demand of vehicle $q^w_{\Delta t}$ in the current time period $t \in T$. The allocation method is based on the path selection with the minimization of the negative effect, so the cost of each part of the negative effect is used as the side weight. When the weights are the same, the allocation is based on the shortest path principle among alternative paths with the same weights. Suppose that the side weight of section $a \in A$ is the sum of the side weight of section $t \in T$ and the weight of the energy supply station, where section $a_j$ is before station $S_j$, that is, the replenishment vehicle passes from section $a_j$ to station $S_j$, then the side weight of section $a \in A$ is:

$$W_{a.t} = \eta(C_{a.t}(x_a) + Fl_a p_v) + (1 - \eta)\mathrm{NE}^{a.t}_{S_j}\delta_a \tag{3.28}$$

where $\eta$ is the weight coefficient and $F$ is the energy consumption cost per unit section length; $\delta_a$ is the variable 0–1, $\delta_a$ is 1 indicates that there is an energy supply site, 0 indicates that there is no energy supply site.

The symbol and meaning of the model are shown in Table 3.6.

#### 3.3.4.2 Hybrid Road Network Equilibrium Model

In order to obtain the flow on the urban traffic network under the special demand flow during holidays, an improved hybrid road network equilibrium model is adopted, which is a time-serial-based dynamic vehicle replenishment allocation solution model, as follows:

$$\min Z_t(x) = \sum_{t\in T}\left(\sum_{a\in A}\int_0^{x_{a.t}} C_{a.t}(x)dx + \sum_{a\in A} F_e l_a x^e_{a.t} p^e_v + \sum_{a\in A} F_g l_a x^g_{a.t} p^g_v\right) \tag{3.29}$$

s.t.

$$\sum_{w\in W}\sum_{p\in P^w} f^w_{p.g} = q^w_{g.t} \tag{3.30}$$

$$\sum_{w\in W}\sum_{p\in P^w} f^w_{p.e} = q^w_{e.t} \tag{3.31}$$

$$\sum_{w\in W}\sum_{p\in P^w} f^{wd}_{p.g} = r\alpha^w_e q^w_{g.t} \tag{3.32}$$

**Table 3.6** Description of model symbols and meanings

| *Gather* | *Meaning* |
|---|---|
| A | A collection of road segments in a network $a \in A$ |
| P | A collection of paths in the network $p \in P$ |
| N | A collection of network nodes |
| S | A collection of energy supply sites in a network node $s \in S, S \in N$ |
| W | A collection of vehicle travel OD pairs in the network $w \in W$ |
| D | There is a collection of vehicles in the network that need to replenish energy $\forall d \in D$ |
| *Parameter* | *Meaning* |
| $\alpha_e^w$ | OD has a charging demand ratio for electric vehicles between $w \in W$ |
| $r$ | The ratio factor of electric vehicles to fuel vehicles |
| $F_e$ | Electricity cost per section length |
| $F_g$ | Fuel cost per section length |
| $p_v^e$ | EV driver unit time value |
| $q_t^w$ | The total traffic flow between OD to $w \in W$ in time period $t \in T$ |
| $\delta_{a.p}^w$ | On the path $p \in P^w$ between OD to $w \in W$, there is no need for replenishment for road section $a \in A$, and the vehicle pass is 1, otherwise it is 0 |
| $\delta_{a.p}^{ws}$ | Between OD to $w \in W$, on path $p \in P^w$ to road section $a \in A$, the vehicle needs to replenish energy at station $s \in S$, otherwise 0 |
| *Variable* | *Meaning* |
| $x_{a.t}$ | The total amount of traffic on path $a \in A$ within time period $t \in T$ |
| $x_{a.t}^g$ | The flow of fuel vehicles on path $a \in A$ within time period $t \in T$ |
| $x_{a.t}^e$ | EV traffic over time period $t \in T$ on Path $a \in A$ |
| $x_{a.t}^d$ | Traffic on path $a \in A$ that needs to be recharged during time period $t \in T$ |
| $x_{a.t}^{e.d}$ | EV traffic on Path $a \in A$ that needs to be refueled during time period $t \in T$ |
| $x_{a.t}^{g.d}$ | The flow of fuel vehicles on Path $a \in A$ that needs to be refueled during time period $t \in T$ |
| $q_{g.t}^w$ | Fuel vehicle traffic over time period $t \in T$ |
| $q_{e.t}^w$ | EV traffic over time period $t \in T$ |
| $f_{p.g}^w$ | The flow of fuel vehicles on path $p \in P^w$ between OD pairs $w \in W$ over time period $t \in T$ |
| $f_{p.e}^w$ | The OD is on path $w \in W$ between path $p \in P^w$ and EV traffic within time period $t \in T$ |
| $f_{p.g}^{wd}$ | The flow of fuel vehicles on path $p \in P^w$ between OD pair $w \in W$ that needs to be refueled during time period $t \in T$ |
| $f_{p.e}^{wd}$ | EV traffic on path $p \in P^w$ between OD pair $w \in W$ that needs to be refueled in time period $t \in T$ |

$$\sum_{w \in W} \sum_{p \in P^w} f_{p.e}^{wd} = \alpha_e^w q_{e.t}^w \tag{3.33}$$

$$q_{g.t}^w + q_{e.t}^w = q_t^w \tag{3.34}$$

$$x_{a.t}^g = \sum_{w \in W} \sum_{s \in S} \sum_{p \in P^w} f_{p.g}^w \delta_{a.p}^w, \ \forall a \in A, \ \forall w \in W \tag{3.35}$$

$$x_{a.t}^e = \sum_{w \in W} \sum_{s \in S} \sum_{p \in P^w} f_{p.e}^w \delta_{a.p}^w, \ \forall a \in A, \ \forall w \in W \tag{3.36}$$

$$x_{a.t}^{e.d} = \sum_{w \in W} \sum_{s \in S} \sum_{p \in P^w} f_{p.e}^{wd} \delta_{a.p}^{ws}, \ \forall d \in D, \ \forall a \in A, \ \forall w \in W \tag{3.37}$$

$$x_{a.t}^{g.d} = \sum_{w \in W} \sum_{s \in S} \sum_{p \in P^w} f_{p.g}^{wd} \delta_{a.p}^{ws}, \ \forall d \in D, \ \forall a \in A, \ \forall w \in W \tag{3.38}$$

$$x_{a.t}^d = x_{a.t}^{e.d} + x_{a.t}^{g.d}, \ \forall a \in A, \ \forall d \in D \tag{3.39}$$

$$x_{a.t} = x_{a.t}^g + x_{a.t}^e + x_{a.t}^d, \ \forall a \in A, \ \forall d \in D \tag{3.40}$$

$$f_{p.g}^w, f_{p.e}^w, f_{p.e}^{wd}, f_{p.g}^{wd} \geq 0, \ \forall w \in W, \ \forall d \in D, \ \forall a \in A, \ \forall s \in S \tag{3.41}$$

$$\delta_{a.p}^w, \delta_{a.p}^{ws} \in \{0, 1\}, \ \forall w \in W, \ \forall a \in A, \ \forall s \in S \tag{3.42}$$

Constraints (3.30) to (3.34) are traffic conservation constraints between OD pairs. The constraints (3.35) to (3.38) are the relational constraints between the path flow and the section flow. The constraints (3.34), (3.39), and (3.40) are quantitative constraints. The constraint (3.41) is a non-negative constraint on the path traffic. The constraint (3.42) is a 0–1 variable. The objective function is the sum of the cumulative travel time cost and the energy consumption cost of the travel path on all road sections, and its main function is to make the solution result of the model meet the user equilibrium condition. The model is solved by the Frank-Wolfe algorithm, and the results of section allocation based on urban transportation energy network under time series are obtained.

## 3.4 Examples of Evaluation of Negative Effects of Urban Transportation Energy Supply Sites and Networks

### *3.4.1 Examples of Negative Effect Evaluation of Urban Transportation Energy Supply Sites*

A CNPC refueling station on the main road of Nanan District of Chongqing City was selected and combined with Baidu Map intelligent transportation, taking every hour as a unit time period, relevant data collection and calculation of established evaluation indicators were carried out to study the negative effects of the unit time period with an example.

#### 3.4.1.1 Standardization of Site Evaluation Index Values

According to the variation law of traffic flow over time, it is divided into morning peak period (7:00–9:00) and evening peak period (17:00–19:00). With the hourly indicator data of this station as the input vector, a total of 30 h of data from morning peak period to the end of evening peak period (7:00–22:00) of two working days are collected and sorted. After standardized processing, the negative effect evaluation value matrix dataset of urban transportation energy supply stations is obtained, as shown in Table 3.7.

#### 3.4.1.2 Evaluation of Negative Effect Coefficient

The weight of each index in Table 3.8 is determined according to the entropy weight method, and the expected value of the negative effect coefficient of urban transportation energy supply station is obtained through calculation. The evaluation index values of dimensionless processing were input into GA-BP algorithm in the form of a $12 \times 30$ evaluation matrix, and the first 20 rows of data were taken as the training set and the last 10 rows of data were taken as the test set to obtain the evaluation output of GA-BP algorithm, as shown in Table 3.8. With the change of time, the negative effect level is also changing, but because the station is located in the urban trunk expressway with large traffic flow, its negative effect level is mainly maintained at levels 3 to 4. The result period was from 12 noon to 10 p.m., the evening peak was concentrated at 5 p.m. and continued to 8 p.m., and the negative effect level was 3. In addition, it can be seen that more vehicles participate in replenishment after 8 o'clock than at noon, which is consistent with the actual research situation.

**Table 3.7** Standardization of evaluation index values

| | $W_1$ | $W_2$ | $W_3$ | $W_4$ | $W_5$ | $W_6$ | $W_7$ | $W_8$ | $W_9$ | $W_{10}$ | $W_{11}$ | $W_{12}$ |
|---|---|---|---|---|---|---|---|---|---|---|---|---|
| $T_1$ | 0.628 | 0.286 | 0.605 | 0.836 | 0.861 | 0.852 | 0.889 | 0.888 | 0.888 | 0.696 | 0.113 | 0.955 |
| $T_2$ | 0.637 | 0.508 | 0.255 | 0.849 | 0.874 | 0.865 | 0.903 | 0.902 | 0.902 | 0.747 | 0.099 | 0.955 |
| $T_3$ | 0.450 | 0.453 | 0.321 | 0.574 | 0.580 | 0.578 | 0.586 | 0.586 | 0.586 | 0.619 | 0.415 | 0.955 |
| $T_4$ | 0.501 | 0.352 | 0.473 | 0.650 | 0.661 | 0.657 | 0.673 | 0.673 | 0.673 | 0.703 | 0.327 | 0.955 |
| $T_5$ | 0.355 | 0.199 | 0.845 | 0.436 | 0.431 | 0.433 | 0.426 | 0.426 | 0.426 | 0.637 | 0.574 | 0.955 |
| $T_6$ | 0.220 | 0.156 | 1.000 | 0.238 | 0.219 | 0.226 | 0.197 | 0.197 | 0.198 | 0.540 | 0.801 | 0.955 |
| $T_7$ | 0.289 | 0.165 | 0.964 | 0.339 | 0.327 | 0.331 | 0.314 | 0.314 | 0.314 | 0.488 | 0.685 | 0.045 |
| $T_8$ | 0.243 | 0.259 | 0.669 | 0.272 | 0.255 | 0.261 | 0.236 | 0.236 | 0.236 | 0.443 | 0.763 | 0.955 |
| $T_9$ | 0.314 | 0.348 | 0.481 | 0.375 | 0.366 | 0.369 | 0.356 | 0.356 | 0.356 | 0.428 | 0.643 | 0.045 |
| $T_{10}$ | 0.567 | 0.424 | 0.359 | 0.746 | 0.765 | 0.758 | 0.785 | 0.785 | 0.784 | 0.340 | 0.217 | 0.500 |
| $T_{11}$ | 0.510 | 0.596 | 0.168 | 0.662 | 0.674 | 0.670 | 0.688 | 0.688 | 0.688 | 0.302 | 0.313 | 0.955 |
| $T_{12}$ | 0.618 | 0.750 | 0.054 | 0.821 | 0.844 | 0.836 | 0.870 | 0.870 | 0.870 | 0.373 | 0.131 | 0.955 |
| $T_{13}$ | 0.286 | 0.485 | 0.281 | 0.335 | 0.323 | 0.327 | 0.309 | 0.310 | 0.310 | 0.624 | 0.690 | 0.955 |
| $T_{14}$ | 0.171 | 0.336 | 0.502 | 0.166 | 0.141 | 0.150 | 0.114 | 0.114 | 0.114 | 0.774 | 0.885 | 0.955 |
| $T_{15}$ | 0.104 | 0.276 | 0.630 | 0.067 | 0.035 | 0.047 | 0.000 | 0.000 | 0.001 | 0.774 | 0.998 | 0.955 |
| $T_{16}$ | 0.969 | 0.331 | 0.512 | 0.933 | 0.965 | 0.953 | 1.000 | 1.000 | 0.999 | 0.807 | 0.002 | 0.500 |
| $T_{17}$ | 0.612 | 0.642 | 0.130 | 0.765 | 0.784 | 0.777 | 0.806 | 0.806 | 0.805 | 0.646 | 0.196 | 0.500 |
| $T_{18}$ | 0.061 | 0.543 | 0.218 | 0.462 | 0.459 | 0.460 | 0.456 | 0.456 | 0.456 | 0.452 | 0.544 | 0.500 |
| $T_{19}$ | 0.076 | 0.364 | 0.453 | 0.305 | 0.291 | 0.296 | 0.275 | 0.275 | 0.275 | 0.202 | 0.724 | 0.500 |
| $T_{20}$ | 0.047 | 0.219 | 0.782 | 0.372 | 0.363 | 0.366 | 0.353 | 0.353 | 0.353 | 0.353 | 0.647 | 0.500 |
| $T_{21}$ | 0.272 | 0.160 | 0.986 | 0.092 | 0.062 | 0.073 | 0.029 | 0.029 | 0.030 | 0.140 | 0.969 | 0.500 |
| $T_{22}$ | 0.162 | 0.196 | 0.853 | 0.148 | 0.122 | 0.132 | 0.094 | 0.094 | 0.094 | 0.108 | 0.904 | 0.500 |
| $T_{23}$ | 0.071 | 0.340 | 0.496 | 0.283 | 0.267 | 0.272 | 0.249 | 0.249 | 0.249 | 0.218 | 0.750 | 0.500 |
| $T_{24}$ | 0.084 | 0.408 | 0.383 | 0.160 | 0.135 | 0.143 | 0.107 | 0.107 | 0.107 | 0.058 | 0.891 | 0.500 |
| $T_{25}$ | 0.097 | 0.471 | 0.298 | 0.350 | 0.339 | 0.343 | 0.327 | 0.327 | 0.327 | 0.000 | 0.672 | 0.500 |
| $T_{26}$ | 0.685 | 0.659 | 0.117 | 0.798 | 0.820 | 0.813 | 0.845 | 0.844 | 0.844 | 0.425 | 0.157 | 0.500 |
| $T_{27}$ | 0.413 | 0.844 | 0.000 | 0.765 | 0.784 | 0.777 | 0.806 | 0.806 | 0.805 | 0.327 | 0.196 | 0.500 |
| $T_{28}$ | 0.092 | 0.544 | 0.217 | 0.574 | 0.580 | 0.578 | 0.586 | 0.586 | 0.586 | 0.925 | 0.415 | 0.955 |
| $T_{29}$ | 0.196 | 0.356 | 0.465 | 0.384 | 0.375 | 0.378 | 0.366 | 0.366 | 0.366 | 0.970 | 0.634 | 0.955 |
| $T_{30}$ | 0.031 | 0.276 | 0.629 | 0.272 | 0.255 | 0.261 | 0.236 | 0.236 | 0.236 | 1.000 | 0.763 | 0.955 |

#### 3.4.1.3 Result Analysis

In GA-BP algorithm, through the training of different algorithms and the result matching, as shown in Table 3.9, trainbfg algorithm was found to have better performance, the training was stopped when the training number reached 26 times, the error convergence was stable, the result was better, as shown in Fig. 3.3, and the value of R correlation coefficient of training fitting regression reached 0.995, as shown in

**Table 3.8** Negative effect coefficient and grade of each time period

| Time period | Negative effect coefficient values | Level |
|---|---|---|
| $A_1$ | 0.186 | 5 |
| $A_2$ | 0.195 | 5 |
| $A_3$ | 0.244 | 4 |
| $A_4$ | 0.192 | 5 |
| $A_5$ | 0.271 | 4 |
| $A_6$ | 0.536 | 3 |
| $A_7$ | 0.506 | 3 |
| $A_8$ | 0.435 | 3 |
| $A_9$ | 0.361 | 4 |
| $A_{10}$ | 0.300 | 4 |

Fig. 3.4. Errors between network output and expected output are shown in Figs. 3.5 and 3.6. Mean Absolute Error (MAE) is 0.017, Root Mean Square Error (RMSE) is 0.020, the minimum relative error is 0.008, the maximum relative error is 0.119, and the average relative error is 0.055. The coefficient of determination R square is 0.975. Population fitness began to stabilize at the 54th generation, as shown in Fig. 3.7.

From the results and convergence of the algorithm, it can be seen that GA-BP algorithm has good applicability to the evaluation of the negative effect of urban transportation energy supply station, and can objectively and reasonably analyze the change of the negative effect of the station over time, which has an important reference role in revealing the influence of urban motor vehicles' energy supplement behavior and optimizing the distribution of energy supply station.

**Table 3.9** Function error comparison of different algorithms

| Algorithmic functions | Mean squared error(MSE) |
|---|---|
| traingdx | 0.00280320 |
| trainlm | 0.00041556 |
| trainbfg | 0.00038434 |
| traingd | 0.00436660 |
| traingda | 0.00278930 |
| traingdm | 0.00422100 |
| trainrp | 0.01189200 |
| trainscg | 0.00040501 |
| trainoss | 0.00115750 |

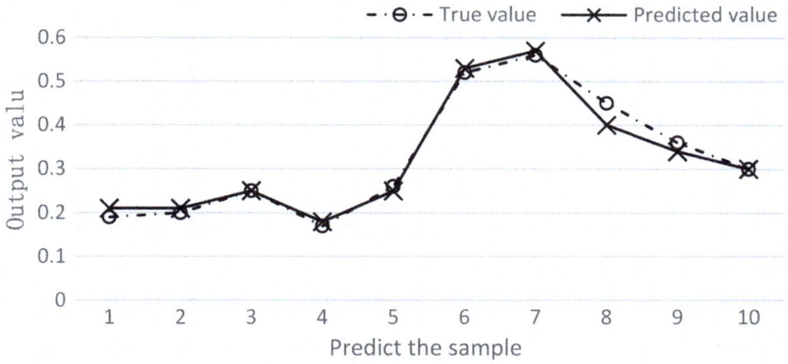


**Fig. 3.3** Comparison between the true value of the evaluation sample and the evaluation value

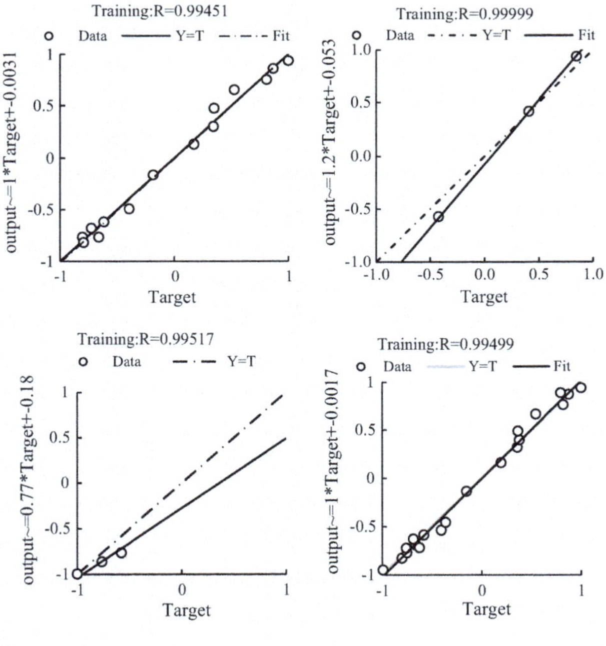


**Fig. 3.4** Linear regression fitting

### *3.4.2 Examples of Negative Effect Evaluation of Urban Transportation Energy Supply Network*

In order to verify the effectiveness of the proposed model and algorithm, the road network in Jiulongpo District of Chongqing City was taken as an example to carry out the verification calculation, as shown in Fig. 3.8. The network has 15 nodes and 20 sections. There are two refueling station nodes and two charging station nodes

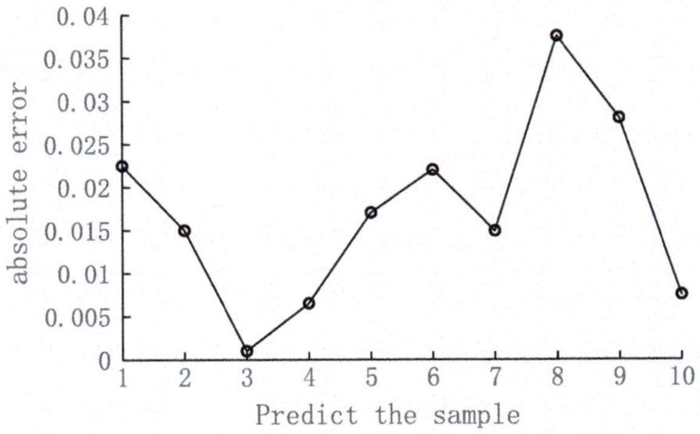


**Fig. 3.5** GA-BP output absolute error

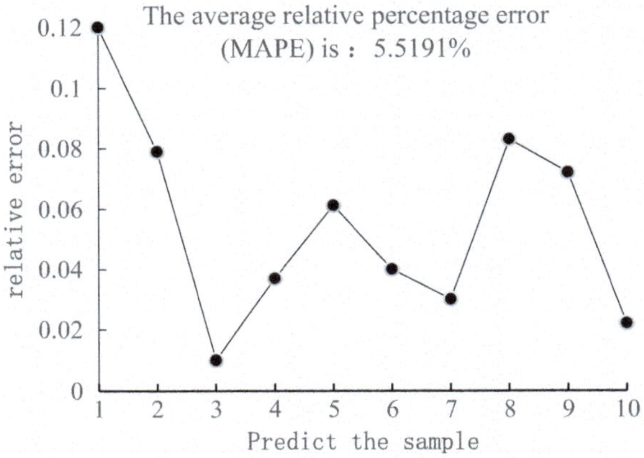


**Fig. 3.6** Absolute relative error of GA-BP output

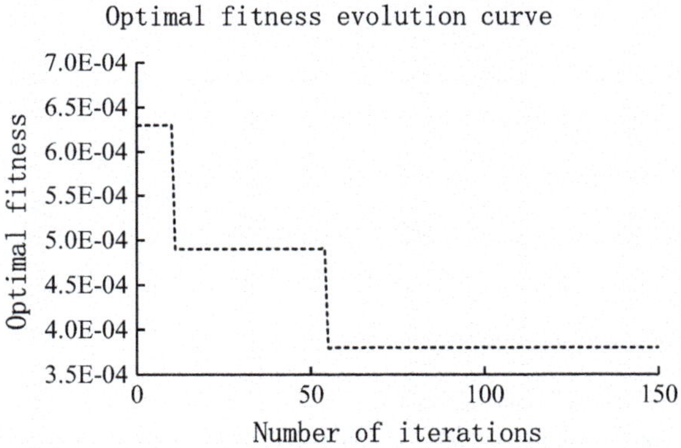


**Fig. 3.7** Optimum fitness curve

in the node, 6, 11 represent the charging station node, 7, 10 represent the refueling station node; There are four OD pairs at the start and end points of the two OD pairs, that is, nodes 1 and 4 are the start points and nodes 2 and 3 are the end points, then the OD pairs are 1→2, 1→3, 4→2 and 4→3. The capacity and flow of different road classes are significantly different, which has an important influence on vehicle routing. The sections are divided into three road grades, which are represented by arrows of different thickness. Among them, sections 1, 2, 5, 6, 7, 10, 12, 13, 14 and 19 are the main trunk roads, sections 3, 9, 11, 12, 15, 17, 18 and 20 are the main roads, and sections 4 and 8 are the secondary roads.

According to the road grade position of each node, the peak period data of the road corresponding to the actual grade is selected for calculation. Figure 3.9 shows the negative effect of the four sites from 6:00 am to 12:00 am from Monday to Thursday. It can be seen that node 6 and node 10 ushered in a higher peak of negative effect in the morning peak, while node 11's peak value was not obvious, showing a high level of negative effect for a long time, and the time points of the four nodes ushered in the highest peak were different. In addition, it can be observed that node 7 has a peak value in a short period of time, while node 6 is on the contrary. Node 6 has a transition time before and after the peak value, and node 10 has a transition time before and after the peak value.

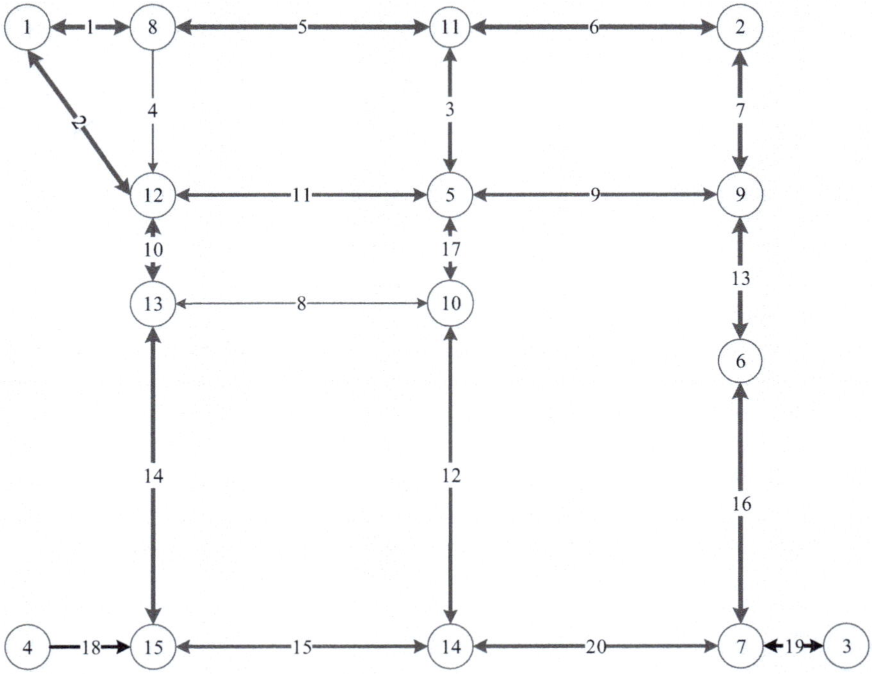


**Fig. 3.8** Actual map and simplified road network map

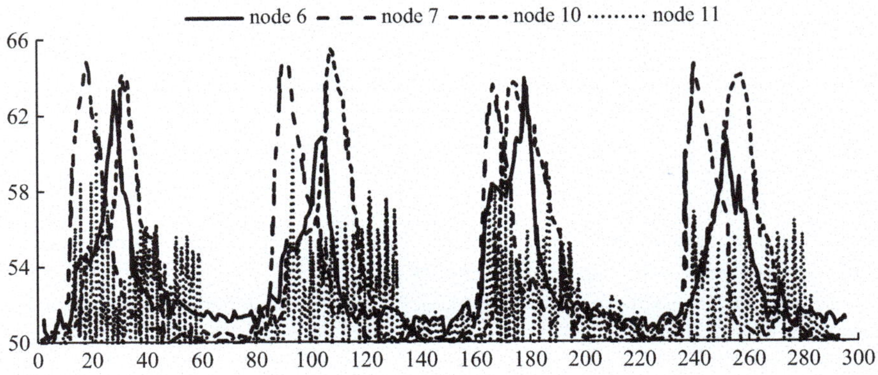


**Fig. 3.9** Changes of negative effects of each node in four days in the morning

To assess the negative cost trend on Friday, we can use the GA-BP algorithm based on the data from the previous four days. The data is measured every five minutes, and each site contains 365 data. The first 292 data are used as the training set, and the last 73 data are used as the test set. The results are shown in Fig. 3.10.

The evaluation error and determination coefficient are shown in Table 3.10. The average relative error is almost below 0.1, and the R-square coefficient is above 0.98. The average relative percentage error is shown in Fig. 3.11. The average relative percentage error of node 6 is 0.50%, that of node 7 is 10.70%, that of node 10 is 3.42%, and that of node 11 is 2.00%. In addition, the linear regression fitting R values of each part of the neural network results are shown in Fig. 3.12, which shows that the fitting results are better.

After getting the evaluation results, we can observe the negative effect trend of each city transportation energy supply station, and also reflect the change trend of the point energy replenishment traffic of the station with time. It can be found that when the energy replenishment vehicle flow increases at each node, the negative effect increases and the peak time prolongs. The reason is that the vehicle flow concentrates on the energy replenishment at a certain node, resulting in a long queuing time, resulting in a corresponding increase in the negative effect generated by energy replenishment. However, the peak value of each node decreases and the peak time is shortened when the supplementary nodes are added. Figure 3.13 shows the negative effect evaluation values of the four transportation energy supply stations obtained. The negative effects of the four nodes on the evaluation day are divided into six time periods according to their incremental trends. It can be seen that the negative effects of the four sites almost rise at the beginning of the moment, and the rising speed and magnitude of node 7 are significantly greater than those of the other three nodes. However, from the beginning of the moment, the negative effects of node 7 decline rapidly, while node 6 and node 10 are still in the rising stage with a small amplitude. Node 11 is in the higher stage of negative effect for a long time. In addition, it can be obviously observed that in the idle period, the negative effect value of node 11 is

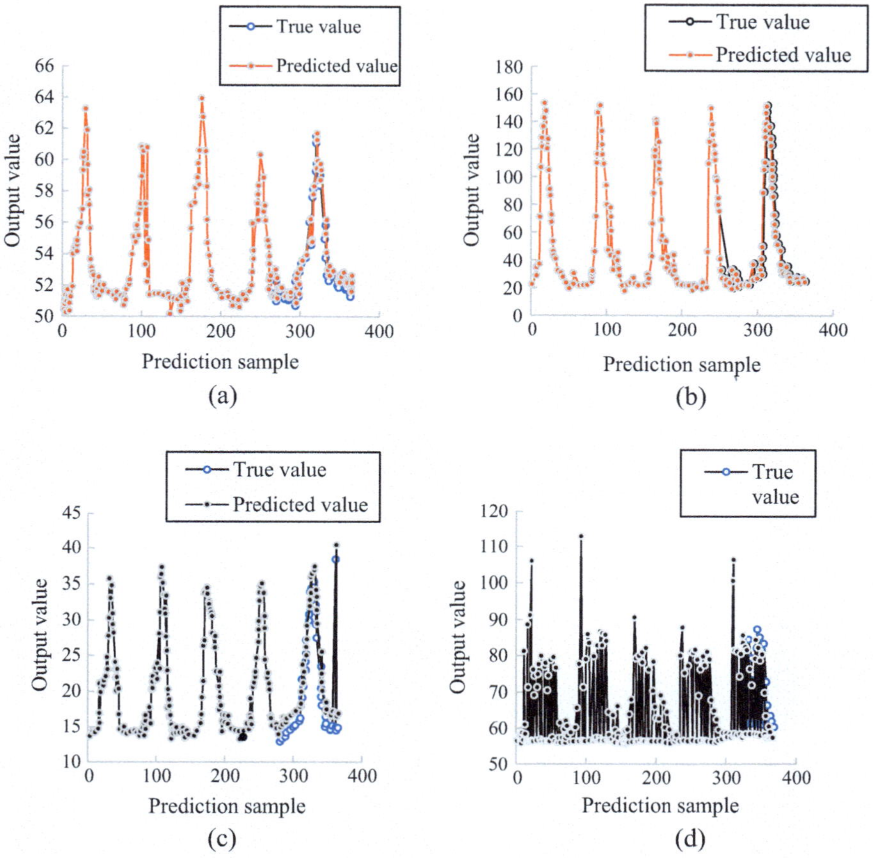


**Fig. 3.10** GA-BP evaluation value and true value of negative effect of each node (**a**) Node 6 Evaluated value and true value, (**b**) Node 7 Evaluated value and true value, (**c**) Estimated value and true value of node 10, (**d**) Estimated value and true value of node 11

**Table 3.10** Evaluation error and determination coefficient of each node

| Node | Average absolute error | Average relative error | Coefficients of determination$R^2$ |
|---|---|---|---|
| 6 | 1.257 | 0.020 | 0.983 |
| 7 | 3.791 | 0.107 | 0.985 |
| 10 | 0.625 | 0.034 | 0.989 |
| 11 | 0.272 | 0.005 | 0.988 |

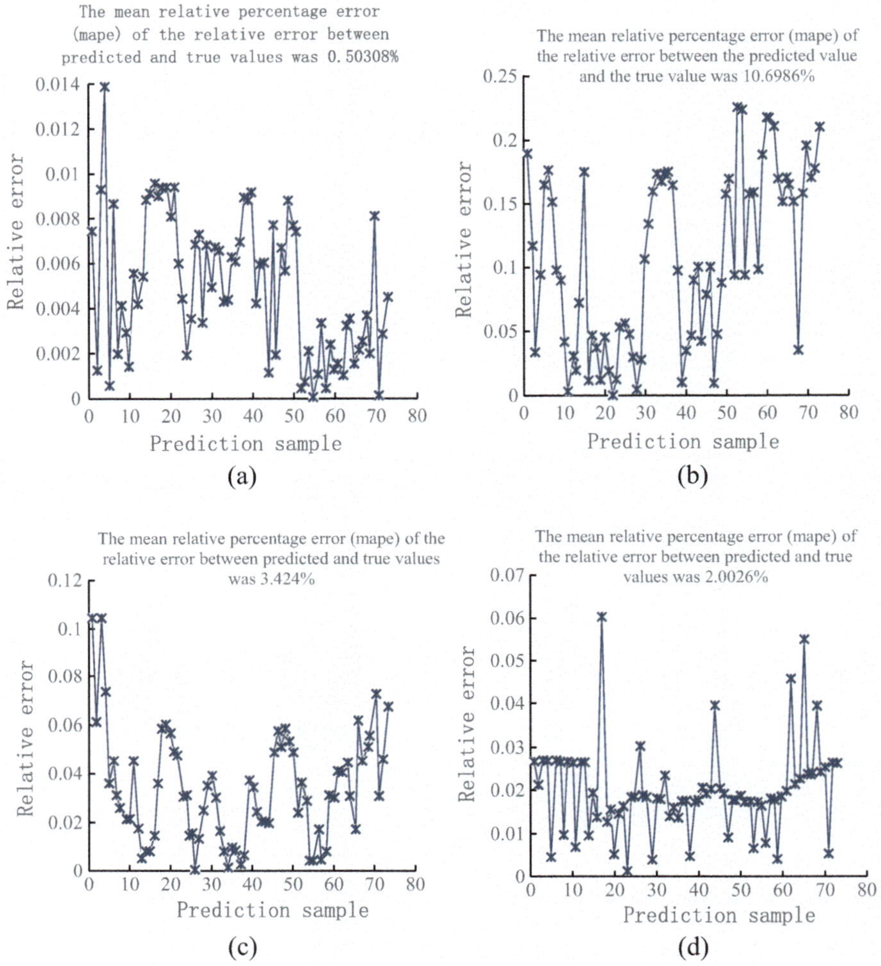


**Fig. 3.11** Error comparison between GA-BP evaluation value and true value of negative effect of each node (**a**) Average relative percentage error of node 6 (**b**) Average relative percentage error of node 7 (**c**) Average relative percentage error of node 10 (**d**) Average relative percentage error of node 11

slightly larger than that of node 6 and much larger than that of node 7 and node 10, and node 10 is basically in the position with the smallest negative effect.

To assess the negative effect of the holiday, it can not be directly assessed according to the traffic flow of the previous few days, an important reason is that in addition to the daily peripheral travel vehicles in the region, the holiday additional travel vehicles. In order to evaluate the negative effect of energy replenishment on holidays, the energy replenishment flow in each time period can be obtained through the distribution model according to the divided time period, and the negative cost of energy replenishment

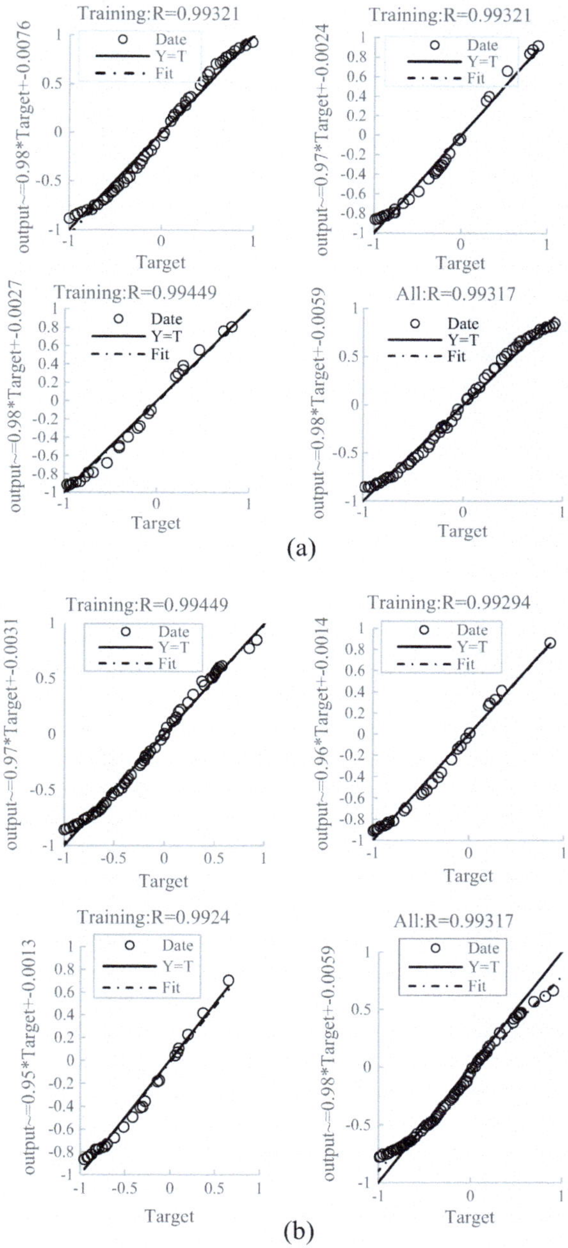


**Fig. 3.12** (**a**) Linear regression fitting of node 6 results (**b**) Linear regression fitting of node 7 results (**c**) Linear regression fitting of node 6 results (**d**) Linear regression fitting of node 7 results

**Fig. 3.12** (continued)

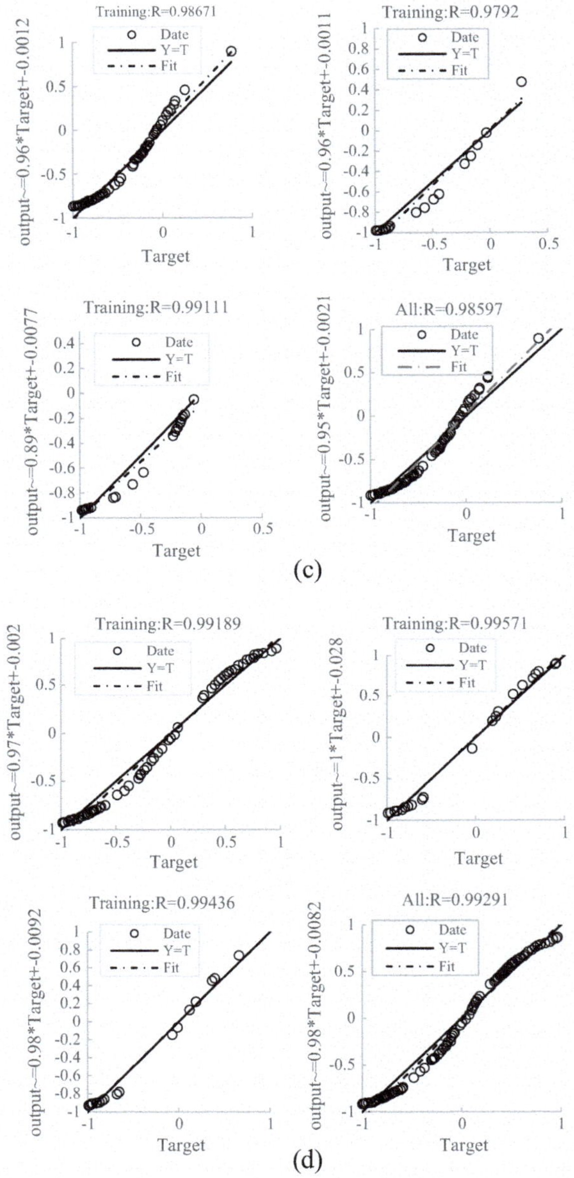


in this time period can be calculated after the energy replenishment flow is obtained. Table 3.11 shows the total demand of each OD point in this period, which is obtained based on the data of the previous Saturday. At present, the market share of electric vehicles and fuel vehicles in Chongqing market is 1:9, the safe driving range of electric vehicles is 170 km, and the anxiety coefficient of power exhaustion is 0.8.

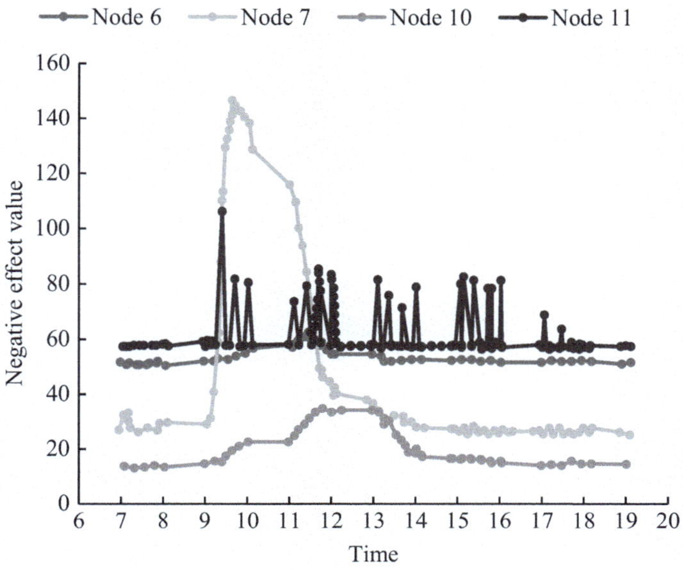


**Fig. 3.13** Relationship between negative effects and time of different transportation energy supply stations in different stages

The calculation of edge weight mainly includes the driving weight of road section and the weight of connected node, in which the BPR function of road section travel time value adopts classical parameters. The power cost per unit section length and the fuel cost per unit section length are 0.15 yuan/km and 1.5 yuan/km respectively. The length of each section of the road network is shown in Table 3.12.

The traffic flow results obtained by using the allocation algorithm are shown in Fig. 3.14. It can be obviously observed that the traffic flow of each section is almost the highest in the time period from 7:10 to 7:45, followed by 7:45–7:55, and the other sections 9 and 12 are basically maintained in the lower traffic flow range. According to the traffic and replenishment of four sites, the replenishment traffic of four nodes in six time periods is summarized and compared. From the time dimension, the replenishment vehicles are also the most in the time period of 7:10–7:45, while there

**Table 3.11** Total demand for corresponding time period

| Time period | 1→2 | 1→3 | 4→2 | 4→3 |
|---|---|---|---|---|
| 6:00–7:10 | 1800 | 4400 | 2500 | 800 |
| 7:10–7:45 | 1100 | 5900 | 2100 | 600 |
| 7:45–7:55 | 350 | 1200 | 500 | 150 |
| 7:55–8:20 | 1100 | 2000 | 1000 | 500 |
| 8:20–9:15 | 1700 | 2300 | 2600 | 1200 |
| 9:15–12:00 | 4100 | 4600 | 6600 | 1900 |

**Table 3.12** Length of each section of the road network

| Sections | Length (m) | Sections | Length (m) |
|---|---|---|---|
| 1 | 362 | 11 | 1106 |
| 2 | 778 | 12 | 1811 |
| 3 | 548 | 13 | 675 |
| 4 | 801 | 14 | 1923 |
| 5 | 1612 | 15 | 1517 |
| 6 | 1109 | 16 | 1825 |
| 7 | 1301 | 17 | 385 |
| 8 | 1210 | 18 | 408 |
| 9 | 2605 | 19 | 388 |
| 10 | 363 | 20 | 952 |

are fewer vehicles in the earliest and noon time period. Compared with the overall traffic flow, the trend is roughly the same, and the flow of node 6 is the smallest, and the volume of node 7 is similar to that of node 10, which is determined by the vehicle flow in the demand for supplementary energy, and the current market share of fuel vehicles still accounts for a large share. For the same charging station, node 11 has a higher flow rate than node 6, which means that node 6 is located in the expressway and Yangjiaping business district, so the flow rate is larger. The reason why node 7 has more traffic than node 10 is determined by the road grade and distance from the end point. Therefore, the share of traffic volume of each road will significantly affect the negative effect of energy replenishment.

The implementation of "dual carbon", the energy side of the change will tend to electric vehicles, to achieve green, the use of real renewable energy, at the same time, new energy vehicles can be connected to the grid to achieve vehicle network interaction. In the next 3–5 years, the use of fuel vehicles will gradually decrease, and the small-scale use of electric vehicles will slowly move to the track of large-scale use, and the energy revolution and the automotive revolution will truly achieve substantial synergy. With the gradual increase in the market penetration rate of electric vehicles, the proportion of charging demand replenishing vehicles in the subsequent road network will also gradually increase. Therefore, the station traffic of different replenishing vehicle types with interest rates is compared, as shown in Fig. 3.15, and the traffic to the charging station node is basically flat when the market share is 30%. When the market share is 50%, the vehicle flow of the charging station node begins to exceed the refueling station node, and it is completely exceeded by 80%, which is even greater than the current gap between the two types of vehicle flow.

After obtaining the traffic flow of the replenished vehicle section, the negative effect on each replenished station and the corresponding path is calculated as shown in Fig. 3.16. The negative effect cost becomes higher during 7:45–7:55 and 7:55–8:20. The total negative effect cost during the total measured time is 16,821.52 yuan, of which the cost for station 6 is 4,464.75 yuan. Site 7 is $7,471.88, site 10 is $7,659.88, and site 11 is $5,408.92. As can be seen from Fig. 3.16, as the market share of electric

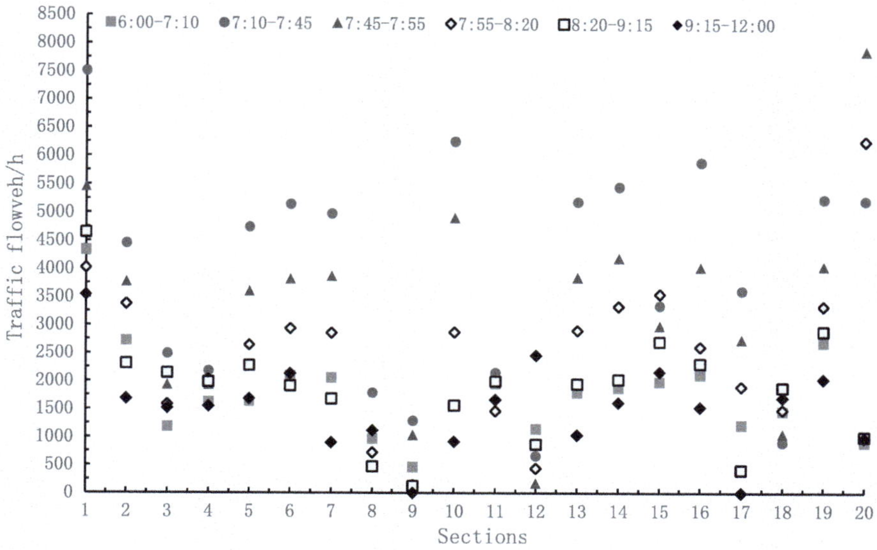


**Fig. 3.14** Balanced distribution results of traffic flow in different time periods of each road section

vehicles increases, the overall negative effect level is also gradually declining. When the market share of electric vehicles is 30%, the negative cost is 12281.23 yuan, down 26.99% year-on-year; When the market share of electric vehicles is 50%, the negative cost is 11197.77 yuan, down 33.43%; When the market share of electric vehicles is 80%, the negative effect cost is 11885.71 yuan, down 29.34% year-on-year. However, with the increase in the number of electric vehicles, charging stations take too long to charge, resulting in increased queues and time delays. How to reduce the charging time, increase the number of charging stations, and speed up the operation efficiency of electric vehicles will be the problems that need to be considered in the future development of electric vehicles.

## 3.5 Summary of This Chapter

This chapter firstly defined the negative effect of urban transportation energy supply network, proposed the specific definition of the specific negative effect, and sorted out the manifestations of different negative effects according to whether the negative effect can be quantified. For each kind of negative supplementary energy effect, the specific quantification method is defined. Starting from the negative effect evaluation of single station, the negative effect evaluation of single recharge station and recharge network is carried out respectively. Considering the characteristics of the surge of travel demand during holidays, an evaluation method of the negative effect of transportation energy supply network during holidays is established. Finally, taking

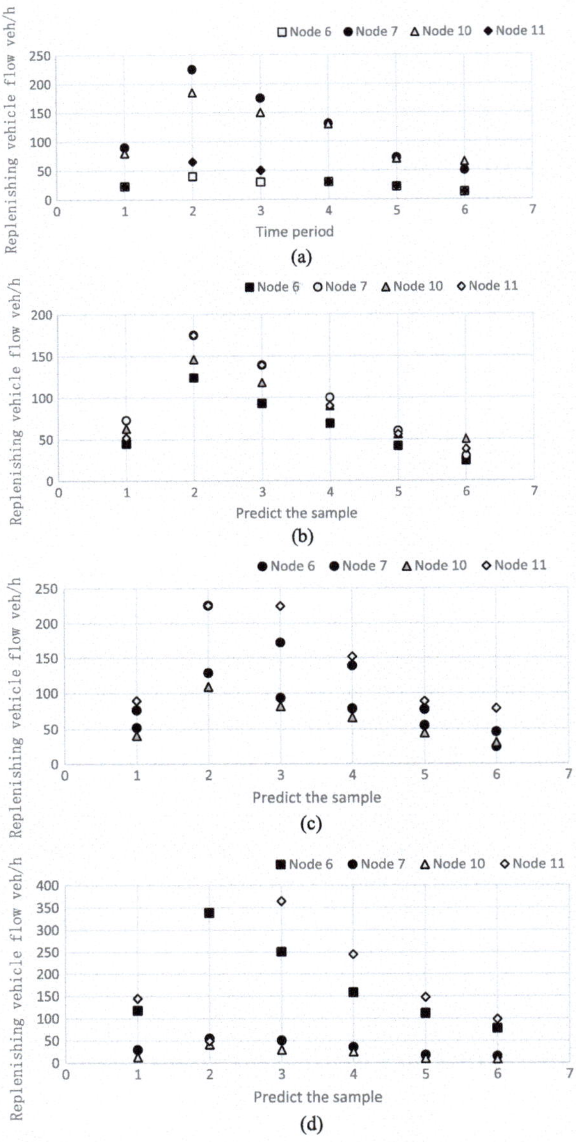


**Fig. 3.15** Balanced distribution results of replenished vehicle flow in each time period (**a**) Market share of electric vehicles 10% (**b**) Market share of electric vehicles 30% (**c**) Market share of electric vehicles 50% (**d**) Market share of electric vehicles 80%

Chongqing regional road network as a case study, a case study on the evaluation of recharge stations and network negative effects is carried out. The evaluation method of the negative effect of the energy supply network in urban transportation proposed in this chapter provided the optimization target for the study of the negative effect reduction mechanism of the energy supply network.

**Fig. 3.16** Negative effect cost of vehicle replenishment during holidays (**a**) Market share of electric vehicles 10% (**b**) Market share of electric vehicles 30% (**c**) Market share of electric vehicles 50% (**d**) Market share of electric vehicles 80%

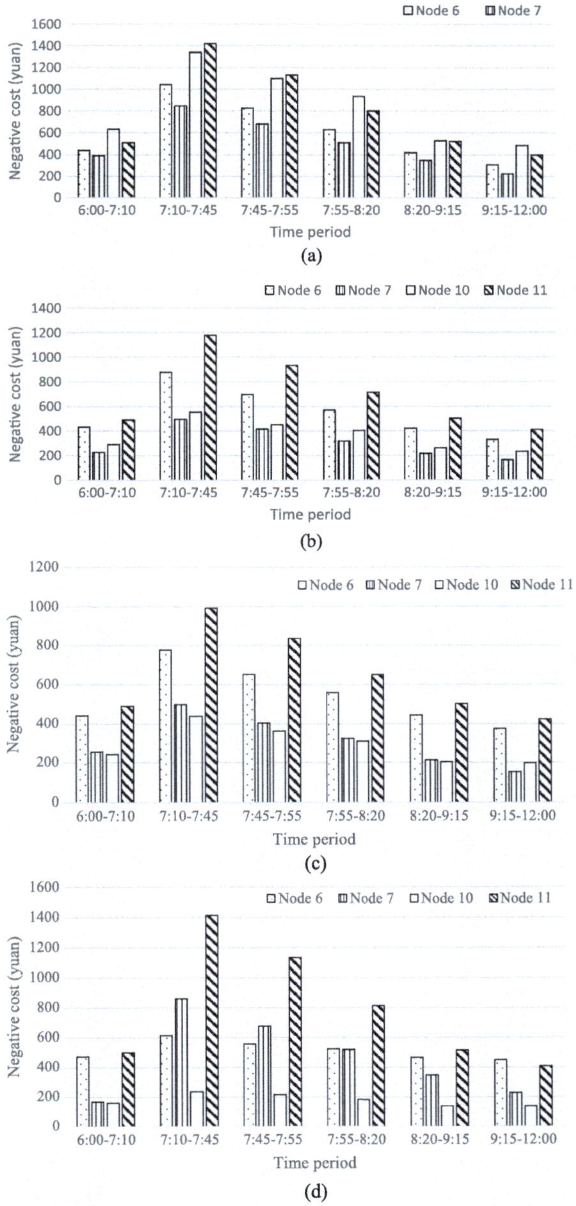

## References

Bowyer DP, Akçelik R, Biggs DC (1984) Guide to fuel consumption analyses for urban traffic management. Australian Road Research Board

Bureau of Public Roads (1964) Traffic assignment manual. Washington DC: Urban Planning Division, US Department of Commerce

Low DW (1974) Optimal dynamic pricing policies for an M/M/s queue. Oper Res 22(3):545–561

Lucas A, Silva CA, Neto RC (2012) Life cycle analysis of energy supply infrastructure for conventional and electric vehicles. Energ Policy 41(1):537–547

Macrina G et al (2019) The green mixed fleet vehicle routing problem with partial battery recharging and time windows. Comput Oper Res 101(1): 183–199

Wang Q, Yuan X, Zhang J et al (2015) Assessment of the sustainable development capacity with the entropy weight coefficient method. Sustainability 7(10):13542–13563

# Chapter 4
# Research on Energy Station Layout Optimization Based on Negative Effect Minimization

## 4.1 Main Influencing Factors of the Site Selection of Supplementary Energy Stations

Due to the function and role of the supplementary energy station and its impact on the traffic network and the surrounding environment, the site selection mainly considers the following factors:

1. Overall urban planning

The overall urban planning highlights the city's politics, economy, and culture, and is the criterion of municipal construction and urban traffic norms, so the layout and construction of the supplementary energy station should comply with the overall urban planning. From the constraints and limitations of urban planning on the layout of supplementary energy sites, the main affected groups are users with supplementary energy needs. Specifically, the urban transportation network and land planning determine the traffic flow in each area of the city, which also affects the demand distribution of transportation energy such as gasoline, diesel, and electric energy. In addition, the city's driving restriction policy will also affect the number of vehicles, which in turn affects the demand for supplementary energy in various areas of the city. In addition, urban land planning also affects the transportation of refined oil products and the construction of the power grid, which will have a direct impact on the investment and site selection of supplementary energy stations. Therefore, the layout and scale of the supplementary energy network should be in line with urban land planning, and a long-term plan should be established to meet the demand for transportation energy for future urban development.

2. Energy supply

Refined oil and electric energy are the main transportation energy sources sold by the supplementary energy station, and their supply capacity will have a key impact on the operation and sales of the supplementary energy station. Therefore, in order



X. Ge, *Urban Transportation Energy Supply Network*, Industrial Ecology and Environmental Management 5, https://doi.org/10.1007/978-3-031-64619-5_4

to ensure the long-term development of the supplementary station, it is necessary to maintain appropriate inventory and durable supply, otherwise the supplementary station will not be able to show the advantages of economy and convenience, and even cause more unnecessary negative effects.

3. Land use and cost

Land use and cost are the factors that must be considered in the layout planning of supplementary energy station. Land use planning carries economic effects such as material exchange, energy flow, and information transmission in the city, and maintains urban economic development and ecological balance. In addition, the construction cost and operating cost of the supplementary energy station are important components of the healthy and sustainable operation of the supplementary energy station. Therefore, the layout planning of the supplementary energy site must consider the factors of land use and construction and operation cost.

4. Environmental safety

Safety, health, environmental protection, and other factors must be placed in a key position in the layout planning of urban replenishment sites. At present, the construction of gas stations, filling stations, and charging stations is already a mature technology, but the production process of gas stations involves high-pressure processes, and the processing objects are flammable and explosive. The charging station should consider the impact of the station area, power supply and outlet corridor, water supply and drainage and flood control facilities, and the road to and from the station. Therefore, when making site selection, safety and environmental protection factors must be considered.

5. Transport network

The location of the supplementary energy station will affect the vehicle driving condition of the nearby regional road network. Factors such as road grade, building type, commercial development and time will affect the road condition, and then cause road congestion, mileage anxiety, redundant detour and even secondary congestion. Therefore, when considering the location selection of the supplementary energy station, it is necessary to fully refer to the traffic network situation in the actual location in order to effectively reduce the burden.

## 4.2 Optimization Method of Replenishment Station Layout

### *4.2.1 Layout Method Based on P-median Value*

The P-median model consists of a set of requirements with a certain number and location and a set of site candidate locations, which plan the $P$ number of sites to a reasonable location and minimize the transportation cost between each demand point and a certain site. P-median model is usually applicable to the site location problem of factories, warehouses, and transfer stations in distribution and service scenarios, and generally requires the least cost between the site and the manufacturer or user. The constraints, objectives, and variables of the P-median model problem are defined as follows, and its objective function is (Hakiml 1964):

$$\text{Min} \sum_{i \in N} \sum_{j \in M} d_i c_{ij} y_{ij} \tag{4.1}$$

The constraints are as follows:

$$\sum_{j \in M} y_{ij} = 1, \ \ i \in N \tag{4.2}$$

$$\sum_{j \in M} x_j = p \tag{4.3}$$

$$y_{ij} \leq x_j, i \in N, j \in M \tag{4.4}$$

$$x_j \in \{0, 1\}, i \in N, j \in M \tag{4.5}$$

Formula:

$N$: $n$ customer points (demand points) in the study scenario, $N = (1, 2, \ldots, n)$;
$d_i$: the demand of the first $i$ customer;
$M$: Candidate sites for m proposed sites in the study object, $M = (1, 2, \ldots, m)$;
$C_{ij}$: Unit transportation cost from $i$ point to $j$ point;
$P$: Total number of proposed facilities ($p < m$);

$$x_j - x_j = \begin{cases} 1, \text{suppose a facility is built in } j \in M \\ 0, \text{otherwise} \end{cases};$$

$$x_j - x_j = \begin{cases} 1, \text{suppose a facility is built in } j \in M \\ 0, \text{otherwise} \end{cases};$$

$$y_{ij} - y_{ij} = \begin{cases} 1, \text{Suppose the customer is in } i \in N, \text{and} \\ \text{the service is provided by facility } j \in M \\ 0, \text{otherwise} \end{cases};$$

$$y_{ij} - y_{ij} = \begin{cases} 1, \text{Assume that the customer is in } i \in N \text{ and} \\ \text{is serviced by facility } j \in M \\ 0, \text{otherwise} \end{cases};$$

Formula (4.1) is the objective function of the model. Formula (4.2) restricts each demand point to allocate only one facility to provide service for it, formula (4.3) limits the maximum number of sites, and formula (4.4) indicates that there is no service for customers where there is no facility. As can be seen from the above, two problems need to be solved to solve a P-median model:

1. Selection and determination of the location of the site facility (variable $x$ in the formula).
2. Whether the customer can reach the site location (variable $y$ in the formula).

Similar to solving the coverage model, in order to minimize the total cost, customer points are assigned to different sites when the location of the site is determined. Besides. For facility location problem, P-median mainly adopts two kinds of methods: heuristic and exact algorithm. Since the P-median model is a non-deterministic polynomial problem, accurate algorithms can usually only solve P-median problems with small data scale or relatively simple scenarios. For large-scale data, heuristic algorithms can be adopted. Here is a greedy takeaway heuristic algorithm, the basic steps of which are as follows:

P-median model greedy take-up algorithm:

Step 1: Set the number $k = m$ of selected sites, that is, select $m$ candidate sites.
Step 2: Assign the customer to the nearest station and calculate the total transportation cost Z.
Step 3: If $k = p$, output the results of $k$ sites and each customer point; Otherwise, go to Step 4.
Step 4: Determine a point in the location of the $k$ candidate sites, the selection condition is: after the point is deleted, its customers are assigned to the nearest site, if the increase in the total cost is minimal, then determine to delete the point.
Step 5: Delete the deleted sites from the collection, let $k = k + 1$ and repeat Step 2.

### 4.2.2 Layout Method Based on P-center

P-center model refers to the selection of a site location in the set of candidate points, so that it satisfies all, and ensures that the distance between the demand point and the nearest site is minimal. The P-center model is widely used in the location of emergency facilities, such as the location of public services such as police stations, fire stations, hospitals, and so on, requiring the service point to be reached as quickly as possible.

The P-center model can be described in mathematical language, and the objective function is (Suzuki and Drezner 1996):

$$\text{MinZ} = \sum_{i=1}^{m}\sum_{j=1}^{n} d_{ij}x_{ij} \tag{4.6}$$

Subject to:

$$\sum_{j=1}^{n} x_{ij} = 1(i = 1, 2, \ldots, m) \tag{4.7}$$

$$\sum_{j=1}^{n} y_j = p \tag{4.8}$$

$$\sum_{i=1}^{m} q_i x_{ij} \leq c_j (j = 1, 2, \ldots, n) \tag{4.9}$$

$$\sum_{i=1}^{m} \sum_{j=1}^{n} x_{ij} y_j = m \tag{4.10}$$

where:

$n$ indicates the site selection candidate node
$c_j$ represents the maximum service capacity of the first $j$ service facility
$p$ represents the total number of facilities
$m$ indicates the number of required points
$d_{ij}$ represents the distance from $i$ node to $j$ node

$$x_{ij} = \begin{cases} 1, \text{The needs in } i \text{ are covered by } j\text{'s range of services} \\ 0, \text{otherwise} \end{cases}$$

$$x_{ij} = \begin{cases} 1, \text{The requirements at node } i \text{ are covered by the scope of service at node } j \\ 0, \text{ otherwise} \end{cases}$$

$$y_j = \begin{cases} 1, \text{the } j\text{th service facility was opened} \\ 0, \text{ otherwise} \end{cases}$$

$$y_j = \begin{cases} 1, \text{The } j\text{th service facility is activated} \\ 0, \text{ otherwise} \end{cases}$$

Formula (4.7) indicates that each demand point is served by only one site, formula (4.8) indicates the number of sites selected, formula (4.9) indicates the service ability constraints of service sites on customer points, and formula (4.10) indicates that there is no demand point where there is no service point.

### 4.2.3 *Layout Method Based on Maximum Coverage*

The coverage model is divided into the maximum coverage problem and the set coverage problem. The set coverage model mainly studies the problem that the total number or construction cost of the site is minimum when the service of all demand points is covered. The maximum coverage model is to study how to select $p$ service site in the alternative location, so that these service points meet the maximum number of demand points or the maximum demand.

The set coverage model takes cost optimization as the research goal, and calculates the minimum total cost when the simulation time is set to meet the demand, so the number of sites is not limited. When the cost of each site is the same, the goal of the set coverage model is to minimize the number of sites. The objective function is (Toregas et al. 1971):

$$\text{Min} \sum_{j \in M} c_j X_j \tag{4.11}$$

Subject to:

$$\sum_{j \in M} Y_{ij} X_j \geq 1 \quad \forall i \in N \tag{4.12}$$

$$Y_{ij} \in \{0, 1\} \quad \forall i \in N, \ \forall j \in M \tag{4.13}$$

$$X_j \in \{0, 1\} \quad \forall j \in M \tag{4.14}$$

where:

$N$: Demand point set
$M$: Candidate set;
$i$: Demand point, $i \in N$;
$j$: Candidate sites, $j \in M$;
$d_{ij}$: Distance from the demand point $i$ to the candidate point $j$;
$c_j$: $j$ Point fixed cost;
$R$: The farthest distance acceptable to the customer;
$X_j$: If you choose to build the station at the point $j$, then $X_j = 1$, otherwise $X_j = 0$;
$Y_{ij}$: If the requirement $i$ can be serviced within a distance $R$ by the candidate site, that is $0 \leq d_{ij} \leq R$, then $Y_{ij} = 1$, otherwise $Y_{ij} = 0$.

The objective function (4.11) indicates that the construction cost of the site is the minimum. If the construction cost c_j of all sites is the same, the objective function is to minimize the number of sites. Formula (4.12) indicates that at least one site can accept and complete the service for any demand point. Equations (4.13) and (4.14) are 0–1 decision variables.

The feature of the set coverage model is to ensure that all requirements are met. In practical applications, some demand points are far away from other demand points, and the demand is small. If this point is taken into account, the revenue obtained will not be able to compensate for the increased service cost, resulting in a decline in total revenue.

Therefore, when the number and distance of sites are relatively fixed, the maximum demand is required to be covered. Therefore, a site selection model with maximum coverage is proposed, which does not require to cover all demand points, but to cover as many demands as possible through the given number of sites and the maximum service range. The objective function is as follows (Vatsa and Jayaswal 2016):

$$\text{Min} \sum_{i \in N} H_i Z_i \tag{4.15}$$

Subject to:

$$\sum_{j \in M} Y_{ij} X_j \geq Z \quad \forall i \in N \tag{4.16}$$

$$\sum_{j \in M} X_j = p \tag{4.17}$$

$$X_j \in \{0, 1\} \quad \forall j \in M \tag{4.18}$$

$$Y_{ij} \in \{0, 1\} \quad \forall i \in N, j \in M \tag{4.19}$$

$$H_i \in \{0, 1\} \quad \forall i \in N \tag{4.20}$$

where:

$N$: Demand point set
$M$: Candidate set;
$i$: Demand point, $i \in N$;
$j$: Candidate sites, $j \in M$;
$d_{ij}$: Distance from the demand point $i$ to the candidate point $j$;
$c_j$: $j$ Point fixed cost of site construction;
$R$: The farthest distance acceptable to the customer;
$X_j$: If you choose to build the station at the point $j$, then $X_j = 1$, otherwise $X_j = 0$;
$Y_{ij}$: If the requirement $i$ can be serviced within a distance $R$ by the candidate site, that is $0 \leq d_{ij} \leq R$, then $Y_{ij} = 1$, otherwise $Y_{ij} = 0$;
$H_i$: If the demand at the demand point $i$ is met, then $H_i = 1$, otherwise $H_i = 0$.

Among them, the objective function (4.15) indicates that the covered customers have the greatest demand; Constraints (4.16) indicate that if the requirements of demand point $i$ are covered, at least one facility site is built within its acceptable limited coverage distance; Constraints (4.17) indicate that the construction number of site facilities is $p$. The constraints (4.18), (4.19), and (4.20) represent three 0–1 decision variables.

The maximum coverage siting model has been proved to be one of the most effective models in facility siting problem. In the model, it is assumed that the coverage of demand points is binary, that is, there are only two cases of "completely covered" and "completely not covered" for users existing on any node. However, in the real 3D map, the site facilities have the actual physical volume, such a binary coverage assumption is flawed. Therefore, the traditional maximum coverage location model has some shortcomings in practical application scenarios.

## *4.2.4 Layout Method Based on Demand Clustering*

### 4.2.4.1 K-Means Clustering

Specific Steps

Step 1: First set some classes/groups, need to predict in advance to determine the number of classes/groups, randomly initialize the respective center points of these groups/classes. The center point exists at a point where each data point vector is the same length. Also determine the number of center points.
Step 2: Calculate the distance between each data point and the center point, and divide the center points closer to the data point into that category.
Step 3: Calculate the center point in each updated class, and use the calculated result as the new center point.
Step 4: Cycle through the above steps until each type of center does not change beyond the expected range after iteration. You can also randomly initialize the center point several times, choosing the one with the most reasonable running result.

The Advantages and Disadvantages of K-Means

Advantages: Fast and easy to calculate.
Disadvantages: Need to determine how many classes/groups of data there are before the experiment.

### 4.2.4.2 Mean Shift Clustering

Mean shift clustering is a sliding window-based clustering algorithm that aims to find dense regions of data points. This is an algorithm based on the center of mass, by updating the candidate points of the center point to the mean of points in the sliding window to locate the center point of each group/class, and then these candidate Windows are screened, similar Windows are deleted, and finally the set of central points and the corresponding grouping are obtained.

Specific Steps

Step 1: Randomly select the center point as the center of the circle, then determine the radius of the circular window, and start sliding.
Step 2: The window will move to a higher density area, each time it slides to a new area, calculate the mean within the sliding window and use it as the center point, the number of points within the sliding window is called the density within the area window.

Step 3: Slide the window, calculate the center point within the window and the density within the window, until the window can no longer accommodate more points in any direction, that is, move until the density inside the circle stops increasing.
Step 4: Steps 1 to 3 will usually produce multiple sliding Windows, when multiple sliding Windows overlap, keep the window covering the most points, and then cluster according to the sliding window covering the data point.

Advantages and Disadvantages of Mean Shift Clustering

Advantages: Unlike the K-Means algorithm, the mean shift clustering algorithm does not need to know how many classes/groups there are in total in advance, and is less affected by the mean than K-Means.
Disadvantages: The selection of window radius has little influence on the experimental results.

#### 4.2.4.3 Density-Based Spatial Clustering of Applications with Noise (DBSCAN)

Similar to the means shift clustering method, DBSCAN is also a density-based clustering algorithm.

Specific Process

Step 1: To understand the radius scale and the minimum number of points, start from any point in the data that has not been visited, and take this point as the center. All the points within the radius of the circle are compared with the minimum number of points as the critical point. If the number of points is greater than or equal to the critical point, this point is marked as the center point; on the contrary, it will be marked as the noise point.
Step 2: If a noise point exists within the radius of a central point, the edge point will be marked by the noise point, and the opposite is still the noise point. Cycle step 1, select all the points in all the data once to cutoff.

Advantages and Disadvantages of Density-Based Clustering

Advantages: The number of classes/groups need not be known in advance.
Disadvantages: Need to know the radius scale and minimum number of points.

#### 4.2.4.4 Clustering with the Expectation Maximization (EM) of Gaussian Mixture Models (GMM)

Gaussian mixture model refers to the linear combination of multiple normal distribution functions. In theory, any type of distribution can be fitted by Gaussian mixture model. This model is often used to solve the same distribution with different parameters, or the distribution with different types of data, such as normal distribution and Bernoulli distribution. Or it can be used to solve data cases that contain multiple different distributions within the same collection.

Specific Process

Step 1: Specify the number of classes and randomly initialize the mean and variance of the normal distribution for each class.
Step 2: Calculate the probability that a data point belongs to a class based on the normal distribution of each class. The closer the data point is to the center of the normal distribution, the more likely it is to belong to this class.
Step 3: Calculate the normal distribution parameters from the probability calculated in the previous step, maximize the probability of the data points, determine the weight of the data by the probability of the data points belonging to this class, you can use the weighted probability of the data points to calculate these new parameters.
Step 4: Repeat iterative steps 2 and 3 until the probability change in the iteration is within a given range.

Advantages and Disadvantages of Gaussian Mixture Model

Advantages of the Gaussian mixture model: The Gaussian mixture model uses the probability of a data point being in a certain location, which means that multiple classes can be contained in the same data point. For example, 30% probability of belonging to class A and 70% probability of belonging to class B are included in point X, which indicates that the model can support mixed coverage, which is more suitable for the actual situation. In addition, the Gaussian mixture model uses mean and variance, which shows that the class can take on an oval shape without being limited to a circle. Therefore, a special type of Gaussian mixture model is K-Means, where the class appears circular if the variance is infinitely close to 0 in all dimensions.

#### 4.2.4.5 Hierarchical Agglomerative Clustering (HAC)

Hierarchical clustering algorithms are divided into two categories according to the order: top-down and bottom-up. HAC is a hierarchical clustering algorithm from top to bottom. First, the aggregation level clustering treats each data point as a single class, and then aggregates the classes by calculating the distance between all the classes. The aggregation continues until all the classes are aggregated into one class.

Specific Process

Step 1: First, we treat all data points as a single class, and second, we choose a metric to measure the distance between the two classes. The average distance between two groups of classes such as the first class and the second class of data points is defined as the distance between the two classes, and this standard is determined.
Step 2: In each iteration update, merge the two classes with the smallest average link into one class.
Step 3: Repeat Step 2 to aggregate all data points into one class, and then select the number of required classes based on the experiment scenario.

Hierarchical Clustering Advantages and Disadvantages

Advantages: The total number of classes does not need to be known in advance; And this algorithm is insensitive to the demarcation of distance metric.
Disadvantages: The program runs inefficiently.

## 4.3 Research on Optimization of Energy Site Layout Based on Demand Clustering Model

### *4.3.1 Analysis of Energy Site Layout Based on Demand Clustering*

When the demand for energy replenishment occurs in urban vehicles, the reasonable distribution of gas stations is an important condition for realizing rapid energy replenishment, reducing the long queue at gas stations, avoiding traffic congestion on main roads, and reducing energy consumption and carbon emissions. Since the demand for supplementary energy is random in time and space, how to accurately describe the distribution of the demand for supplementary energy is the basis of the location of supplementary energy under random demand. Different from previous studies on discrete random variables representing uncertain data, this paper breaks down the limitations of imprecision of graph theory and difficult statistics of quantity and location distribution law, and uses Gaussian cluster mixed model to describe vehicle replenishment distribution, clustering urban vehicle replenishment demand points into multiple demand regions that obey Gaussian distribution. According to the regional distribution of energy replenishment demand, site selection is carried out to reduce the time of vehicle replenishment, rationally optimize and allocate energy replenishment vehicles, reduce carbon emissions in the process of vehicle replenishment, and reduce the negative effect caused by vehicle replenishment aggregate.

### *4.3.2 Random Replenishment Demand Analysis Based on GMM Clustering*

In this book, the advantage of GMM method to analyze the uncertainty of supplementary energy demand is that: ① Any complex distribution can be truly represented by the mixed Gaussian distribution. The hybrid Gaussian distribution function is used to predict the occurrence of replenishment demand, and the vehicle distribution probability at different locations is obtained. ② The model can not only accurately reflect the distribution of vehicle replenishment demand, but also predict the occurrence law of future demand, and make optimization in advance to avoid the accumulation and queuing of replenishment vehicles, which will cause negative effects.

The theory of digital signal processing shows that multiform and complex signal waves can be converted into a number of sine wave combinations by Fourier transform. Similarly, the replenishment demand of random distribution in two-dimensional geographic space can also be transformed into several combinations of standard distributions by relevant mathematical processing. In this book, the energy demand of fuel vehicle is described by $K$ hybrid Gaussian model. Assuming that the hybrid model consists of a Gaussian distribution, the probability density function of the model is (Toral and Chakrabarti 1993):

$$p(x) = \sum_{k=1}^{K} \pi_k \bullet N(x|\mu_k, \sum_k) \tag{4.21}$$

$$N(x|\mu_k, \sum_k) = \frac{1}{2\pi\left|\sum_k\right|^{\frac{1}{2}}} \bullet \exp\left[-\frac{(x-\mu_k)^T \sum_k^{-1} (x-\mu_k)}{2}\right] \tag{4.22}$$

where, $x$ represents a random variable, which is related to the longitude and latitude coordinates; $\pi_k$ is the mixing coefficient, which represents the probability of the supplementary energy demand appearing in each region; $N(x|\mu_k, \sum_k)$ is the first $k$ component in the mixed model, and $\mu_k$ represents the mean value of longitude and latitude coordinates of the first $k$ region where the demand for supplementary energy exists. $\sum_k$ is the covariance matrix.

It is difficult to obtain the probability density $p(x)$ of the mixed model directly from the given data, so introducing a new $K$ latitude random variable $z$, $z_k(1 \leq k \leq K)$ can only be 0 or 1, the elements $z_k$ in this vector satisfy $z_k \in (0, 1)$, when $\sum_k z_k = 1$, representing the probability of being selected first $k$, for any demand point $x_k$, there is a $z_k$ corresponding to it, proving that the demand point $x_k$ is in the Gaussian demand, that is $p(z_k = 1) = \pi_k$; Otherwise, $\sum_k z_k = 0$ indicates that the fist $k$ is not selected. Let's assume that $z_k$ is independent and equally distributed, then the joint probability distribution of $z$ is: Thus defining a joint distribution $p(x, z)$, its edge distribution $p(z)$ and conditional probability $p(x|z)$, corresponds to the formula medium $\pi_k$ and $g(x|\mu_k, \sum_k)$.

$$p(z) = p(z_1)p(z_2)\ldots p(z_K)=\prod_{k=1}^{K} \pi_k^{zk} \tag{4.23}$$

Since the value of $z_k$ is 0 or 1, and only one $z_k$ of $z$ is 1, others $z_j(j \neq k)$=0, the above formula holds.

Because each class of data follows a normal distribution, the $k$ class can be represented as:

$$p(x|z) = \prod_{k=1}^{K} N\left(x\middle|\mu_k, \sum_k\right)^{z_k} \tag{4.24}$$

The edge distribution $p(z)$ and conditional probability $p(x|z)$ are respectively given above. According to the multiplication rule, the mixed Gaussian distribution density is obtained:

$$\begin{aligned} p(x) &= \sum_z p(z)p(x|z) \\ &= \sum_z \left(\prod_{k=1}^{K} N\left(x\middle|\mu_k, \sum_k\right)^{z_k}\right) = \sum_{k=1}^{K} \pi_k N\left(x\middle|\mu_k, \sum_k\right) \end{aligned} \tag{4.25}$$

It can be seen that formula (4.21) of the GMM model is similar to formula (4.25), but new variables $z$ are introduced into formula (4.25), which are called implicit variables here. For the location distribution data of vehicle replenishment, the location of each demand point is known, but a demand point is randomly selected, and the specific gas station to which this demand point should belong is not known in advance. Therefore, an implicit variable $z$ is introduced to describe this phenomenon.

When clustering the random locations of the replenishing vehicles, the probability of randomly replenishing vehicles choosing a certain gas station is mainly considered, and then the replenishing vehicles are divided into regions and assigned gas stations, and the mean value and covariance of longitude and latitude coordinates of each station are estimated. Calculated under Bayesian theory:

$$\begin{aligned}\gamma(z_k) = p(z_k = 1|x) &= \frac{p(z_k = 1)p(x|z_k = 1)}{p(x|z_k = 1)} \\ &= \frac{p(z_k = 1)p(x|z_k = 1)}{\sum_{j=1}^{K} p(z_j = 1)p(x\big|z_j = 1)} \\ &= \frac{\pi_k N(x\big|\mu_k, \sum_k)}{\sum_{j=1}^{K} \pi_j N(x\big|\mu_j, \sum_j)} \end{aligned} \tag{4.26}$$

By introducing hidden variables z, $\gamma(z_k)$ can be calculated by hidden variables z and known x, each submodel has unknown parameters $\pi_k$, $\mu_k$, $\sum_k$, direct derivation can not be calculated, so the EM algorithm is used to estimate.

## 4.3.3 Site Selection Model Considering Random Needs

### 4.3.3.1 Problem Description and Model Hypothesis

Vehicles in the city have random energy replenishment needs. After accurately describing the uncertain distribution of vehicles, how to plan the location of gas stations according to the demand distribution is the core of realizing rapid energy replenishment, avoiding traffic congestion, and reducing carbon emissions. In order to solve this problem, this book establishes an integer programming site location model considering random demand, adopts GMM clustering algorithm and K-means algorithm to determine the demand area, takes the minimum site as the goal, and applies the "double coverage" model to site selection to ensure that the total recharge time is minimized. Therefore, this book proposes a site-planning model that takes service priority into account, which assumes:

1. There is only one main gas station and one standby gas station in each region to provide priority replenishment service and backup replenishment service for vehicles;
2. The main gas station and the standby gas station in a replenishment demand area are not the same service point; However, the main gas station in this area can also be a backup gas station in other demand areas, and vice versa;
3. The total replenishment time to the main gas station of each replenishment demand area is less than the total replenishment time to other stations;
4. The demand in each replenishment demand area is independent of each other.

### 4.3.3.2 Symbol Description

The model notation comments are shown in Table 4.1.

**Table 4.1** Parameter description

| Symbol | Definition |
|---|---|
| $K$ | Total number of replenishment demand areas (unit: units) |
| $M$ | Total number of optional filling stations (unit: units) |
| $W$ | A collection of enabled gas stations, $m \in W$ |
| $A_i$ | Number of vehicles generating replenishment demand in Area $i$ (unit: units) |
| $t_{ij}$ | The total replenishment time from the replenishment demand area $i$ to the station $j$, when considering the spatial randomness of the demand, $t$ is a random variable; Otherwise, it is the mean, that is $\bar{t}_{ij}$ |
| $t_{sr}$ | Ideal total time for refueling from the refueling demand area to the main refueling station (in minutes) |
| $t_{s1}$ | Total replenishment time from the replenishment demand area to the primary site (unit: minutes) |
| $t_{s2}$ | Total replenishment time from the replenishment demand area to the standby site (unit: minutes) |
| $V_i$ | Refueling vehicle $i$ single refueling volume (unit: ml) |
| $a_l, b_l$ | They are the upper and lower limits of the service capacity load of the level $l$ gas station |
| $w_j$ | Supply capacity of gas station $j$ |
| $y_j$ | 0–1 variable, gas station $j$ set to station $y_j$=1, otherwise, $y_j$= 0 |
| $\alpha_{ij}$ | 0–1 variable, Refueling demand area $i$ to the main gas station $j$ for refueling, $\alpha_{ij}$=1; Otherwise, $\alpha_{ij}$=0 |
| $\beta_{ij}$ | 0–1 variable, Replenishment demand area $i$ to the standby gas station $j$ for refueling, $\beta_{ij}$=1; Otherwise, $\beta_{ij}$=0 |
| $x_{ij}$ | 0–1 variable, Demand point $i$ to gas station $j$ replenishment, $x_{ij}$=1, Otherwise $x_{ij}$= 0 |

#### 4.3.3.3 Model Building

The objective function is to minimize the number of enabled sites:

$$\min f = \sum_{j=1}^{M} y_j \tag{4.27}$$

Model constraints:

$$\sum_{i=1}^{K} (A_i \cdot \sum_{j=1}^{M} (\Pr(t_{ij} \leq t_{sr}) \cdot \alpha_{ij})) \geq \varphi \sum_{i=1}^{K} A_i \tag{4.28}$$

$$\sum_{j=1}^{M} t_{ij}\alpha_{ij} \leq t_{s1}, \forall i \tag{4.29}$$

$$\sum_{j=1}^{M} t_{ij}\beta_{ij} \leq t_{s2}, \forall i \tag{4.30}$$

$$\sum_{j=1}^{M} \alpha_{ij} = 1, \forall i \tag{4.31}$$

$$\sum_{j=1}^{M} \beta_{ij} = 1, \forall i \tag{4.32}$$

$$\alpha_{ij} + \beta_{ij} \leq y_j, \forall i \tag{4.33}$$

$$\sum_{i=1}^{K} (\alpha_{ij} + \beta_{ij}) \geq y_j \tag{4.34}$$

$$\sum_{j \in W} \alpha_{ij}\bar{t}_{ij} \leq \bar{t}_{im}, \ \forall m \in W, \forall i \tag{4.35}$$

$$\sum_{i \in A_i} V_i x_{ij} \leq a_l w_j y_j \tag{4.36}$$

$$\sum_{i \in A_i} V_i x_{ij} \leq b_l w_j y_j \tag{4.37}$$

$$t_{ij} = t_{ij}^1 + t_{ij}^2 + t_{ij}^3 \tag{4.38}$$

The objective function (4.27) follows the economic principle of the location problem: in the case of fixed market demand, the fewer the number of gas stations, the greater the probability of service demand of each station, and the higher the revenue of each station. Constraint (4.28) is a classical constraint in the site location problem, that is, the probability of the demand area completing the replenishment at the site within the total replenishment time $t_{sr}$ is $\varphi$, where, $P(t_{ij} \leq t_{sr})$ represents the probability that the total complement energy of the demand point $i$ at the site $j$ is less than a certain standard $t_{sr}$. The parameters obtained by GMM clustering can be used to calculate its specific value, and the constraint conditions can be used as an important index of site benefit evaluation. Constraints (4.29) and (4.30) require that the total energy replenishment time at the primary site and secondary site of the demand area be no higher than the standard time $t_{s1}$ and time $t_{s2}$ respectively. The constraint conditions (4.31) and (4.32) indicate that there is only one primary station and one standby station in each demand area. It should be noted that in practical applications, when a certain main gas station can provide service in time, the standby station can be selected to complete the replenishment. Of course, if the constraint is relaxed, there are multiple standby stations for one demand. Constraints (4.33) and (4.34) ensure that when Site $j$ is used as a primary or secondary site, $y_j = 1$. Constraint (4.35) ensure the total recharge time for each demand area to its primary site is less than the average total recharge time to other vehicle sites. The constraint conditions (4.36) and (4.37) indicate that the total demand of the demand point allocated to the gas station cannot exceed the maximum load of the gas station. $a_l$, $b_l$ are respectively the upper and lower limits of service capacity load of grade (total three) gas stations; The constraint condition (4.38) represents the total replenishment time from the demand point $i$ to the gas station $j$ including: supplementary driving time $t_{ij}^1$, supplementary queuing time $t_{ij}^2$, service time $t_{ij}^3$. The replenishment travel time is the time taken by the vehicle to travel from the demand point to the gas station, and the queuing time represents the waiting time for the replenishment vehicle to arrive at the replenishment. Service time refers to the time taken by the staff to replenish the energy of the vehicle from entering the refueling station to the end of the replenishment and leaving the refueling station. Note that the queuing time is not included. There are many factors that affect the vehicle replenishment time, such as: the number of refueling guns, refueling methods, the business proficiency of service personnel, the volume of vehicle fuel tank, etc., but it can be summarized as service efficiency. After field investigation of CNPC (5 km gas station, 7 km gas station, Qinglong Road gas station), Sinopec (junction station, Daojiao gas station) and Shell gas station in Nanan District of Chongqing, the average energy replenishment time of each vehicle is 4 min. Therefore, $t_{s1} = t_{s1}^1 + t_{s1}^2 + 4$, $t_{s2} = t_{s2}^1 + t_{s2}^2 + 4$.

Rewrite random variables

$$\mathrm{P}(t_{s1} - \sum_{j=1}^{M} t_{ij}\alpha_{ij} \geq 0) \geq 1 - \theta_1 \tag{4.39}$$

$$\mathrm{P}(t_{s2} - \sum_{j=1}^{M} t_{ij}\beta_{ij} \geq 0) \geq 1 - \theta_2 \tag{4.40}$$

where, $\theta_1$, $\theta_2$ is the control parameter, formula (4.39) represents the probability of the total replenishment time $t_{ij}$ from any demand point in the replenishment demand area to the main site within time $t_{s1}$ is $1 - \theta_1$, and formula (4.40) is the same.

The distribution of replenished vehicles is random, so it can be represented by Gaussian mixture distribution. Let the longitude and latitude coordinates of the demand point $i$ be $(e_i, n_i)$. It follows Gaussian mixture distribution $f(e_i, n_i)$; let the latitude and longitude coordinates of the gas station $j$ be $(e_j, n_j)$, the speed of the vehicle to recharge is $v$. The model is established to:

$$\sum_{j=1}^{M} \iint\limits_{(e_i-e_j)^2+(n_i-n_j)^2 \leq (vt_{s1}^1)} f(e_i, n_i) de_i dn_i * \alpha_{ij} \geq 1 - \theta_1 \tag{4.41}$$

$$\sum_{j=1}^{M} \iint\limits_{(e_i-e_j)^2+(n_i-n_j)^2 \leq (vt_{s2}^1)} f(e_i, n_i) de_i dn_i * \beta_{ij} \geq 1 - \theta_2 \tag{4.42}$$

Among them,

$$f(e_i, n_i) = \frac{1}{2\pi\sigma_{ie}\sigma_{in}} * \exp\left\{-\frac{1}{2}\left[\frac{(e_i - \overline{e}_i)^2}{\sigma_{ie}^2} + \frac{(n_i - \overline{n}_i)^2}{\sigma_{in}^2} - \frac{2(e_i - \overline{e}_i)(n_i - \bar{n}_i)}{\sigma_{ie}\sigma_{in}}\right]\right\} \quad (4.43)$$

$\overline{e}_i$, $\overline{n}_i$ denotes the horizontal and vertical coordinates of demand $i$, $\sigma_{ie}$, $\sigma_{in}$ is the standard deviation. By using the Gaussian distribution constraints (4.39), (4.40), (4.41), and (4.42) can be obtained, and formula (4.43) is converted from formula (4.22).

## 4.3.4 Design of Two-Stage Demand Clustering Algorithm

A two-stage algorithm was adopted to solve the model, and the algorithm flow is shown in Fig. 4.1.

### 4.3.4.1 Clustering Stage

Clustering algorithms are generally divided into partition clustering method, hierarchical clustering method, density-based clustering method, grid-based clustering method, and integrated clustering algorithm. EM algorithm, as the most popular algorithm among clustering algorithms, was proposed by Dempster et al. in 1977 for the maximum likelihood estimation of the parameters of a probability model containing Hidden variables (Myung 2003). The main steps are as follows:

Step 1: Define the number of components $K$, set the initial value of each $k$ parameter $\pi_k$, $\mu_k$, $\sum_k$, and calculate the log-likelihood function.

Step 2: E process, according to $\pi_k$, $\mu_k$, $\sum_k$, calculate the probability $\gamma(z_{nk})$.

$$\gamma(z_{nk}) = \frac{\pi_k N(x|\mu_k, \sum_k)}{\sum_{j=1}^{K} \pi_j N(x|\mu_j, \Sigma_j)} \quad (4.44)$$

Step 3: M process, update $\pi_k$, $\mu_k$, $\sum_k$ according to the calculation $\gamma(z_{nk})$ in E process.

$$\mu_k^{new} = \frac{1}{N_k} \sum_{n=1}^{K} \gamma(z_{nk}) x_n \quad (4.45)$$

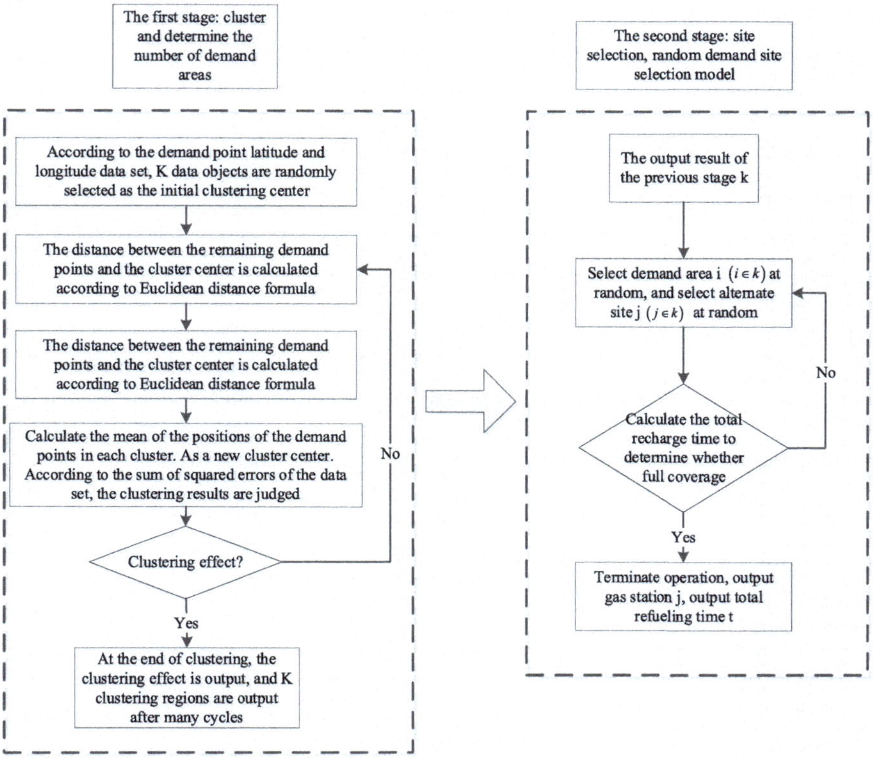


**Fig. 4.1** Two-stage algorithm flow

$$\sum_k^{new} = \frac{1}{N_k}\sum_{n=1}^{K}\gamma(z_{nk})(x_n - \mu_k^{new})(x_n - \mu_k^{new})^T \tag{4.46}$$

$$\pi_k^{new} = \frac{N_k}{N} \tag{4.47}$$

where, $N$ represents the number of points, $N_k$ can represent the number of points belonging to the first $k$ cluster.

$$N_k = \sum_{n=1}^{N}\gamma(z_{nk}) \tag{4.48}$$

Step 4: Calculate the logarithmic likelihood function.

$$\ln p(x|\pi, \mu, \Sigma) = \sum_{n=1}^{N}\ln\left\{\sum_{k=1}^{K}\pi_k N(x|\mu_k, \sum_k)\right\} \tag{4.49}$$

Step 5: Check whether the parameters converge or whether the likelihood function of the logarithm converges; if not, return to step 2.

#### 4.3.4.2 Site Selection Phase

Based on the clustering results of the distribution of replenished vehicles, with the goal of enabling the minimum number of stations, design the location steps:

Step 1: Obtain $k$ location of a demand area and $m$ location of an alternative gas station as feasible points.

Step 2: Select a demand area $i, i \in k$ and an alternative gas station $j, j \in m$, calculate its total recharge time, and meet all the demand areas are covered.

Step 3: Return to Step 2 to find the optimal exchange combination, reduce the value of the objective function, and update the status. When the optimal position can no longer be found, the operation is terminated and the result is printed.

### 4.3.5 *Comparative Analysis of Examples*

Based on the distribution location and quantity data of vehicles in a street in Nanan District of Chongqing, the probability of urban spatial random recharge demand is described. According to the rationality of the application of Gaussian mixture model in the field of random distribution, the urban recharge demand is clustered.

In order to realize the solution of the model and algorithm, the values of relevant parameters in the location selection process are shown in Table 4.2.

#### 4.3.5.1 Analysis of Site Selection Results

By analyzing the data of replenishment vehicles in Chongqing, the spatial distribution law of replenishment vehicles is obtained. The replenishment vehicles occur randomly at the starting place (residential area, etc.), during the trip, and at the

**Table 4.2** Values of related parameters

| Parameters | Values |
|---|---|
| $K$ | 12 |
| $M$ | 4 |
| $t_{s1}$ | 20 (min) |
| $t_{s2}$ | 22 (min) |
| $\theta_1$ | 0.1 |
| $\theta_2$ | 0.1 |

destination (company, etc.), and their geographical location reflects strong randomness. Figure 4.2 shows the current geographical distribution of vehicle replenishment demand in Nanping Street, Nanan District, Chongqing, indicating the possible points of replenishment demand. Through the distribution of replenishment data, it is found that some areas of replenishment demand are more intensive, and some areas are more dispersed. GMM clustering algorithm and K-means algorithm are used for clustering, and the algorithm is a continuous iterative process. According to the distribution characteristics of the original data, the horizontal and vertical coordinates in the figure represent the latitude and longitude, and 12 demand regions as shown in Fig. 4.3 are obtained after the algorithm is used for clustering.

Through Baidu map search and field investigation, it is found that there are currently two gas stations in Nanping Street, Nanan District, Chongqing. According to the site selection requirements of gas stations, they are far away from schools, business districts, and other areas, and they are on the right side of the road with large traffic flow and convenient transportation. Therefore, there are four alternative stations in this area except the existing gas stations. In Fig. 4.4, the hollow rectangle represents an established gas station, the hollow circle represents an alternate station, and ® represents an isolated demand point.

In order to detect the impact of random energy demand on site selection, comparison is made between sites with random demand and sites without random energy demand. The mean value is used instead of the random distance and quantity, i.e. the mean value model.

$$\sum_{i=1}^{K}(\lambda_i \sum_{j}^{M}(I_{\bar{t}_{ij} \le t_{sr}} \alpha_{ij})) \ge \varphi \sum_{i=1}^{K} \lambda_i \tag{4.50}$$

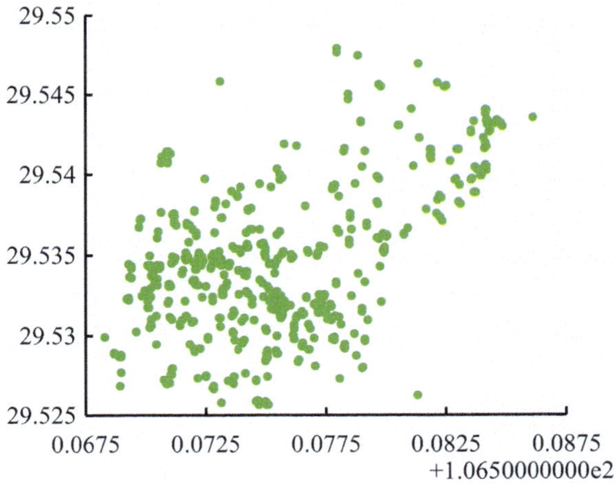


**Fig. 4.2** Original demand distribution

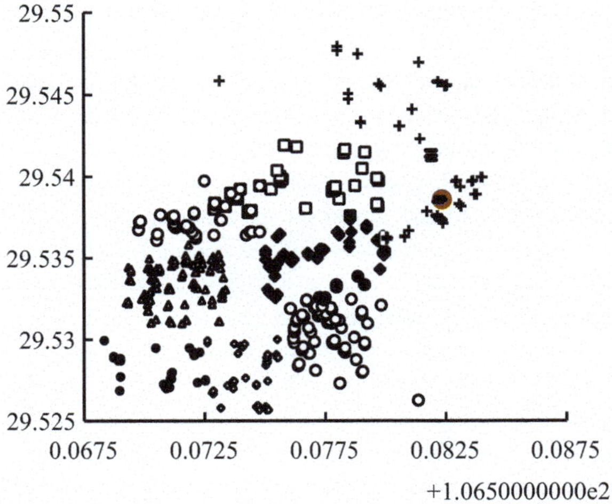


**Fig. 4.3** Algorithm iteration results

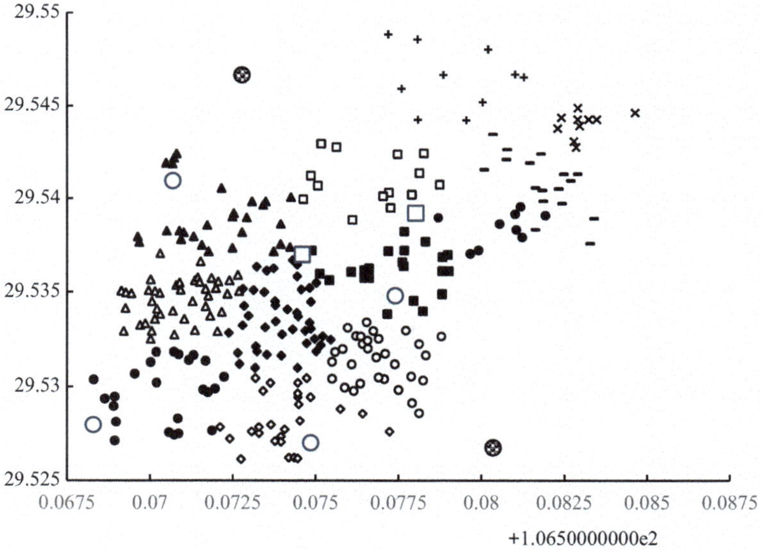


**Fig. 4.4** Clustering results of Gaussian mixture model

$$\sum_{j=1}^{M} \bar{t}_{ij}\alpha_{ij} \leq t_{s1}, \forall i \tag{4.51}$$

$$\sum_{j=1}^{M} \bar{t}_{ij}\beta_{ij} \leq t_{s2}, \forall i \tag{4.52}$$

$I_{\bar{t}_{ij} \leq t_{sr}}$ is the indicator function, when $\bar{t_{ij}}\, t_{ij} \leq t_{sr} t_{sr}$, $I_{\bar{t}_{ij} \leq t_{sr}} = 1$, or $I_{\bar{t}_{ij} \leq t_{sr}} = 0$.

As can be seen from Table 4.3, the four candidate sites in the random replenishment demand all provide replenishment service, which reduces the replenishment time of each replenishment demand area, and the average replenishment time is 8.75 min. Among them, there are a total of 5 recharge areas for selecting alternative site 3, accounting for almost half. However, the mean value model does not consider the randomness of the energy replenishment demand, so the total energy replenishment time is longer, which proves that the driving time of vehicles increases during the energy replenishment process or the vehicles gather seriously during the energy replenishment process, resulting in an increase in the queuing time. From the view of the travel time of the figure, the random recharge model can save the travel time in most demand areas, except for a long time to the gas station in some demand areas. It is found in the figure that the queuing time generated by the random replenishment model is all below the mean value model, indicating that the random replenishment model can significantly optimize vehicle replenishment distribution, effectively avoid vehicle replenishment aggregation, and shorten vehicle replenishment queuing time. Obviously, from the perspective of energy replenishment travel time, energy replenishment queuing time, and total energy replenishment time, the location selection results obtained by the random energy replenishment model are better than the mean value model, indicating the effectiveness of the scheme.

As can be seen from Figs. 4.5, 4.6, and 4.7, in the site selection considering spatial random demand, under the constraint of total energy replenishment time, all vehicles can complete energy replenishment within 20 min when the randomness of demand is considered; without the randomness of energy replenishment demand, there is

**Table 4.3** Analysis of results in each region

| | Random replenishment site selection | | | | | Mean complement energy siting | | | | |
|---|---|---|---|---|---|---|---|---|---|---|
| Demand area | Travel time | Queue time | Service time | Total time | Site | Travel time | Queue time | Service time | Total time | Site |
| 1 | 3 | 2 | 4 | 9 | 4 | 2.75 | 5 | 4 | 11.75 | 1 |
| 2 | 1 | 2 | 4 | 7 | 1 | 3.25 | 10.75 | 4 | 18 | 3 |
| 3 | 2 | 2 | 4 | 8 | 2 | 2 | 6.5 | 4 | 12.5 | 3 |
| 4 | 3 | 1 | 4 | 8 | 3 | 3.5 | 11.25 | 4 | 18.75 | 1 |
| 5 | 2 | 7 | 4 | 13 | 3 | 4 | 13.5 | 4 | 21.5 | 3 |
| 6 | 1 | 1 | 4 | 6 | 3 | 1.25 | 11.75 | 4 | 17 | 4 |
| 7 | 2 | 0 | 4 | 6 | 1 | 3 | 3.75 | 4 | 10.75 | 3 |
| 8 | 1 | 5 | 4 | 10 | 2 | 2.75 | 8.5 | 4 | 15.25 | 1 |
| 9 | 3 | 2 | 4 | 9 | 3 | 3.5 | 8.5 | 4 | 16 | 4 |
| 10 | 1 | 1 | 4 | 6 | 3 | 3.5 | 8.75 | 4 | 16.25 | 1 |
| 11 | 2 | 6 | 4 | 12 | 4 | 3.5 | 15.5 | 4 | 23 | 3 |
| 12 | 3 | 4 | 4 | 11 | 1 | 3 | 8 | 4 | 15 | 4 |

only a probability of 16.6% to complete energy replenishment service within the specified time, so the randomness of energy demand is adopted. It can significantly improve the vehicle replenishment efficiency and reduce the negative effect caused by the vehicle replenishment period.

This book focuses on the randomness of the supplementary energy demand, shows the description method of supplementary energy demand and the location planning model of gas station, and uses the mixed Gaussian model to quantitatively cluster the supplementary energy demand from the actual supplementary energy demand of Nanan District of Chongqing. On this basis, a siting model of gas station considering the random demand of energy replenishment is established to improve the efficiency

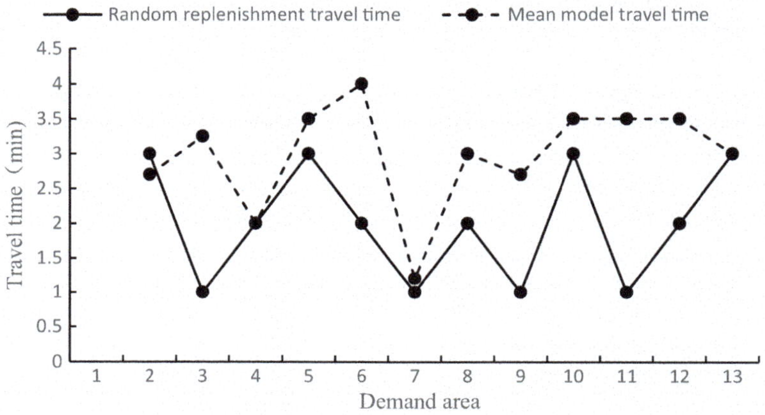


**Fig. 4.5** Analysis of travel time under different models

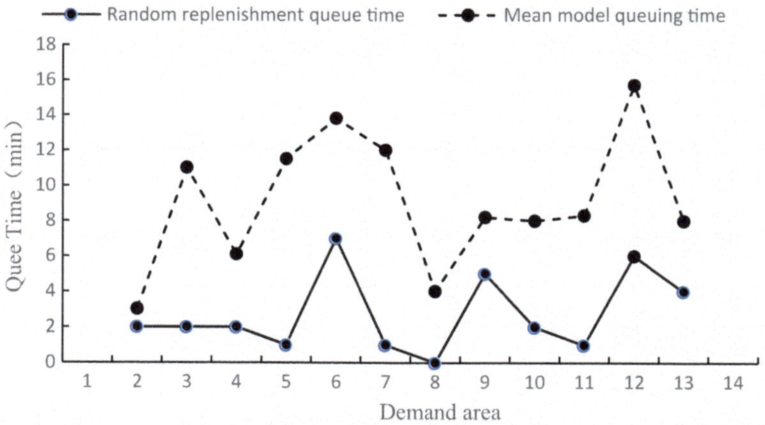


**Fig. 4.6** Analysis of queuing time under different models

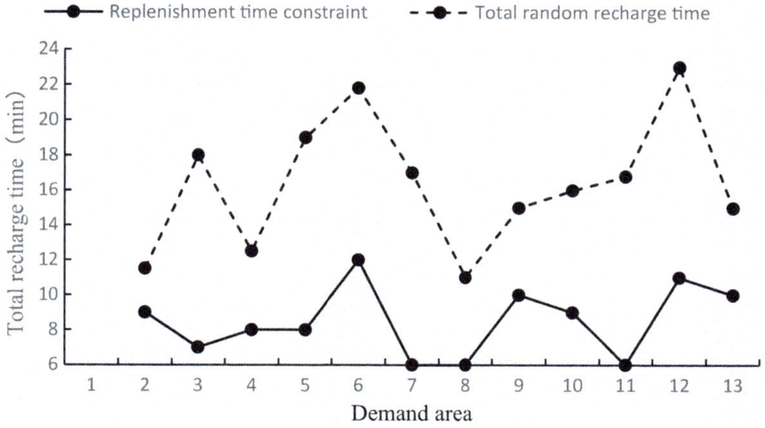


**Fig. 4.7** Analysis of total replenishment time under different models

of vehicle energy replenishment and reduce the negative effect of energy replenishment. In the case analysis, the energy demand of the whole area of Nanping Street in Nanan District of Chongqing is divided into 12 demand areas by Gaussian clustering, and then the gas station layout is obtained by random energy model. The results show that randomness has obvious effect on gas station planning. Compared with the equilibrium model, the average replenishment time is shortened from 21.81 min to 8.75 min, and the total replenishment time is reduced from 195.75 min to 105 min. The energy replenishment waiting time in all demand areas is reduced, the energy replenishment driving time is partially reduced, and the energy replenishment is completed within a reasonable time. The experimental results prove the effectiveness and superiority of this method.

## 4.4 Suggestions on Spatial Layout Planning of Urban Transportation Energy Supply Stations

### *4.4.1 Strategic Suggestions on Urban Transportation Energy Supply Network Level*

With the continuous advancement of urban construction, the sustainable and efficient development of economy and society, the number of motor vehicles has increased year by year, and the increase in the demand density of supplementary energy has put forward higher requirements for the convenience, service, and efficiency of supplementary energy station. Based on the current investigation and research of the entire urban transportation energy supply station, the scientific, systematic, and dynamic correlation of the layout of energy supply station is analyzed, and on the basis of combining qualitative and quantitative indicators, reliability and feasibility and other principles, the optimized deployment of energy supply station in the road network area is carried out to reduce road network congestion and station queuing and other negative effects. Improve the transportation efficiency of urban transportation network and the income of energy supply stations, and promote the urban economic, efficient and green development.

Gas station is an important part of urban energy supply network. The quality of its location directly affects the vehicle fuel replenishment efficiency and road operation in the relevant road network area, and also affects the effective operation and operation of related enterprises. With the development and construction of the city, the transportation network has a new pattern, and the distribution of gas stations has a non-negligible impact on people's travel activities, enterprise operations, and social and economic development. The reasonable location and layout of gas stations can support the transfer of people and the transportation of goods, facilitate people's travel, and make the economic activities of the city more reasonable and orderly. Different from the general site selection, gas stations mainly deal with gasoline and diesel, which are inflammable, explosive, and toxic dangerous goods storage facilities. Gas stations in cities with high personnel density have high safety requirements, so they should meet the national norms during construction, and the safe arrangement distance between oil storage tanks should be fully considered during site selection. Determine the safety distance according to the grade of the gas station and the fire resistance grade and building nature of the surrounding buildings, and its site selection requirements and standards can not be sloppy. In addition to considering safety factors, road distribution, population density, and vehicle type in the area where the gas station is located directly determine the actual influencing variables such as vehicle flow, inbound and outbound vehicles and single vehicle refueling volume of the gas station. Based on the deployment of the existing urban energy supply network, the number of stations and the number of refueling guns that meet the actual demand for energy replenishment are proposed, and a more reasonable location strategy is proposed. Develop a more scientific location of station facilities to reduce carbon emissions and other negative effects in the road network.

Green energy is an important part of future new energy vehicle technology. Already, other countries are focusing on new energy vehicles and charging stations to provide good alternatives to internal combustion engine vehicles. The popularity of electric vehicles not only needs to increase the number of charging stations and charging piles, but also needs to consider the characteristics of the charging station itself to solve the problems of electricity transportation, charging capacity, battery compatibility, charging time, charging cost, and so on. In the planning process of the replenishment network of new energy vehicles, it is necessary to carefully analyze the needs and actual conditions of the research area, such as the annual increase of electric vehicle ownership, the future development of battery technology, and the innovation of charging and replacing technology, the popularity of electric vehicle sharing and other factors, and formulate reasonable deployment strategies to reduce the range anxiety of electric vehicle drivers and improve the penetration rate of electric vehicles. Promote reasonable competition and cooperation between enterprises at different charging stations, improve the operational income and efficiency of the replenishment network, and accelerate the development of green transportation.

### *4.4.2 Strategic Suggestions on the Supply and Demand Level of Urban Transportation Energy*

If there is no scientific prediction of the construction number of gas stations, the planning and construction of gas stations will lag behind the development of motor vehicles and the urbanization process, and the negative effect phenomenon of the existing urban energy supply network cannot be alleviated, and the pressure of the supplementary energy load of some gas stations and the possible traffic congestion on the road where the gas stations are located during the peak period of supplementary energy can be reduced. At the same time, the number of gas stations can not be over-expanded, in the actual construction of gas stations, construction costs and operating costs must also be considered as a factor. Excessive deployment of gas stations will reduce the competitiveness of a single station, increase the land cost of initial construction, reduce operating income, and fail to effectively use land resources and protect the ecological environment, so as to achieve the goal of reducing the burden of urban energy supply network.

At present, the number of electric vehicles in China is increasing year by year, and the deployment of a reasonable number of charging stations can effectively slow the mileage anxiety of electric vehicle drivers and reduce the detour cost. At present, the utilization rate of public charging piles for electric vehicles in China is very low, and the phenomenon of "zombie piles" still exists. The lack of reasonable charging station layout and accurate charging load prediction has become the main reason for the low efficiency of the use of charging piles. Solving the low utilization rate of charging piles is the premise of whether electric vehicles can accelerate the deployment efficiency of electric vehicles. Therefore, the assessment of the demand for replenishment of electric vehicles and the prediction of the actual deployment number of charging stations, in other words, the calculation of the number of charging stations and charging piles to be deployed is also of great significance for rational layout of charging facilities, improvement of equipment utilization rate, and reduction of construction cost and management overhead.

### *4.4.3 Strategies and Suggestions for Urban Traffic Replenishment at Individual Level*

As far as the current development trend is concerned, the unreasonable deployment of gas stations in China is caused by factors such as unreasonable urban development and pre-construction planning, ineffective supervision by government departments, competition between enterprises and self-employed people and blind pursuit of interests. Due to the unscientific layout of some gas stations at present, it affects the construction of the entire transportation system, reduces the efficiency of vehicle replenishment, and forms a series of negative effects. Therefore, the analysis of factors such as population size, car ownership, and geographical location meets the requirements of urban planning, environmental protection, and fire safety, meets the specifications of Urban Road Traffic Planning and Design, estimates the demand for replenishment of different types of fuel vehicles in the road network area, and considers the economic factors such as land cost and refined oil transportation cost. Based on the driver's personal habits and external factors in the replenishment process involving braking, steering, inbound, stopping, outbound, and importing to road vehicles, and the actual situation of existing stations, optimize the deployment of gas stations, withdraw, merge, transform and build new stations, so as to improve the replenishment efficiency, service capacity and operating income of gas stations in the road network. And meet the supply needs of future urban development.

In terms of charging station deployment, electric vehicles and fuel vehicles have great differences in energy consumption. Electric vehicles are limited by the size of their battery capacity, and even with recent improvements in battery capacity, the actual range is still generally lower than that of conventional fuel vehicles. Under the same driving range, the charging frequency of electric vehicle drivers is higher than the refueling frequency of fuel vehicles, so travelers need to consider factors such as road network characteristics and traffic status, but also need to consider the charging state of electric vehicles, uncertain charging time and charging location. Considering the future expansion of the number of electric vehicles and the development trend of battery capacity, in the deployment of charging stations and charging piles in the road network, the charging characteristics of electric vehicles and the corresponding user characteristics of drivers, such as users' travel habits and energy replenishing needs, are studied according to the operating status, characteristics and future trends of charging stations. Combined with the distribution of existing charging facilities, battery performance, road structure, and other factors, the purpose of reducing urban road congestion and charging station operating costs.

## 4.5 Summary of This Chapter

From the perspective of urban overall energy supply network planning, considering that the unreasonable network layout of urban energy supply stations is one of the main reasons causing the negative effect of supplementary energy, based on the existing layout, the layout optimization of urban energy supply stations is studied to minimize the negative effect of supplementary energy network. This chapter first makes clear the principle of site layout optimization considering the minimization of negative effects; Secondly, the traditional research methods of layout planning are analyzed, and the optimization model of energy site layout based on demand clustering is established. Finally, relevant suggestions are put forward for the spatial layout planning of urban transportation energy supply stations.

## References

Hakiml SL (1964) Optimum locations of switching centers and the absolute centers and medians of a graph. Oper Res 12(3):379–517

Myung IJ (2003) Tutorial on maximum likelihood estimation. J Math Psychol 47(1):90–100

Suzuki A, Drezner Z (1996) The p-center location problem in an area. Locat Sci 4(1–2):69–82

Toral R, Chakrabarti A (1993) Generation of Gaussian distributed random numbers by using a numerical inversion method. Comput Phys Commun 74(3):327–334

Toregas C, Swain R, ReVelle C et al (1971) The location of emergency service facilities. Oper Res 19(13):1363–1373

Vatsa AK, Jayaswal S (2016) A new formulation and benders decomposition for the mult-period maximal covering facility location problem with server uncertainty. Eur J Oper Res 251(2):404–418

# Chapter 5
# Research on Cooperative Game of Urban Transportation Energy Supply Network Based on Minimal Negative Effects

## 5.1 Principle Analysis of Multi-Subject Cooperative Game of Urban Transportation Energy Supply Network

### *5.1.1 Cooperative Game Theory*

Depending on whether a binding agreement can be reached, games are divided into cooperative games and non-cooperative games. Cooperative game studies how players distribute the benefits of cooperation when they reach cooperation, that is, the problem of income distribution; Emphasize group rationality, and do not discuss how rational individuals reach the process of cooperation, but directly discuss the results of cooperation and the distribution of benefits. Cooperative games take a cooperative approach, or compromise. The reason why compromise can increase the interests of both sides and the interests of society as a whole is because cooperative games can produce a cooperative surplus, and this surplus is generated from this relationship and method. As for how the surplus of cooperation is distributed among the parties to the game, it depends on the strength and contribution of the parties to the game. Therefore, compromise must be negotiated by all parties to the game, reaching consensus, and cooperating. According to the number of people participating in the game, cooperative games can be divided into two-person cooperative games and n-player ($n > 2$) cooperative games.

The general representation of cooperative games is that let $N$ be a non-empty finite set of participants ( all possibilities for cooperation need to be considered), and each subset $S \subset N$ can be thought of as a union, set $N$ is called a grand alliance, and set $\phi$ is called an empty alliance. For the set of alliances, there are $2^N$, and for each $S \subset 2^N$, $|S|$ is used to represent the number of elements in $S$, and $e^S$ is used to represent the eigenvectors of $S$ (Branzei et al. 2008), where



X. Ge, *Urban Transportation Energy Supply Network*, Industrial Ecology and Environmental Management 5, https://doi.org/10.1007/978-3-031-64619-5_5

$$\begin{cases} (e^S)^i = 1, i \in S \\ (e^S)^i = 0, i \in N \backslash S \end{cases} \tag{5.1}$$

Here $N = \{1, \ldots, n\}$.

Thus, the cooperative game defined as having the form of a feature function is an ordinal $\langle N, v \rangle$ where $N$ is the set of participants and $v : 2^N \to \mathbb{R}$ is the feature function satisfying $v(\phi) = 0$. The real-valued function $v(S)$ can be interpreted as the maximum benefit or the maximum cost savings that can be saved by the alliance members when $S$ cooperates, and it is usually assumed that the game $\langle N, v \rangle$ has a characteristic function $v$.. Cooperative games with the form of characteristic functions are usually interpreted as transferable payment games (TU-games).

Cooperative games can be divided into two stages. The first stage is the stage of profit struggle, that is, collective rationality. The alliance as a whole play with the outside of the alliance or the alliance to strive for cooperation surplus; It is also possible that all the intellectuals are in a major league, then it is a problem of integration optimization. The second stage is the stage of benefit distribution, that is, individual rationality. Members of the alliance negotiate and bargain over the remaining cooperation and finally reach a distribution plan. The most common cooperative game distribution solutions are the Nash bargaining solution, the Core and Shapley values, and so on.

First of all, we will introduce the Nash bargaining game, that is, the bargaining game between two people, which is the basic problem of cooperative game, and the essence is the division and distribution of specific interests between two subjects.

The Nash bargaining game needs to satisfy three basic elements. The first is the distribution protocol, which is a binary group $s = (s_1, s_2)$ that describes the benefits received by each person, and the distribution agreement is a central element of Nash's bargaining game. The second is the utility couple, which is a binary group $u = (u_1, u_2)$ that describes the subjective utility evaluation of the allocation agreement by both parties, respectively. Where $u_i$ is the expected utility of Person $i$ in the bureau, and is a real-valued function of the feasible allocation protocol set $S$ to the set of real numbers, $u : S \to R$. Finally, there is the breakdown point of negotiations, in binary $d = (d_1, d_2)$, any negotiation has the possibility of breaking down, $d_i$ indicates that when the negotiation breaks down, the benefits that person $i$ can get in the situation. Thus, a two-person bargaining problem can be expressed in terms of $B(S, d, U)$.

To take into account the stability of the alliance, the concept of nuclear is introduced here. For game $G = (N, v)$, if the allocation vector $x$ satisfies $\forall S \subset N, \sum\limits_{i \in S} x_i \geq v(S)$, $x$ is said to be in the kernel of $G = (N, v)$. That is, the core is made up of a number of distributions that guarantee that the big league is better than any kind of small league, at least no worse. If an alliance structure prevents all participants from benefiting from the alliance reorganization, the alliance structure is the nucleus of the cooperative game. That is, in this alliance structure, all participants have no incentive to change the current structure. The nucleus does not necessarily exist, and if the nucleus does, it must be a valid solution to the game. Finally, Lloyd Shapley proposed a new solution concept of n-player cooperative game proposed in

1953, the core idea of which is distributed according to the contribution of the people in the game, the greater the contribution, the more you get.

The Shapley value is calculated as $\phi(v) = (\phi_1(v), \phi_2(v), \ldots, \phi_{|C|}(v))$ (Shapley 1952), where

$$\varphi_c(v) = \sum\nolimits_{c \in S} \frac{(|C| - |S|)!(|S| - 1)!}{|C|!}[v(S) - v(S - \{c\})],$$
$$c = 1, 2, \ldots, |C| \tag{5.2}$$

$S$ is any collaboration including Company $c$; $v(S) - v(S - \{c\})$ is the marginal contribution of Company $c$ to the cooperation $S$. Therefore, the Shapley value allocates the weighted sum of its contributions to each business. The Shapley value describes fairness, and the kernel describes stability. If the Shapley value is not in the kernel, and we take the Shapley value as the allocation scheme, in fact, because of the instability of the alliance, the alliance cannot be formed at all, if the Shapley value is in the kernel, it is best to take the Shapley value as the distribution scheme.

### *5.1.2 Analysis of the Multi-Subject Cooperation Effect of Urban Transportation Energy Supply Stations*

Urban energy supply station is the necessary foundation for the normal operation of the city, with the improvement of urbanization level, the scale of the city gradually expands, but the current urban energy supply station has uncoordinated expansion, facilities supply and demand imbalance, and other problems, so the key to solving the problem is whether the layout of the energy supply station is reasonable or not, reasonable energy supply layout can ensure the balance of supply and demand. The current study analyzes the layout of energy supply stations, evaluates the contradiction between supply and demand at different stations, and provides corresponding management insights. However, such evaluations do not take into account the cost of site renovation and the competition between sites, and although the recommendations are effective, the actual implementation may encounter corresponding difficulties, such as increasing the service capacity of the site, requiring a large amount of cost investment, and whether the benefits can cover the costs. This introduces the concept of cooperative game, cooperation can bring many benefits, first of all, in terms of benefits, only need to increase a small amount of information cost, you can obtain more benefits; Secondly, cooperation can reduce the negative impact of the site, such as the phenomenon of vehicle queuing during peak hours, which can effectively avoid the impact of long queues on the main road; Finally, cooperation can improve the competitiveness of the site and increase customer satisfaction.

### 5.1.3 *Influencing Factors of Cooperation Between Urban Transportation Energy Supply Stations*

In general, cooperation faces two problems: how to cooperate and how to distribute the benefits. For the problem of income distribution, the existing theoretical research is very mature, so this part is easy to solve, mainly considering the problem of how to cooperate. Energy supply stations are jointly operated by a number of different subjects, there is a competitive relationship between the sites, and the benefits and risks of all parties need to be considered when taking cooperation, and the main influencing factors are brand influence, site management mode, and revenue. For gas stations, PetroChina and Sinopec are the most familiar brands to consumers, while charging stations are Tesla and China Southern Power Grid, and brands can give products a wider influence. For gas stations, different brands of oil come from different sources, resulting in differences in quality, and for unfamiliar consumers, choosing a branded site will make them more at ease. It mainly refers to the management system within the station and the management of customer service quality. If the service of the site does not reach the level of customer satisfaction, then the user will have directional thinking, thinking that the service situation of the cooperation point is similar, so it will choose other sites, thus bringing negative effects to the cooperation parties. Finally, there are the benefits. Taking charging stations as an example, time-of-use pricing is prone to differences in electricity prices at different stations at the same time, and after cooperation, electricity prices need to be reformulated to ensure the benefits of cooperation.

## 5.2 Behavior Analysis of the Main Body of Urban Transportation Energy Supply Network

### 5.2.1 *Analysis of Government Behavior*

With the vigorous promotion of new energy vehicles in the country, a series of relevant policies have been introduced. In December 2021, the Ministry of Finance issued the Notice on the Financial Subsidy Policy for the Promotion and Application of New Energy Vehicles in 2022, extending the subsidy for new energy vehicles to the end of 2022, and the subsidy for new energy vehicles in 2022 will decline by 30% on the basis of 2021. In January 2022, the National Energy Administration proposed to speed up electric vehicle charging base stations, and made it clear that by the end of the "14th Five-Year Plan", a balanced and intelligent charging infrastructure system could meet the charging needs of 20 million electric vehicles. In January 2022, the Ministry of Finance issued the "Notice on Adjusting the Oil Price Subsidy Policy for Rural Passenger Transport and Taxi", requiring continued subsidies for taxi and urban public transport; Among them, the subsidies for changing fees to taxes are directly issued to taxi drivers, and the subsidies for raising prices are coordinated by local governments. The subsidy is an urban transportation development incentive fund to support the operation of new energy vehicles in the field of urban transportation and energy. The cause of the negative effect of energy supply station is that the short-time refueling vehicles exceed the service capacity of the station, and the station queue easily leads to traffic congestion. The reasons for the queues at different stations are analyzed, and the following three are summarized: the NDRC's oil price adjustment, the normal peak period, and the promotion activities of energy supply stations.

#### 5.2.1.1 Adjust Oil Prices

The National Development and Reform Commission adjusts the domestic oil price according to the international market crude oil price, and adjusts the price every 10 working days. Car owners are the most sensitive to the rise and fall of oil price, and when the oil price rises the next day, the number of vehicles refuels will surge; When the oil price drops, the total sales volume of refined oil products at the station will increase. At this time, the stations with queue congestion are random and cannot be effectively predicted, which shows the necessity of station cooperation. The queuing of stations caused by the adjustment of oil prices is the impact of the government's macro-control, so more incentives should be given to the processing of stations.

#### 5.2.1.2 Normal Peak

Affected by commuting, commuting time is the peak period of energy supply stations, and the sections with large traffic flow are prone to queue congestion. Due to space restrictions, the increase of vehicles is easy to cause a disorderly queue-jumping phenomenon, affecting the normal service in the station, if the energy supply station is not properly managed, it will affect the main road. This kind of queue is basically fixed, and the appropriate partner can be found according to the historical situation.

#### 5.2.1.3 Promotional Activity

Promotion is the most common way to attract customers to the site, when the site is in the area of the maximum discount, the driver is affected by the strength of the discount and attracted to the station to replenish energy, easy to cause queuing phenomenon, affecting the normal vehicle driving. Such queues are caused by the stations themselves, so the public Security Bureau can set stricter fines according to the extent of the impact on the traffic on the main road, and let the stations take the initiative to shoulder the work of traffic removal.

### 5.2.2 Behavior Analysis of Energy Supply Stations

Energy supply stations are usually operated by different entities, such as CNPC, Sinopec, Shell, etc.; Charging station special calls, national grid, star charging, etc. Due to the large gross profit space of the refined oil market, charging stations have also been strongly supported by the state in recent years, and each business entity wants to occupy a larger market share, which makes the market competition more intense. The energy supply station is to serve the replenishment vehicle, and the station will also consider its own income. Due to the large capacity of the energy station market, the demand is also increasing year by year, and each business entity will take corresponding measures in order to attract users, such as choosing a lot with a large traffic flow to build sites, increasing sites, and taking price promotion methods to attract more customers. Cao Ying analyzed the factors that affect the sales volume of gas stations, including location, traffic flow, and brand. In the actual investigation, it was found that the sales volume of gas stations in urban areas and highway intersections was the highest. Secondly, for the same location, different traffic flow has a great impact on the sales of the site. Finally, the appeal of different brands to customers is different, through the 5-min node, the station sales rate of 5% to 10% is at an average level. Li Xia also added the internal and external environment of oil products and gas stations. In the external environment, she suggested the visibility of the appearance and the patency of the intersection, and in the internal management system, mainly aimed at customers, to improve the quality of customer service.

At this stage, the research only considers the sales of the station, but does not consider whether the establishment of the station or the promotion activities will have a negative impact on the road, such as the traffic occupying the road queue. In addition, the blind promotional activities of the station will easily lead to queuing and congestion at the gas station. For the situation that the station cannot handle, the degree of traffic congestion will be more serious, which not only affects the normal vehicles, but also increases the management cost of the government. With the increasing energy demand, solving or optimizing this problem has become an urgent problem in the field of urban transportation energy.

### 5.2.3 Behavior Analysis of Replenishment Vehicles

For drivers, when choosing a site, they pay more attention to the three factors of price, location, and brand, and in the service experience of the site, they are more concerned about the queuing situation of the site. When the queuing time is greater than the detour time, the driver may choose other stations to refuel, that is, the goal of the vehicle is to minimize the replenishment time. The model is as follows:

$$\min(t_i, t_{ij}) \tag{5.3}$$

$$t_i = t_q^0(\tau_i + 1) \tag{5.4}$$

$$t_{ij} = t_{cong}[1 + \alpha(\frac{v_c}{c_{net}})^{\beta}] \tag{5.5}$$

Formula (5.3) indicates that the vehicle evaluates its own queuing time and detour time; Formula (5.4) represents the queuing time of vehicles at the station, which is determined by the current number of queuing vehicles $\tau_i$ and the average service time $t_q^0$; Formula (5.5) is the BRP function (Bureau of Public Roads 1964), which represents the time for the vehicle to arrive at the free station.

## 5.3 Cooperation Model Between Urban Transportation Energy Supply Stations

### *5.3.1 Analysis of Cooperation Problems of Urban Transportation Energy Supply Stations*

In the time-varying road network, each time period will generate energy replenishment demand, due to the fixed location and limited number of energy supply stations in the road network, if the station is not properly distributed, it will cause some stations in a specific period of time to have more refueling vehicles than the service capacity of the station, at the same time, another part of the energy supply station in the road network may be located in a more remote location, fewer vehicles entering the station. Therefore, the phenomenon of large differences in the number of supplementary vehicles in the same period of time and between different stations in the road network is called energy supply service imbalance.

Based on the above phenomenon, a cooperative game model of energy supply station is proposed. As shown in Fig. 5.1, there are two energy supply stations in an area, assuming that the maximum number of vehicles that can be accommodated at the station is $P$, and at a certain time, $n$ vehicles go to station $S_1$ for refueling, $m$ vehicles go to site $S_2$ for refueling, and $n > P > m > 0$, at this time, site $S_1$ has queue congestion due to the refueling vehicles exceeding the upper limit. Introducing government incentives, the two stations cooperate to introduce some queuing vehicles from $S_1$ to $S_2$ and compensate them for their costs, so as to ease the operational pressure of site $S_1$, keep the queue length within an acceptable range, and reduce the queuing time of users to achieve service balance at the site.

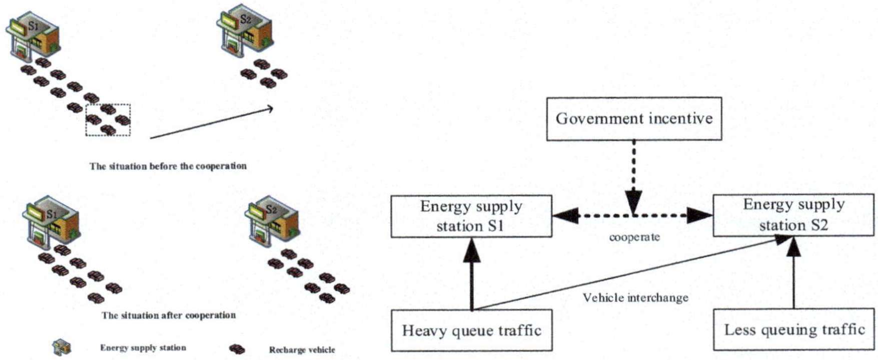


**Fig. 5.1** Cooperative game model of energy supply stations

## 5.3.2 *Problem Assumptions and Symbolic Illustrations of the Total Revenue Model of Urban Transportation Energy Supply Stations*

### 5.3.2.1 Problem Hypothesis

To articulate the issue more clearly, make the following assumptions:

1. Consider only the average yield per vehicle
2. Persuaded vehicles agree to go to the partner site
3. Traffic flow at idle stations is always below maximum service capacity
4. The maximum service capacity of the partner site is the same

### 5.3.2.2 Symbol Description

Based on the above analysis, the following symbols are involved in the question:

$G_1, G_2$: Government subsidies to support site cooperation;
$C_1, C_2$: The operating costs of the site;
$Q_1, Q_2$: The additional cost of establishing an information platform when the site cooperates;
$R_1, R_2$: The revenue of the site after the cooperation;
$T$: The calculation period is divided into $n$ periods on average, with an interval of $\triangle t = T/n$ in each period.
$p$: The average profit per vehicle to the site;
$f$: Compensation paid to the changing vehicle;
$\eta$: Maximum service capacity of the site (vehicles/hour);
$x_t^1, x_t^2$: Number of vehicles at the station in each period, and $x_{st}^2 < \eta, \forall t \in T$;
$\varepsilon, \gamma$: The compensation coefficient of stations $S_1$ and $S_2$ for the vehicle, and $\varepsilon + \gamma = 1$;

$\lambda_i^k, \mu_i^k$: Determine the number of vehicles and service capacity of the station in each period;

#### 5.3.2.3 Energy Supply Station Cooperation Model

Define the characteristic function $V(S)$ as the total revenue of the alliance, in order to maximize the profit of the alliance, the following calculation is required:

$$\max V(S) = R_1 + R_2 \tag{5.6}$$

$$R_1 = \triangle t \sum_{i=1}^{m} \lambda_i^k p - \triangle t \sum_{i=1}^{m} \alpha f(\lambda_i^k - \eta) - C_1 - Q_1 + G_1 \tag{5.7}$$

$$R_2 = r_2 - C_2 - Q_2 + G_2 \tag{5.8}$$

$$r2 = \triangle t \sum_{i=1}^{n} x_t^2 p + \triangle t \sum_{i=1}^{n} \mu_i^k (x_t^1 - \eta) p - \triangle t \sum_{i=1}^{n} \beta \mu_i^k f(x_t^1 - \eta) \tag{5.9}$$

$$T = \{[0, \triangle t], \ldots, [(i-1) \triangle t, i \triangle t], \ldots, [(n-1) \triangle t, n \triangle t]\} \tag{5.10}$$

$$\lambda_i^k = \begin{cases} x_t^1, & x_t^1 < \eta \\ \eta, & x_t^1 \geq \eta \end{cases} \tag{5.11}$$

$$\mu_i^k = \begin{cases} 0, & x_t^1 \leq \eta \\ 1, & x_t^1 > \eta \end{cases} \tag{5.12}$$

$$x_t^2 < \eta, \forall t \in T \tag{5.13}$$

Equations (5.7) and (5.8) indicate the respective benefits of site cooperation. Equations (5.9) are the sales revenue of the $S_2$ site. Equations (5.10) indicate the time period. Equation (5.11) indicates the vehicle served by station $S_1$ at each moment. Equation (5.12) is the 0–1 variable, when the vehicle at station $S_1$ exceeds the service limit, $\lambda_i^k = 1$, otherwise $\lambda_i^k = 0$. Equation (5.13) indicates the limit on the number of vehicles entering station $S_2$.

### *5.3.3 Revenue Distribution Model Based on Shapley Values*

Once the benefits of the alliance of energy supply stations have been determined, the next step is to consider a reasonable and equitable distribution method. Shapley value is a more classic benefit distribution method of cooperative games, which is distributed according to the marginal contribution of each member, avoiding the average distribution situation to a certain extent. Let $I = \{1, 2, \ldots, n\}$ be $n$ energy supply station participants, and if any subset of $s \in I$ has a characteristic function $v(s)$, the following prerequisites need to be met (Branzei et al. 2008):

$$v(\emptyset) = 0 \tag{5.14}$$

$$v(s_1 \cup s_2) \geq v(s_1) + v(s_2) \quad s_1 \cap_2 s = \emptyset \tag{5.15}$$

Obviously Eqs. (5.14) satisfy the conditions. For formula (5.15), according to the previous model, the cooperative site can absorb more vehicles, and the alliance revenue is greater than the sum of the benefits of the site's separate operation, so the energy supply station cooperation can meet the requirements of Eq. (5.15).

Shapley gives the calculation method for the alliance allocation, also known as the Shapley value (Shapley 1952).

$$\varphi_i(v) = \sum_{|s \subseteq S_i|} \frac{(|s| - 1)!(|n| - |s|)!}{|n|!}(v(s \cup i) - v(s)) \tag{5.16}$$

where $S_i$ represents the alliance of all members $i$, $|s|$ is the number of alliance members, and $(|s| - 1)!(n - |s|)!/n!$ is the probability of each situation after cooperation. $v(s \cup i) - v(s)$ is the contribution of member $i$ to the participation in the cooperation process.

### 5.3.4 Revenue Distribution Model Based on Shapley's Modified Value

The Shapley value considers the projects of all consortiums of the participants and calculates the marginal contribution of each participant in the project. In the case of only two site federations, the consortium scheme has only {A, B} and {B, A}, as shown in Fig. 5.2, which defaults to the two possibilities being equal in the distribution, but this equalization situation is obviously not in line with the status quo. The alliance of energy supply stations is based on the fact that there are too many vehicles at a certain station, and it is necessary to cooperate with nearby idle stations to eliminate queue congestion, and the number of vehicles in idle stations also affects the effectiveness of cooperation. Therefore, the congestion site has the initiative to cooperate, so there is only one situation left for the alliance, and the Shapley value needs to be improved.

On the basis of the Shapley value, add the influencing factor $\theta_j(0 < \theta_j < 1), j = 1, 2, \ldots, n$ of the income distribution. The factors affecting the revenue distribution of the energy supply station alliance were analyzed, and the three indicators of brand influence, contribution, and risk bearing of the site were selected.

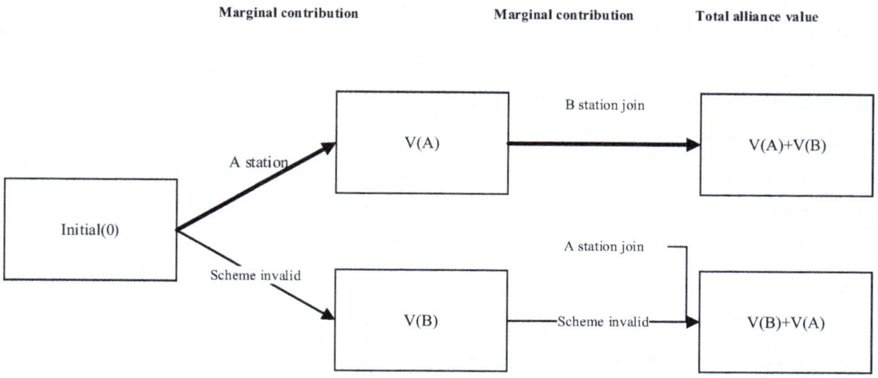


**Fig. 5.2** Site consortium formation diagram

#### 5.3.4.1 Brand Influence $\theta_1$

Most consumers know more about PetroChina and Sinopec and are less familiar with other private brands. In the list of domestic energy supply station brands in 2021, Sinopec and PetroChina ranked the top two, and Shell ranked third, which shows that consumers recognize these three brands more. Then in the cooperation, the site with a high brand can improve the trust of consumers and have a greater influence on the customer's decision-making. Therefore, brand effect plays an important role in site cooperation.

In order to mobilize the enthusiasm of vehicle owners to participate, several groomers should be arranged at the congestion site, introduce the location of the idle site, etc., persuade customers to go to the idle station, and distribute coupons. Idle stations can set up a sign at the entrance of the congested station to introduce the location and distance, which is convenient for the owner to go.

#### 5.3.4.2 Risk-taking $\theta_3$

Benefit and risk coexist, and the higher the risk-taking site, the greater the proportion of its return. In the site alliance, the risk mainly comes from the site's business data, employees' appraisal performance, customer losses, etc. The Alliance shall distribute the benefits on a risk-return basis.

The factors influencing the distribution of revenue in the construction alliance are shown in Table 5.1.

The table is normalized, and the values of different influencing factors $j$ are calculated and set to $b_{ij}$.

**Table 5.1** Influencing factors of income distribution

| Influencing factor | Brand influence ($\theta_1$) | Contribution degree ($\theta_2$) | Risk taking ($\theta_3$) |
|---|---|---|---|
| Energy supply station 1 | $p_{11}$ | $p_{12}$ | $p_{13}$ |
| Energy supply station 2 | $p_{21}$ | $p_{22}$ | $p_{23}$ |
| … | … | … | … |
| Energy supply station $n$ | $p_{n1}$ | $p_{n2}$ | $p_{n3}$ |

$$b_{ij} = \frac{p_{ij}}{\sum_{1}^{n} p_{ij}} \text{ and } \sum_{1}^{n} b_{ij} = 1, (j = 1, 2, 3) \tag{5.17}$$

After normalization, a new matrix is obtained:

$$B = \begin{Bmatrix} b_{11} & b_{12} & b_{13} \\ b_{21} & b_{22} & b_{23} \\ \vdots & \vdots & \vdots \\ b_{n1} & b_{n2} & b_{n3} \end{Bmatrix} \tag{5.18}$$

Due to the different degrees of influence of different factors, the final weight factor is determined by the AHP method (Podvezko 2009). If $\beta_j (0 \leq \beta_j \leq 1)$ is the weight coefficient of factor $j$, the weight vector is:

$$\beta = (\beta_1, \beta_2, \beta_3)^T \tag{5.19}$$

After adjustment, the comprehensive vector is $\lambda$

$$\lambda = (\lambda_1, \lambda_2, \lambda_3)^T = B * \beta \tag{5.20}$$

Then the model of the improved Shapley value is (Liu et al. 2020):

$$\varphi_i'(V) = \varphi_i(V) + V(I) * (\lambda_i - 1/n), \ i = 1, 2, \ldots, I \tag{5.21}$$

The improved Shapley value fully considers the influencing factors of different site cooperation, and improves the rationality and objectivity of distribution.

**Table 5.2** Traffic volume per time (vehicle/hour)

| Time frame | Number of vehicles at $S_1$ station | Number of vehicles at $S_2$ station | Maximum number of services per hour | Number of vehicles changing stations |
|---|---|---|---|---|
| 7:00–8:00 | 79 | 27 | 60 | 19 |
| 8:00–9:00 | 85 | 32 | 60 | 25 |
| 11:00–12:00 | 69 | 28 | 60 | 9 |
| 12:00–13:00 | 72 | 31 | 60 | 12 |
| 17:00–18:00 | 76 | 34 | 60 | 16 |
| 18:00–19:00 | 83 | 33 | 60 | 23 |

## 5.3.5 *Analysis of the Cooperation Between the Two Sides of Urban Transportation Energy Supply Stations*

### 5.3.5.1 Study Parameters

Given a traffic network, there are two energy supply stations. Among them, site $S_1$ is often queued during peak hours, and site $S_2$ is generally idle. The calculation period is from 6 am to 10 pm, $t = 1$ hour, based on the peak travel period of the city, the peak time of the energy supply station is 7:00–9:00, 11:00–13:00, 17:00–19:00, so the alliance can be reached in these three time periods.

Assuming that each vehicle can make a profit of 130 yuan, the operating cost of the $S_1$ station is 700 yuan/day, and the operating cost of the $S_2$ station is 500 yuan/day. After the alliance, the additional cost of the information platform is 130 yuan/day, the government subsidy is 1000 yuan/day, the compensation for changing vehicles is 20 yuan/vehicle, and the maximum service volume of the two stations is 60 vehicles/hour. The specific traffic flow is shown in Table 5.2.

In the road network, when the traffic exceeds the service level, subsequent vehicles will choose to leave because they are not served. In cooperative games, the site has two optional strategies, operating separately and cooperating in pairs. $R_1$ and $R_2$ represent the benefits of non cooperative state between Site 1 and Site 2, while V(S) represents the benefits of cooperation between the two sites. It can be found that the cooperation between sites can increase the revenue, this cooperation is a convex game, and the Shapley value is within the kernel (Table 5.3).

### 5.3.5.2 Profit Distribution

Through the previous revenue calculation, it can be known that the benefits of cooperation between the two parties are greater than the separate operation, so the site is more willing to reach cooperation, and the next step is the issue of revenue sharing, and the specific distribution process is as follows.

**Table 5.3** Returns of different strategies in cooperative games

| Time frame | $R_1$/yuan | $R_2$/ yuan | $R_1 + R_2$/ yuan | $V(S)$/ yuan |
|---|---|---|---|---|
| 7:00–8:00 | 7800 | 3510 | 11,310 | 13,400 |
| 8:00–9:00 | 7800 | 4160 | 11,960 | 14,820 |
| 11:00–12:00 | 7800 | 3640 | 11,440 | 12,430 |
| 12:00–13:00 | 7800 | 4030 | 11,830 | 13,150 |
| 17:00–18:00 | 7800 | 4420 | 12,220 | 13,980 |
| 18:00–19:00 | 7800 | 4290 | 12,090 | 14,620 |
| Total revenue | 46,100 | 23,550 | 69,650 | 82,070 |

**Table 5.4** Comparison of the two allocation methods (yuan)

| Allocation method | station $S_1$ | station $S_2$ |
|---|---|---|
| Independent operation | 46,100 | 23,550 |
| Value of Shapley | 52,310 | 29,760 |
| The correction of Shapley | 53,179.4 | 28,890.6 |

Step 1: According to the evaluation of experts and site-related personnel, give the correction parameters and weights.

$(\theta_1, \theta_2, \theta_3)_{S_1} = (0.686, 0.367, 0.573)$, $(\theta_1, \theta_2, \theta_3)_{S_2} = (0.314, 0.633, 0.427)$; The weight coefficients of the three influencing factors are $\beta = (0.5, 0.3, 0.2)^T$, from which the correction factors $\lambda_1 = 0.57$, $\lambda_2 = 0.43$ are obtained. For vehicles, the compensation coefficients $\varepsilon = \lambda_1$ and $\gamma = \lambda_2$.

Step 2: Calculate the revenue distribution of the site after cooperation.

After the cooperation, the additional income generated includes the income brought by the changing vehicle, as well as the government subsidy, and the additional cost incurred is the information platform fee and the compensation of the changing vehicle, which results in the increased income after the cooperation of 12,420 yuan. Therefore, the initial distribution of the alliance can be calculated according to Eq. (5.16), and the Shapley correction can be obtained according to Eq. (5.21), as shown in Table 5.4, the total revenue of site S1 is 53179.4 yuan, and the total revenue of site S2 is 28890.6 yuan.

#### 5.3.5.3 Analysis of the Results

From the table, it can be seen that the revenue of site $S_1$ increased by 15.4%, and the revenue of site $S_2$ increased by 22.7%, and both sites have increased to a certain extent compared to when they operated independently, which is consistent with the original intention of the cooperative alliance. Secondly, in terms of distribution rationality, the Shapley correction value is used, in which the size of the correction factor will affect the income of the alliance members, so it will be more reasonable for the partners to jointly formulate the weight and impact factor.

In order to solve the imbalance of energy supply station services, a cooperative game model of energy supply stations is proposed, which transforms the situation handled within a single site into multi-site cooperation in the past. In order to ensure the reasonable distribution of benefits, a correction factor is introduced on the basis of the Shapley value. The examples show that the revised distribution model can not only increase the benefits of both partners, but also make the number of vehicles at the station more balanced, which provides a theoretical reference for reducing the burden of urban energy supply stations.

## 5.4 Tripartite Cooperative Game Model Between Government, Energy Supply Station, and User

### *5.4.1 Analysis of Tripartite Cooperation Between Government, Energy Supply Stations, and Users*

It is believed that users and local governments are very rational and that they will also consider environmental protection in the pursuit of profit maximization. Assuming that the user (P) and the local government (G) are the two sides of the game, in a complete information environment, the user's strategy set is {queuing to replenish energy, changing stations to replenish energy}, where queuing replenishment means that the queue is too long, which affects the normal vehicle driving, and the replacement station replenishment is that the vehicle chooses a less idle station to replenish energy, which does not affect normal traffic. The policy set of local governments is $Sg$ ={regulated, non-regulated}. When the user wants to save the time of detour (recorded as $Rpt$) and queue up at the current station to replenish energy, the local government will impose a fine (recorded as $F$, $F \geq Rpt > 0$); When the user chooses to change stations for refueling, it is necessary to find an idle station, especially when it is far from the destination, and there is a higher time and monetary cost (denoted as $Cpt$), and of course the local government will reward these vehicles (marked $Rpm$). Regardless of whether users are queuing for energy replenishment or changing stations, local governments have the obligation to supervise the queuing of energy supply stations. The regulatory cost here is recorded as $Cgm$, and when the user is found to be queuing up on the main road to replenish energy, a fine can be charged (denoted as $F$); If there is no regulation, local governments will be punished by the central government, denoted as $Cgn$. Therefore, the choice of actions and payments of users and local governments is shown in Table 5.5.

**Table 5.5** Action options and payments by users and local governments

| Participant | Pure strategy selection | Cost | Income |
|---|---|---|---|
| User | Queue recharge | fine($F$) | Save detour time($Rpt$) |
| | Change stations for replenishment | Time and expense($Cpt$) | Rewards($Rpm$) |
| Local government | Regulatory | Regulatory cost($Cgm$) | Fines and positive impact($F+E$) |
| | Non- regulatory | Negative impact($Cgn$) | 0 |

### 5.4.2 *Game Analysis Between Users and Governments*

A payment matrix for the game between users and local governments is established, as shown in Table 5.6.

Table 5.6 shows that when local governments regulate it, the user's choice of action depends on the value of $Rpt - F$ and $Rpm - Cpt$ When local governments do not regulate, the choice of user behavior depends on the value of $Rpt$ and $Rpm - Cpt$. The process of the game is shown in Fig. 5.3.

Obviously, there is no pure strategic Nash equilibrium in the game, it belongs to the mixed strategy Nash equilibrium. Suppose $\gamma$ represents the probability of users queuing up for energy replenishment, and $\theta$ represents the probability of local government supervision. Then, the probability of changing stations is $(1-\gamma)$, and the probability of non-supervision is $(1-\theta)$. Therefore, the expected revenue of local governments is:

$$\begin{aligned} Eg = \theta[(\gamma(F+E-Cgm)+(1-\gamma)(E-Cgm)] \\ +(1-\theta)[\gamma(-Cgn)+(1-\gamma)(-Cgn)] \end{aligned} \tag{5.22}$$

The expected return for users is:

$$\begin{aligned} Ep = \gamma[(\theta(Rpt-F)+(1-\theta)Rpt] \\ +(1-\gamma)[\theta(Rpm-Cpt)+(1-\theta)(Rpm-Cpt)] \end{aligned} \tag{5.23}$$

**Table 5.6** Game matrix for users and local governments

| | | Queue recharge | Change stations for replenishment |
|---|---|---|---|
| Local government | Regulatory | $(F+E-Cgm, Rpt-F)$ | $(E-Cgm, Rpm-Cpt)$, |
| | Non- regulatory | $(-Cgn, Rpt)$ | $(-Cgn, Rpm-Cpt)$, |

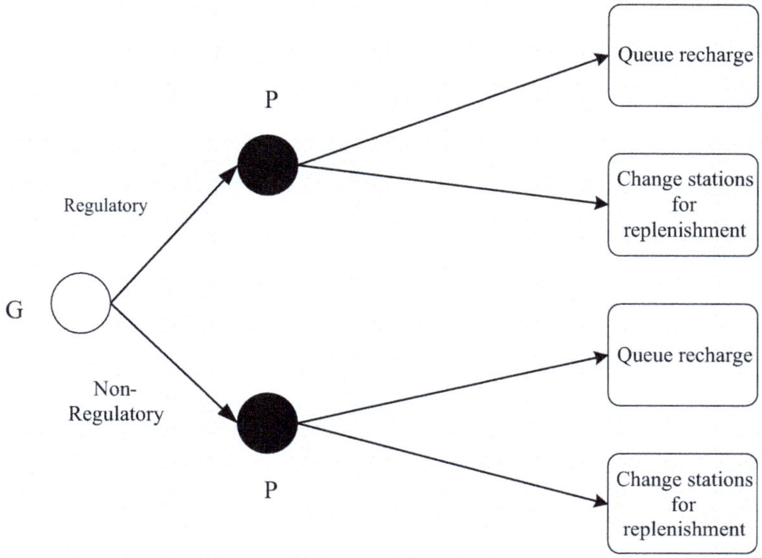


**Fig. 5.3** Game process between users and local governments

In Eqs. (5.22) and (5.23), the partial derivatives of $\theta$ and $\gamma$ are calculated and equal to zero.

$$\frac{\partial Eg}{\partial \theta} = \gamma(F + E) - Cgm + Cgn = 0 \tag{5.24}$$

$$\frac{\partial Ep}{\partial \gamma} = -\theta F + Rpt - (Rpm - Cpt) = 0 \tag{5.25}$$

After solving the system of equations, the values of A and B can be obtained:

$$\gamma* = (Cgm - Cgn)/(F + E) \tag{5.26}$$

$$\theta* = (Rpt - (Rpm - Cpt))/F \tag{5.27}$$

Therefore, it can be concluded that the Nash equilibrium of the hybrid strategy is:

$$(\theta*, \gamma*) = ((Rpt - (Rpm - Cpt))/(F + E)), (Cgm - Cgn)/F) \tag{5.28}$$

It can be found that when $Rpt - (Rpm - Cpt) < 0$, regardless of whether the local government has supervision or not, the user will choose to go to the idle site for replenishment, and when $Rpt - (Rpm - Cpt) > 0$, if the local government chooses to supervise, the user will detour to the idle station, once not supervised, it will be regarded as illegal. In order to achieve the former ideal goal, $Cpt < Rpm - Rpt$ must be met, that is, if the user needs to change stations consciously, then the cost must be controlled in the difference between the return on changing stations and the time and cost saved by queuing. Further analysis, government incentives for changing vehicles are usually fixed, and it takes some time to find an idle energy supply station, then the resulting revenue and other revenues are almost collected by the supply station and the local government. Therefore, in order to increase the motivation of users to bypass idle stations, it is necessary to distribute benefits between energy supply stations, users, and local governments.

### *5.4.3 Multi-Party Mixed Game Analysis*

Further analysis of the problem of users queuing at energy supply stations shows that the goals that all parties want to achieve can be divided into two levels, the first basic goal is that when the government adopts strict supervision, users can consciously go to stations with fewer vehicles; However, the ideal goal is that the government does not regulate and users still choose to change stations to detour. Here's a simple analysis.

#### 5.4.3.1 Basic Objectives

In order to get users to agree to change stations, then the subsidy to users is greater than or equal to the cost increase caused by changing stations, and who will bear this part of the subsidy, considering the huge user group in the country, if all borne by the government, it will be a huge financial expenditure, and if the subsidy is not standardized, it will also produce new problems.

#### 5.4.3.2 Ideal Goal

In order to meet the requirements of $Rpt - (Rpm - Cpt) < 0$, from the user's point of view, its income is determined by the government subsidy $Rpm$, which is generally fixed. In addition, in the actual energy supply station queue, the queuing time is different in each time period, especially when the station queue overlaps with the peak period of the road network, the user will give up the change station because of the difficulty of changing stations, therefore, in order to maximize the enthusiasm of users, the government and the transportation energy supply station need to participate in the processing, improve the subsidies to users, and at the same time the station sends a guide to let the vehicle go out smoothly from the exit gate.

### 5.4.4 Cooperative Game Model Between Governments, Energy Supply Stations, and Users

Suppose that users, enterprises, and local governments form a set $N = \{1, 2, 3\}$, the non-empty subset $S(S \in N)$ is a consortium, and $N$ is an overall consortium. When three parties participate in cooperation at the same time, a game alliance $G = (N, v)$ is formed.

Suppose $V(p) = Cp$, $V(f) = Cf$, $V(g) = Cg$ is the cost of independent processing by users, sites, and local governments, $V(p, f) = Cpf$ is the cost of cooperative processing between users and enterprises, $V(p, g) = Cpg$ is the cost of cooperative processing between users and local governments, $V(f, g) = Cfg$ represents the cost of cooperative processing between enterprises and local governments, and $V(N) = Cn$ represents the cost of three-party cooperative governance. These costs satisfy $V(N) \leq \sum i \in N$.

To simplify the study, game $G = (N, v)$ can be solved using the Shapley method (Shapley 1952). According to the Shapley method, the average marginal contribution of each member in each distribution scheme of all games can be expressed as:

$$\Phi i(v) = \sum_{S \in N} \frac{(n - |S|)!(|S| - 1)!}{n!}[v(S) - v(S/i)]\, i = 1, 2, \ldots, n \tag{5.29}$$

where: $|S|$ indicates the number of members in the alliance;

| | |
|---|---|
| $(n - \|S\|)!$ | Indicates the number of arranged alliances; |
| $(\|S\| - 1)!$ | Represents the number of alliances that exclude players; |
| $n!$ | Represents the number of permutations of n players; |
| $v(S/i)$ | Represents the contribution of other players after removing the $i$ in $S$; |
| $v(S) - v(S/i)$ | Represents player $i$'s marginal contribution to the alliance; |

Different cooperation situations and marginal contributions under the cooperation game of the three parties are shown in Table 5.7.

So, if the three parties cooperate on governance, then $n = 3$, the cost they bear is:

Cost to the user:

**Table 5.7** Marginal contribution

| Possibility | Order of cooperation | Government/$g$ | Corporation /$f$ | User/$p$ |
|---|---|---|---|---|
| 1/6 | $(g, f, p)$ | $Cg$ | $Cfg - Cg$ | $Cn - Cfg$ |
| 1/6 | $(g, p, f)$ | $Cg$ | $Cn - Cpg$ | $Cpg - Cg$ |
| 1/6 | $(f, g, p)$ | $Cfg - Cf$ | $Cf$ | $Cn - Cfg$ |
| 1/6 | $(f, p, g)$ | $Cn - Cpf$ | $Cf$ | $Cpf - Cf$ |
| 1/6 | $(p, g, f)$ | $Cpg - Cp$ | $Cn - Cpg$ | $Cp$ |
| 1/6 | $(p, f, g)$ | $Cn - Cpf$ | $Cpf - Cp$ | $Cp$ |

$$\varphi p(v) = \frac{2Cp + (Cpf - Cf) + (Cpg - Cg) + 2(Cn - Cfg)}{6} \tag{5.30}$$

Cost of sites:

$$\varphi f(v) = \frac{2Cf + (Cpf - Cp) + (Cfg - Cg) + 2(Cn - Cpg)}{6} \tag{5.31}$$

Cost to the government:

$$\varphi g(v) = \frac{2Cg + (Cpg - Cp) + (Cfg - Cf) + 2(Cn - Cpf)}{6} \tag{5.32}$$

$V(N) \leq \sum_{i \in N} V(i)$ is the premise. In Eq. (5.8), $Cpf - Cf$,$Cpg - Cg$, $Cn - Cfg$ are all less than zero and $\varphi p(v) < 1/3$. It is found that in the three-party cooperative game, the cost borne by the user is only less than 1/3 of the original cost. Similarly, costs for businesses and local governments have fallen to less than a third of their original costs. The cost is greatly reduced, indicating that cooperation can achieve win-win results. In order to control the cost of the interchange detour within a certain range of $Cpt < Rpm - Rpt$ and achieve the desired goal, the cost structure of the interchange detour should be analyzed, including the time and expense required to find an idle station in the supervised area. Therefore, it is necessary to create a collaborative governance situation in which local governments encourage and guide, enterprises respond to the government, and users actively participate.

### *5.4.5 Analysis of Cooperation Game Examples Among Government, Energy Supply Stations, and Users*

Suppose that the utility of the user queuing at the site is 10 yuan, the total cost of changing the station is 15 yuan, and the subsidy received by the user is 20 yuan. The cost of government supervision is 300 yuan/day, the average penalty of superiors is 200/ day, and the positive impact of government supervision is 150 yuan/day. It can be concluded that there are two transportation energy supply stations in the system, and only considering the vehicles that need to change stations, there are 104 vehicles. Under non-cooperation, the cost of the government is $V(g) = 200$, the cost of the transportation energy supply station is $V(f) = 1200$, and the cost of the user is $V(p) = 1040$. The total cost of cooperation between the user and the energy supply station $V(p, f) = 2200$, the total cost of cooperation between the user and the government is $V(p, g) = 1190$, the cost of cooperation between the enterprise and the local government is $V(f, g) = 1480$, and the cost of tripartite cooperation governance is $V(N) = 2350$. Therefore, it can be concluded that the costs borne by the three parties are respectively

Cost to the user:

$$\varphi p(v) = \frac{2Cp + (Cpf - Cf) + (Cpg - Cg) + 2(Cn - Cfg)}{6} = 968.3 < 1040 \quad (5.33)$$

Cost of site:

$$\varphi f(v) = \frac{2Cf + (Cpf - Cp) + (Cfg - Cg) + 2(Cn - Cpg)}{6} = 1193.3 < 1200 \quad (5.34)$$

Cost to the government:

$$\varphi g(v) = \frac{2Cg + (Cpg - Cp) + (Cfg - Cf) + 2(Cn - Cpf)}{6} = 188.3 < 200 \quad (5.35)$$

Through calculation, it can be found that the costs of the three have been reduced, but the amplitude is not high, and it has not reached the ideal of 1/3. There are two main reasons: First, the cost of the three-party cooperation is lower than that of the two-party cooperation can only be theoretical, and can only be lower than the cost before the cooperation; The second is that cooperation is only a theory, there is no actual data for reference, so the value of subjectivity is larger, but the effect of tripartite cooperation is better than that of incompatibility.

### 5.4.6 Research Conclusion of Cooperation Game Among Government, Energy Supply Stations, and Users

First of all, by analyzing the game between the public and local governments, the conclusion is drawn: if the cost can be controlled through the difference between the compensation of switching stations and the time and income saved by the current queuing, then the goal of the public consciously switching stations is likely to be achieved regardless of whether the government will supervise. Secondly, using the cooperative game model, it is found that when the public, enterprises, and local governments try their best to cooperate, the cost of all parties is less than 1/3 of the original, which proves the necessity and internal mechanism of tripartite cooperation.

## 5.5 Suggestions on Multi-Subject Cooperative Game Strategies of Urban Transportation and Energy Supply Stations

### *5.5.1 Strategic Suggestions at the Level of Policy and Institutional Support*

Local governments should bring the problem of long queues at energy supply stations into the scope of market regulation. Local governments can regulate the behavior of energy stations from the aspects of regional layout and road facilities renovation, and standardize public behavior by strengthening education and introducing punishment mechanisms. For energy supply stations and local governments, it is necessary to negotiate the expansion or addition of sites. For the stations with more idle hours, they can be abolished, and at the same time, new stations can be added to the sections with high demand, so as to avoid the queue of replenishment vehicles in the peak period being too long, occupying the motor lane and causing traffic congestion. In addition, in the case of reduced traffic at stations caused by the reconstruction of the existing road network, local governments can allow sites to re-select land and reduce the corresponding costs. For the public, local governments can make full use of propaganda walls and online columns to enhance citizens' awareness of law, morality, and sharing. It is possible that the phenomenon of uncivilized behavior will gradually disappear by strictly investigating the vehicles that queue up to affect road traffic and publicizing the vehicles that refuse to listen.

### *5.5.2 Policy Suggestions at the Energy Supply Site Level*

The station should introduce new technology and management system to regulate the vehicle replenishment behavior. The station should cooperate with the local government to implement various measures, rationally regulate the layout of the station, guide the orderly queue of vehicles, and improve service efficiency to solve the problem of excessive vehicles in the station. Establish a standardized scoring system for the site, cooperate with the government to establish a regional monitoring system to determine whether the vehicle leader of the energy supply station is in a reasonable area, if the occurrence of excessively long queues repeatedly, the site will be interviewed, and the score will be deducted for the site, if the score is lower than a certain threshold, the site will be internal rectification. In addition, through government-enterprise cooperation, the establishment of an information platform, real-time release of site status information, so that users can understand the situation in advance, and choose the right site. Finally, the site can first evaluate its own state, arrange guides at the entrance, and set up eye-catching information boards to remind users to reduce the waste of time generated by irregular queuing.

### 5.5.3 *Strategies and Suggestions at the Individual Participation Level*

In addition to the publicity and constraints of local governments and energy stations, the most important thing is to guide the public to actively participate in the entire decision-making process. Not only can they detour to an idle station, but they can also actively guide other users to follow the rules, and electric car users can use mobile phone software and phone calls to report energy supply stations with extremely long queues, sharing alerts to people around them to pass on positive energy. When local governments solicit opinions, the public should give some suggestions so as to provide solutions for participating in government decision-making.

## 5.6 Chapter Summary

Firstly, the relevant knowledge of cooperative game theory and the positive effect of urban energy supply station cooperation are introduced, and the influencing factors of urban energy supply station are analyzed. Among them, many subjects include the government, urban energy supply stations, and energy replenishment vehicles, and explain the positioning of different subjects in the transportation energy supply network and the current problems, due to the lack of corresponding regulations of the government, there is no good management and vehicles within the energy supply station and more attention to their own interests, the balance of supply and demand of urban transportation energy supply needs to be solved urgently.

Secondly, from a horizontal perspective, cooperation between transportation energy supply stations is proposed. For stations with a large number of vehicles, some vehicles are induced to free stations to replenish energy to balance the supply and demand of transportation energy, and the benefits generated by cooperation are distributed using Shapley's modified value, and the balance between supply and demand of transportation energy can be achieved through numerical example analysis, and the income of all partners can be improved.

Finally, from a vertical perspective, the cooperation between the government, transportation energy supply stations, and vehicles is proposed. Assuming that all parties are bounded rationality, the government has two strategies of supervision and non-supervision, and users have the strategy of choosing to change stations and not changing stations, the benefits of all parties are analyzed, and the revenue matrix is constructed, and the cost of the three can be reduced to 1/3 of the original through the distribution of Shapley value, which shows the effectiveness of tripartite cooperation and provides countermeasures and suggestions to solve problems.

The future energy market will gradually develop toward electric energy, and in the current vigorous development of the new energy vehicle industry, the construction of supporting facilities has not kept up, especially the charging pile industry. In addition, the battery life problem of new energy vehicles has also aggravated the "range anxiety" of car owners, further triggering the "competition" for charging piles. Due to the long charging time of electric vehicles, it is easy to queue up at charging stations. Therefore, similar to the solution at gas stations, cooperation between charging stations can be adopted, which can both sell electricity to each other and direct some queued electric vehicles to idle charging stations. At the same time, private pile sharing is encouraged, and private charging piles in the community are connected to the sharing platform for other users to use to reduce charging anxiety.

## References

Branzei R, Dimitrov D et al (2008) Models in cooperative game theory. Springer Science & Business Media, Cham

Bureau of Public Roads (1964) Traffic assignment manual. Urban planning division, US Department of Commerce, Washington, DC

Liu Y et al (2020) A review of fuzzy AHP methods for decision-making with subjective judgements. Expert Syst Appl 161(5):113738

Podvezko V (2009) Application of AHP technique. J Bus Econ Manage 10(2):181–189

Shapley LS (1952) A value for N-person games. RAND Corporation, Santa Monica, CA

# Chapter 6
# Research on Vehicle Scheduling Based on Negative Effect Minimization

## 6.1 Theoretical Analysis of Supplementary Vehicle Scheduling

There are many kinds of replenishment vehicles in the urban road network, such as traditional fuel vehicles, hybrid electric vehicles, pure electric vehicles, and natural gas vehicles. Due to the characteristics of the vehicle itself and the transportation energy supply station, the replenishment time of traditional fuel vehicles and natural gas vehicles is short, and the replenishment time can be completed in 2–5 min. Hybrid electric vehicle is to combine the traditional gasoline car engine with the electric motor of a pure electric vehicle at the same time to operate force, at low speed, you can use the motor to provide power; When the speed is high, it can be switched to provide power to the internal combustion engine, so in the process of replenishment, you can choose to refuel or charge, and the replenishment can be completed in a short time, and the negative effect of the above models in the process of replenishment is relatively small. For pure electric vehicles, the range is limited, it takes at least 30 min of replenishment time in fast charging mode, and at least 6–8 h in slow charging mode to full the battery, and the negative effect of pure electric vehicles in the replenishment process is relatively large. Therefore, the research of replenishment vehicle scheduling based on negative effect minimization mainly focuses on the replenishment scheduling of traditional fuel vehicles and pure electric vehicles.



X. Ge, *Urban Transportation Energy Supply Network*, Industrial Ecology and Environmental Management 5, https://doi.org/10.1007/978-3-031-64619-5_6

**Table 6.1** Basic elements of supplementary energy vehicle scheduling

| Element | Element specification |
|---|---|
| Vehicle | Vehicles on the road |
| Recharge vehicle | Driving a vehicle on the road with a replenishment requirement |
| Recharge station | A station that supplies vehicles with transportation energy, such as oil, electricity, or natural gas |
| Recharge network | The replenishment network consists of vehicle starting point, destination, and replenishment station |
| Recharge path | The path that the vehicle travels from the place where the energy demand is generated to the energy station |
| Vehicle remaining range | The distance a vehicle can travel before it runs out of fuel or power |
| Restriction condition | The mileage constraint and the maximum capacity of the replenishment station should be taken into account when the vehicle scheduling is carried out |
| Goal | Under the condition of meeting the restriction, the goals of minimizing negative effect, minimizing cost, and minimizing time need to be achieved |

### *6.1.1 Related Theories of Vehicle Scheduling with Supplementary Energy*

Energy replenishment vehicle scheduling refers to allocating energy replenishment stations and formulating reasonable driving routes for drivers with energy replenishment needs according to internal and external factors, so that the vehicle can efficiently and reasonably complete the in-route charging plan under certain constraints (such as vehicle mileage limit), and achieve the goals such as minimum energy replenishment cost and shortest time. The scheduling work of replenishment vehicles can fully coordinate the utilization rate of equipment at each replenishment station, and ensure that the vehicle can successfully complete the formation of replenishment behavior under the low endurance state, so as to minimize the negative effect brought by the replenishment behavior. The basic elements of supplementary vehicle scheduling are shown in Table 6.1.

### *6.1.2 The Role of Replenishment Vehicle Scheduling*

#### 6.1.2.1 Promote Orderly Development of Urban Transportation

Vehicle replenishment is a necessary condition to ensure the smooth completion of the journey when the vehicle is in the low endurance state, so the replenishment behavior is very important for the development of urban traffic. The route selection of replenishing vehicles is quite different from that of non-replenishing vehicles. During the peak period of replenishing, the disorderly replenishing behavior of a large number of replenishing vehicles will bring huge disturbance to the traffic network operation, and easily cause local congestion around the replenishing stations. Therefore, the effective scheduling mechanism of energy replenishment vehicles can orderly dispatch energy replenishment vehicles to the energy replenishment station, minimize the negative effect of energy replenishment behavior on the traffic network, and promote the orderly development of urban traffic.

#### 6.1.2.2 Promote the Coordinated Development of Regional Supplementary Energy Stations

Due to the randomness and heterogeneity of vehicles in selecting recharge stations, large-scale recharge behavior is easy to cause congestion of local charging stations and unbalanced utilization of equipment at regional recharge stations. This phenomenon is most obvious in electric vehicle replenishment, electric vehicle replenishment time is longer, in the peak of replenishment, local charging station time in a high load state, will cause serious damage to the power grid. Therefore, effective energy replenishment vehicle scheduling can promote the coordinated development of energy replenishment stations in the region, avoid the phenomenon of long queues of individual energy replenishment stations, and reduce the burden on the transportation energy supply network.

#### 6.1.2.3 Save the Driver's Supplementary Energy Cost

Since the driver cannot fully grasp the road information, when going to the recharge station, the user may increase the mileage and travel time due to detour, and the driver does not know the current queuing situation in the charging station, there is a great possibility to choose the station with high recharge traffic, so the need to queue up, increasing the time cost of the user. Therefore, effective replenishment vehicle scheduling can recommend replenishment stations and plan replenishment routes for replenishment vehicles according to the current road information and service information of each replenishment station, so as to reduce the driver's replenishment time cost and replenishment trip cost, and improve driver satisfaction.

### *6.1.3 Factors That Complement Vehicle Response Scheduling*

The following factors are taken into account when the replenishment vehicle decides whether to respond to the replenishment scheduling scheme.

#### 6.1.3.1 Whether the Replenishment Time Is in the Peak Replenishment Period

When the driver triggers the demand for replenishment at the peak of replenishment, there are more replenishment vehicles on the road network, and the replenishment flow of each replenishment station is also large. Since the driver is in the incomplete information state of road traffic and replenishment station service, the driver is more inclined to respond to the scheduling scheme and go to the replenishment station for replenishment according to the scheduling scheme. At this time, if the driver's supplementary utility in response to scheduling is higher than that in self-supplementary, the driver's satisfaction with the scheduling scheme will be improved, and the driver's responsiveness to the scheduling scheme will also be improved. If the driver triggers the replenishment demand in the peak period of replenishment, there are fewer vehicles with replenishment demand at this time, and there are surplus at the service desks of each replenishment station, the driver is more inclined to replenish energy by himself and choose the replenishment station according to his own travel or preference.

#### 6.1.3.2 Urgency of the Trip

When the driver's travel time is relatively urgent, the driver is more sensitive to time. If the total travel time of the scheduling scheme is less than the time of self-replenishment, the driver tends to respond to the scheduling strategy. Instead, be a driver when the travel time is relatively abundant, time is not the main factor concerned by the driver, and some economic compensation can be given to the driver through the incentive mechanism to stimulate the driver to respond to the scheduling strategy.

#### 6.1.3.3 Remaining Range

When the remaining driving range of the vehicle is low, the driver is more sensitive to the distance to the recharge station. Because as the driving range continues to decrease, the driver will have range anxiety, and electric vehicles are more obvious in this regard. Due to drivers incomplete access to traffic network information, it is easy to spend more travel time and mileage in the trip, which increases the psychological anxiety of drivers. At this time, the driver prefers to respond to the scheduling strategy and arrives at the replenishment station for replenishment before the remaining mileage is exhausted.

#### 6.1.3.4 The Replenishment Cost of Response Scheduling

Drivers need to spend extra mileage, extra travel time, mileage anxiety cost, purchase energy cost, etc., when they go to the replenishment station by themselves. When the driver responds to the scheduling, it may get the incentive benefit given by the scheduling mechanism, but it may also increase some other supplementary energy costs due to scheduling. Therefore, when the total cost of the driver to go to the replenishment station for replenishment is greater than the cost of replenishment by the response scheduling mechanism, the driver will prefer the response scheduling for the consideration of his own interests.

## 6.2 Analysis of Vehicle Networking Communication Technology

### *6.2.1 Overview of Vehicle-Connected Communication Technology*

With the development of Internet and wireless communication technology, Internet of Things technology has gradually matured and has been applied to many fields, among which the Internet of Things is an extension of the application of the Internet of Things in the field of transportation. The Internet of vehicles is the key technology to build "intelligent transportation system", and it is also an important part of smart city in the field of transportation, and it is of great significance to alleviate urban road traffic congestion in the future. The application of vehicle networking technology is first reflected in the car, with the rapid increase of electric vehicle ownership, the integration of electric vehicles and vehicle networking technology has become a hot topic, compared with cars, electric vehicles have the characteristics of power constraints, slow speed and so on, it is easier to achieve accurate control; In addition, electric vehicles also have the advantages of energy saving, economy, and environmental protection. However, at present, the domestic research on electric vehicle networking is still in its infancy, only a small number of models are equipped with some intelligent services, but with the support of policies and advanced technology, some domestic automobile companies such as BYD, Changan, etc., have now invested a lot of money and technical strength to conduct research.

In the vehicle networking system, the vehicle is the main communication entity, and a number of vehicle units form a wireless Ad hoc network, which is intended to connect the vehicle to the network through wireless communication technology, realize intelligent traffic management and scheduling, and better meet the network service needs of drivers and users. Because the vehicles move on the road network, the road topology changes frequently and the link state is unstable in the vehicle-connected system, which is an inevitable challenge to design the wireless communication technology of vehicle-connected.

### 6.2.2 Vehicle Networking System Architecture

Because the international Internet of Things has not yet set a unified structural standard, the Internet of vehicles as an extension of the application of the Internet of Things in the field of transportation, its structural standards are not unified. By analyzing the current needs of residents for vehicles, searching and researching related literature on reading networking and vehicle networking, combining the seven-layer Open System Interconnect (OSI) model and the definition of vehicle networking, the vehicle networking system can be divided into three layers from low to high, namely the perception layer, the network layer and the application layer. Figure 6.1 shows.

The bottom layer is the perception layer, which is the basis of the entire vehicle networking system. The sensing layer is composed of sensors, RFID, controllers, and other hardware devices deployed in the electric vehicle and connected with the central control chip, which is responsible for collecting relevant information about people, vehicles, and the environment, and can realize the control of the vehicle by receiving and executing relevant instructions. The middle layer is the network layer, which is mainly used for data transmission. The network layer provides a unified interface for the perception layer. Due to the complexity and diversity of devices in the perception layer, the network layer includes various communication network technologies, such as 3G/4G mobile telecommunications network, wireless local area network, Bluetooth RF, and other short-range wireless communication technologies as well as the Internet. The upper layer is the application layer, the main function for the driver to provide various types of services, to achieve the calculation, processing, monitoring, and management of the acquired data. In this scheme, the functions of the application layer are realized by the Android client, whose function is to provide users with visual services and reduce the complexity of operations. The application layer can also be divided into three sub-layers from bottom to top: data layer, which

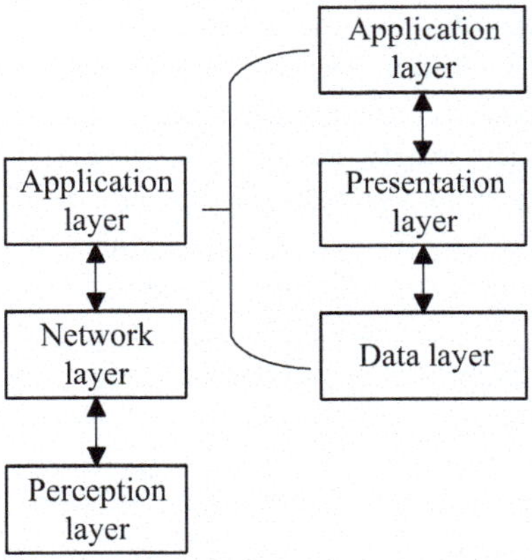


**Fig. 6.1** Architecture of vehicle networking system

is used to organize and store data; Presentation layer, mainly for data processing; The application layer can provide corresponding services according to the needs of users.

### *6.2.3 Vehicle Networking Replenishment Scheduling*

Vehicle network replenishment scheduling is composed of four entities: transportation energy supply network, replenishment station infrastructure, road network, and replenishment vehicle. It aims to provide more comprehensive services for replenishment vehicle drivers through reliable communication among each entity, and at the same time, real-time monitoring of the vehicle's driving range (such as the status of electric vehicle batteries). Help urban transportation energy network to make more reasonable energy scheduling. It can be seen that the real-time and reliable communication technology between entities is the key to ensure the quality of supplementary scheduling service. The technology of vehicle networking can realize the effective communication between vehicle and vehicle, vehicle and road, and vehicle and station by using the form of vehicle self-organizing network, thus solving the problem of information transmission in vehicle replenishing network.

Based on the vehicle networking cloud platform system, the wireless communication network is used to realize the information transmission between the cloud platform and the transportation energy supply network, the transportation network, and the vehicle. Among them, the information of the transportation energy supply network is mainly energy price, geographical location information, queuing situation in the station, and the number of available equipment. Network information includes real-time road conditions; Vehicle information includes vehicle mileage status, replenishment requirements, location information, etc. The cloud platform of the Internet of vehicles will analyze and calculate the collected information, and reduce a series of negative effects caused by the supplementary energy through the scheduling of supplementary energy vehicles.

Taking electric vehicle networking replenishment scheduling as an example, the vehicle networking platform first collects information such as grid load, electric vehicle replenishment, charging station service, etc., classifies electric vehicles according to electricity, and optimizes vehicle charging time before the day. Then the charging information of the vehicle is released to the driver. If the driver accepts and makes an appointment to charge, the charging load required by the electric vehicle is superimposed on the base load of the grid. If the driver does not accept, the driver will re-select the charging time. Then, in the corresponding recharge time, real-time optimization of vehicle charging is carried out. If the driver chooses to accept, the driver is guided to the charging station that meets the requirements; otherwise, the driver re-uploads the geographical location or chooses the charging station independently, and the planning is re-made by the vehicle networking vehicle scheduling platform.

## 6.3 Analysis of Matching Matrix of Replenishment Vehicle Scheduling

In the process of vehicle replenishment, due to the variability of road traffic information and vehicle flow at replenishment station, and the driver cannot fully grasp the service information of road information and replenishment station, the extra mileage and extra travel time of the driver will be increased in the process of replenishment. Therefore, based on the implementation of the network of vehicles scheduling system, the matching matrix of vehicle replenishment selection in terms of expected arrival time, queuing time, estimated service time, driving distance, and replenishment cost can be designed to quickly assist drivers in selecting replenishment schemes and make adaptive adjustments to the schemes according to travel needs, effectively improving the speediness and accuracy of replenishment behaviors. Reduce a series of negative effects brought by the supplement energy.

### *6.3.1 Supplementary Vehicle Scheduling Matching Matrix Principle*

The supplementary vehicle matching principle based on special matrix is shown in Fig. 6.2. Firstly, the key elements are extracted according to the vehicle replenishment demand and energy supply network. Secondly, on the basis of extracting key elements, the matching plan is constructed by using a formatted expression. The matching scheme can be preset in advance, for example, in the process of vehicle driving, according to various situations and travel plans, multiple matching schemes can be planned and stored in the pre-plan database, or can be obtained from the historical replenishment behavior; Thirdly, according to the real-time road traffic information and the supplementary station service information, the feature matrix is generated, compared with the pre-scheme feature matrix and sorted to select the best matching plan. Finally, according to the new or modified in the network, the matching plan is adjusted to determine the final matching scheme, including the expected arrival time, queuing time, estimated service time, driving distance, and replenishment cost of the replenishment vehicle.

### *6.3.2 Steps of Replenishment Vehicle Scheduling Matching Based on Eigenmatrix Pre-plan*

According to the demand of vehicle replenishment in the road network, the scheme matching of vehicle replenishment based on the feature matrix is proposed. The main idea is to calculate the deviation between the real-time information feature matrix

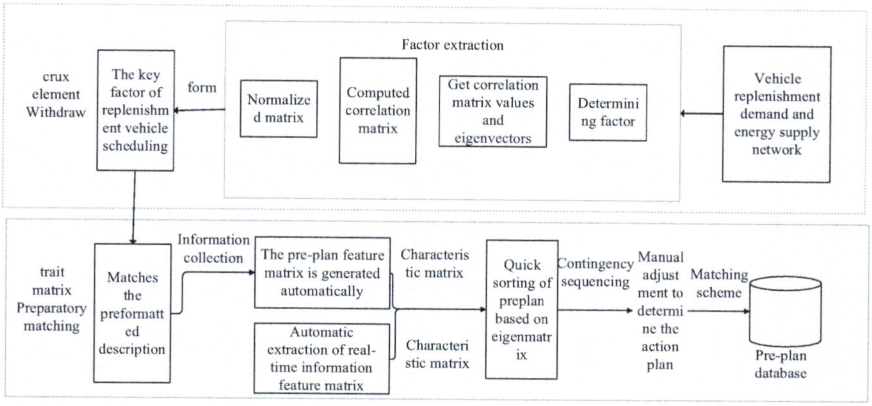


**Fig. 6.2** Principle of matching matrix for replenished vehicle scheduling

and the pre-plan feature matrix, and the pre-plan with the smallest deviation is the best matching scheme. The specific steps are as follows:

#### 6.3.2.1 Matching Plan Formatting

The matching plan is formatted, as shown in Fig. 6.3. The matching plan mainly consists of two parts: the feature area and the scheme area. Among them, the feature area consists of the information of the replenishing vehicle and the information of the energy supply network. The replenishment vehicle information refers to the state information of the vehicle after the replenishment demand is generated, which mainly includes the remaining vehicle mileage, vehicle location information, and vehicle travel plan. The information of the energy supply network includes the location information of the supplementary energy station, the energy price information, and the service information of the station. The scheme area includes matching schemes for vehicle replenishment, such as selection of replenishment station, expected arrival time, queuing time, estimated service time, driving distance, driving route, replenishment cost, etc.

#### 6.3.2.2 Feature Matrix Is Automatically Generated and Extracted

The automatic generation and extraction of feature matrix mainly includes the automatic generation of pre-plan feature matrix and the automatic extraction of real-time information feature matrix. Among them, the pre-plan matrix is automatically generated according to the data in the feature area. The real-time information feature matrix assigns values to the elements in the feature region according to the real-time information of road traffic conditions and energy supply network. Each pre-plan generated corresponds to an eigenmatrix.

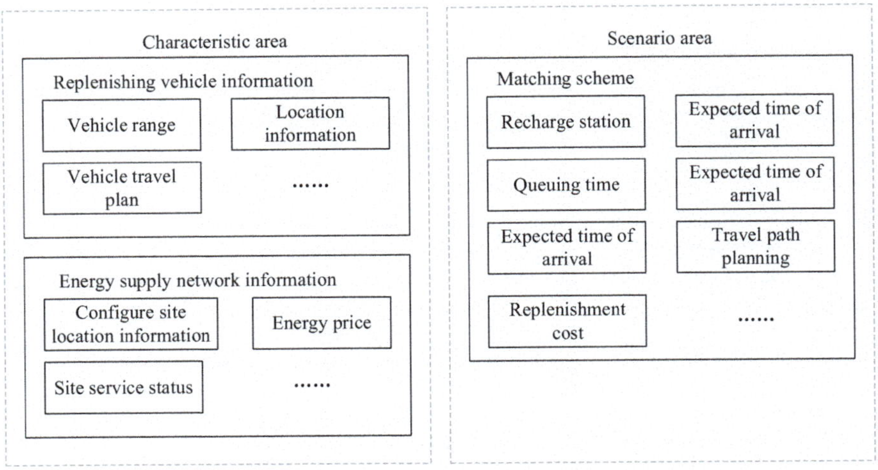


**Fig. 6.3** Composition of matching plan

#### 6.3.2.3 Quick Sorting of Pre-plan Based on Eigenmatrix

Firstly, the deviation between the pre-plan eigenmatrix and the real-time information eigenmatrix is calculated using the norm formula, and then traversed, and all the deviation values are sorted. The pre-plan with the smallest deviation value is the best matching pre-plan. The deviation value is calculated as follows:

$$D = \|\mathrm{A} - \mathrm{B}_i\|^2 \quad i = 1, 2, \ldots j \tag{6.1}$$

where, $D$ is the deviation value; A is the real-time information feature matrix; $\mathrm{B}_i$ is the pre-plan characteristic matrix; $i$ is the plan sequence, $j$ plans in total.

#### 6.3.2.4 The Driver Determines the Matching Scheme

According to the driver's replenishment intention, the plan is selected by combining the real-time traffic information and the transportation energy supply network, and some contents of the matching plan can be added or modified according to the actual situation, so as to make adaptive adjustments and determine the final matching plan.

## 6.4 Research on Balanced Distribution of Random Vehicle Replenished Energy Based on Two-layer Programming

### *6.4.1 Analysis of Random Vehicle Replenishment Balance Problem*

From the perspective of road network system and supplementary drivers, the negative effect of supplementary energy congestion can be alleviated through the scheduling and allocation of supplementary energy vehicles. Travel vehicles are divided into two categories: with and without supplementary energy demand. Based on real-time traffic information, each vehicle can timely understand the flow condition of each section, the oil price of the station, and the queuing situation, and make the supplementary energy distribution, so that the single vehicle's supplementary energy is relatively optimal and the total network's supplementary energy impedance is reduced. Two supplementary energy scheduling methods are considered to make the supplementary vehicles evenly distributed at each station to avoid agglomeration. The first is the oil price scheduling, by adjusting the oil price, or adding additional service scheduling vehicles to select the station with the lowest energy impedance. The second is the queuing degree scheduling, relying on the vehicle self-organizing network, to achieve real-time information exchange between vehicles, roads, and energy supplement stations, drivers timely access to the congestion of routes and energy supplement stations, and make reasonable choices. When the congestion degree is serious, the driver will actively choose other modes (time, energy station) for energy replenishment, thereby reducing the waiting time for energy replenishment. The scheduling mechanism is used to allocate the vehicle to satisfy the minimum total impedance.

As shown in Fig. 6.4, travelers have multiple travel paths $r$ ($r$ = 1,2,3) from starting point 1 to ending point 7, and each route passes through multiple replenishment stations $j$ ($j$ = 2,3,…,6), due to travel habits, vehicles generally choose route 1 as the driving route, and only some vehicles choose route 2 and 3. The replenishing impedance of the vehicle from the starting point 1 to the end point 7 when choosing the path 1 is:

$$C_{17}^{1} = C_{12} + C_{23} + C_{34} + C_{47} + \sum_{2,3,4\hat{I}O} (x_2^v C_2 + x_3^v C_3 + x_4^v C_4) \tag{6.2}$$

Among them, the recharging impedance represents the difficulty of the recharging process; $C_{j-1,j}$ is the travel impedance of the distance, representing the sum of travel cost and travel time; $C_j$ represents the replenishment impedance faced by the vehicle at the station $j$, and represents the sum of waiting time and waiting cost; $x_j^v$ represents the variable 0–1, or 1 if the vehicle $v$ chooses the recharge station $j$; Otherwise, it is 0.

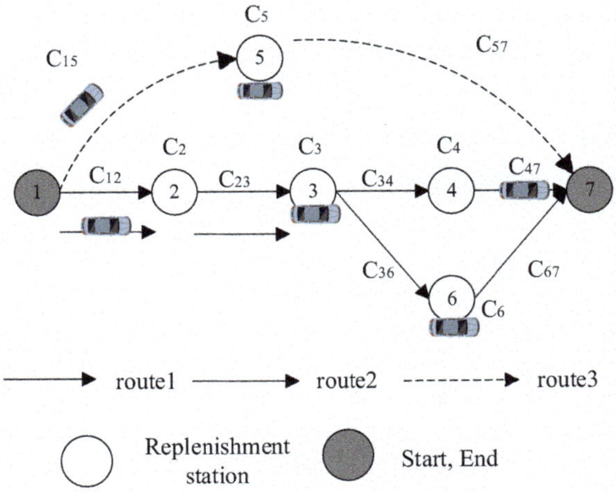


**Fig. 6.4** Joint selection of vehicle travel routes and stations

## 6.4.2 *Construction of Random Vehicle Replenishment Equilibrium Model*

### 6.4.2.1 Modelsumptions

To clarify the scope of use of this book, the following assumptions are made:

(1) There are two types of traffic in the road network, namely, energy replenishment vehicles and vehicles that do not need energy replenishment, and vehicles judge whether they need energy replenishment according to the remaining fuel; And the two types of vehicles are of the same model.
(2) In the case of complete information, all drivers of vehicles in need of replenishment will choose or change the driving path according to the queuing situation at the supply station.
(3) Considering the size limit of the supply station, when the number of vehicles waiting for replenishment is greater than the number of refueling guns in a supply station, the vehicles need to queue for service according to the order. The fuel consumption of the car depends on the distance and flow. While the vehicle is waiting at the supply station and replenishing energy, the engine is turned off, producing no fuel consumption and carbon emissions.
(4) Each car only needs to be replenished once in a single trip and filled up. After the vehicle completes the replenishment, it travels along the designated path without stopping in the middle until the end.

### 6.4.2.2 Symbol Definition

Build a digraph $G = (O, A)$ to represent the entire city energy supply network. Where, $O$ is a node set, $A$ represents a set of all paths in the urban energy replenishment

**Table 6.2** Symbol definition

| Symbol | Definition | Symbol | Definition |
|---|---|---|---|
| $R$ | A collection of all the routes in the network, $r \in R$, $R \in W$ | $T_o$ | the travel time of free flow (min) |
| $V$ | Assemble vehicles on the network | $v_{a.m_1}$ | For vehicles (vehicles) on route a that do not need replenishment |
| $f_w^t$ | The conversion ratio between time and impedance | $v_{a,m_2}$ | Recharge vehicle flow on Route a (vehicles) |
| $f_w^c$ | Conversion ratio between cost and impedance | $\alpha$, $\beta$ | Is a constant. The values are respectively 0.15, 4 |
| $T$ | Travel time (min) | $T_{\text{n}}$ | Waiting time for replenishmentmin) |
| $F$ | Travel cost (yuan) | $\mu$ | Unit cost for vehicle use (yuan/min) |
| $L$ | The maximum acceptable detour distance for the modifier (km) | $\eta$ | Labor cost (元/min) |
| $d$ | Original driving distance (km) | $d_a$ | Section length $a$ |
| $C_{ab}$ | The selection $ab$ induced impedance | $c_v$ | Carbon emission rate per kilometer (kg/km) |
| $C_j$ | Impedance for selecting a gas station $j$ | $\lambda_e$ | Unit carbon emission cost (yuan /kg) |
| $C_a$ | A section $a$ capacity | $f_v$ | Fuel consumption per kilometer (L/km) |
| $T_\omega^r$ | $OD\ \omega pathr$ Travel time (min) | $\lambda_f$ | Unit fuel consumption cost (yuan/L) |
| $x_j^v$ | Represents the variable 0–1, or 1 if the vehicle $v$ selects the recharge station $j$ | | |
| $x_a^v$ | Represents the variable 0–1, or 1 if the vehicle $v$ selects the road station $j$ | | |
| $\zeta_v$ | Represents the variable 0–1. If vehicle recharge is selected in the road network, it is 1 | | |

network, and $a$ is one of the sections, $a \in A$; $W$ indicates the set of OD pairs on the network. $w$ indicates one OD pair in $W$. The model notation is shown in Table 6.2.

#### 6.4.2.3 Model Construction

When there are multiple vehicles in the road network with the demand of replenishing energy, the vehicles are evenly distributed to meet the dual goals of the optimal individual replenishing energy and the road network replenishing energy system. A two-layer programming model is established to combine the global optimization of replenishment system with the individual optimization of replenishment. At the same time, the upper level target will also affect the lower level individual decision. Through the mutual influence and mutual restriction between the upper level and the lower level, the relative optimal of each other is achieved.

Upper-level Planning

From the perspective of road network system, a road network vehicle equilibrium model is established, taking into account the replenishment efficiency, replenishment cost, replenishment time, and negative effects of all replenishment vehicles in the entire urban road network. With the goal of minimizing the total impedance of energy replenishment on the road network, under the constraints of FIFO, detour distance, energy replenishment frequency, route and energy replenishment station joint selection, the two scheduling mechanisms of oil price and queuing degree were set to rationally allocate vehicle energy replenishment, and reduce the travel cost and travel time of vehicle replenishment. From the perspective of the operation efficiency of the urban energy supply network and the entire road network, the following optimization model is proposed, taking travel time and travel cost as variables to meet the goal of minimum total impedance:

$$\min c_{\omega}^{r} = \sum_{r=1}^{R} \sum_{v=1}^{V} (f_{\omega}^{t} T_{r}^{v} + f_{\omega}^{c} F_{r}^{v}) \tag{6.3}$$

where, $T_r^v$ represents the travel time of vehicles on the path $r$, $F_r^v$ represents the travel cost of vehicles $r$ on the path $r$. This objective function is subject to the following conditions:

Fifo constraint indicates that the replenishment vehicles have a certain service order when they arrive at the replenishment station, that is, the vehicles that arrive at the replenishment station first in a certain period of time are replenished before the vehicles that arrive later. Assume that the replenishment $k$ impedance corresponding to time is $c(k)$, and the replenishment $k+ \triangle k$ impedance corresponding to in is $c(k+ \triangle k)$. According to this definition, vehicles entering the replenishment station at time k will leave at time $k + c(k)$, and vehicles entering the replenishment station at time $k+ \triangle k$ will leave at time $k+ \triangle k + c(k+ \triangle k)$. Available expression:

$$k + \Delta k + c(k + \Delta k) > k + c(k), \quad \forall \triangle k, k > 0 \tag{6.4}$$

In the formula, k, $k+ \triangle k$—represents the time to enter the recharge station; $c(k)$, $c(k+ \triangle k)$ represents the replenishment impedance at the corresponding time, that is, while waiting.

From the above formula, it can be deduced:

$$1+\frac{c(k + \Delta k)-c(k)}{\triangle k} \geq 0 \tag{6.5}$$

If $c(k)$ is differentiable and $\triangle k \to 0$, then

$$\begin{aligned} &1+\tau_a'(k) \geq 0 \\ &\tau_a'(k) \geq -1 \end{aligned} \tag{6.6}$$

In the formula, the impedance differential of the time $k$ complement station $a$ is expressed $\tau_a'(k)$. Formula (6.5) indicates that when the decline rate of the replenishment impedance is greater than or equal to $-1$, the vehicle satisfies the first-come first-replenishment principle in the replenishment process. On the contrary, it means that the vehicle will jump the queue in the process of replenishing energy, which does not meet the principle of priority.

Detour distance constraint refers to the increase of driving distance when the replenishment vehicle selects the station. If the replenishment station is far away from the destination, the driver generally will not choose it. According to the minimum cost, the driver chooses to detour the total distance not more than 3 times the original distance. The minimum total impedance obtained should meet the detour distance constraint:

$$\frac{d+L}{d} \le 3, \quad d, L \ge 0 \tag{6.7}$$

According to hypothesis (4), the number of replenishment times is only replenished once:

$$\sum_{j \in O} x_j^v = 1, \quad \forall j \in O \tag{6.8}$$

The constraint of joint selection of travel path and recharge station means that in the process of travel, when there are multiple travel paths and multiple recharge stations in the same OD pair, the joint selection of the recharge station and travel path can obtain the recharge impedance of the travel path:

$$C_w^r = \sum_{a,b \in A}^{n} C_{ab} + \sum_{j=1, j \in O}^{m} x_j^v C_j, \quad \sum_{j \in O}^{m} x_j^v = 1 \tag{6.9}$$

Lower-level Decision Model

Due to the factors such as time, cost, service capacity of gas station, and driver's personal replenishment habit, it is a big problem to choose the replenishment station. The strategy model is based on the minimum recharging impedance of travelers, combined with the driver's travel habits and the randomness of recharging energy, to build a decision model, so that the vehicle satisfies the minimum total recharging impedance of the road network and the optimal individual replenishment energy. The model mainly considers the travel time and travel cost of the vehicle in the process of recharge.

The travel time of the vehicle in the process of travel is an important basis for judging the travel efficiency. Therefore, it is reasonable to use the minimum travel time as one of the indicators in the process of vehicle replenishment for joint selection

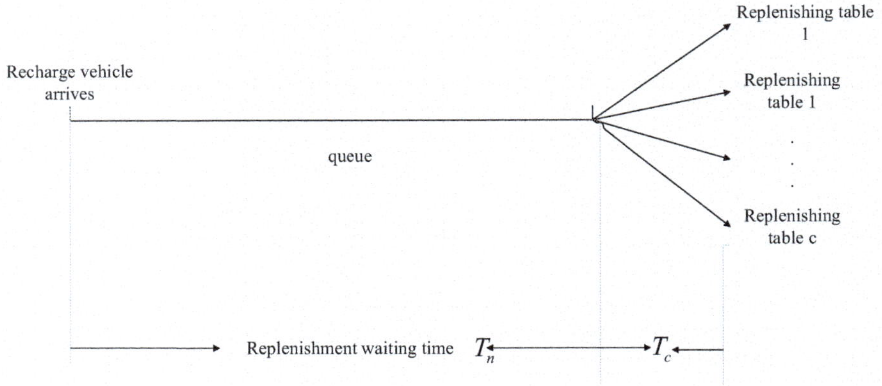


**Fig. 6.5** Schematic diagram of multi-service desk queue replenishment

of supply station and route. Vehicle travel time mainly considers vehicle travel time, vehicle waiting time, and vehicle recharge time.

Travel time indicates that the replenishment vehicle driving on the road network will be limited by the traffic flow and road network capacity. The travel time of the vehicle on the path $r$ can be obtained by the BPR function Bureau of Public Roads (1964):

$$T_{\omega}^{r} = \sum_{a \in A} t_a = \sum_{a \in A} T_o \left[ 1 + \alpha \left( \frac{v_{a,m_1} + v_{a,m_2}}{C_a} \right)^{\beta} \right] x_a^v \tag{6.10}$$

Replenishment waiting time represents the time from entering the replenishment queue to entering the replenishment station when the vehicle arrives at the replenishment station. Specific calculations are as follows:

As shown in Fig. 6.5, in the queuing system $M/M/C$ ($C \geq 2$), the waiting recharge time of the replenishing vehicle $i$ is set $t_i$. If $C = 2$ the queuing system is $M/M/2$, it means that the first two vehicles arriving at the replenishment station do not have to wait in line and can directly replenish energy, that is, $t_1 = t_2 = 0$. When the number of vehicles arriving at the same replenishment station to replenish energy is $n \geq 3$, the remaining vehicles need to queue up to replenish energy, that is, $t_i > 0$ ($i \geq 3$), and

$$t_3 = \min\{t_1', t_2'\} \tag{6.11}$$

$t_i'$, $i = 1, 2, 3...$ represents the replenishment time of the vehicle $i$. The text assumes that the replenishment time of all vehicles is equal to 3.5 min.

When the replenishment vehicle arrives at the replenishment station, the vehicle is about to face replenishment waiting $t_i$ and replenishment service. The waiting factor is assumed $t_c$ to be a constant, which represents the waiting time between the replenishment vehicle being the first in the queue and entering the replenishment service desk. By mathematical induction:

$$T_n = \overbrace{t_c + t_c + \cdots + t_c}^{n-2} = (n-2) * t_c, \quad n \geq 3 \tag{6.12}$$

Therefore, when there is $C$ refueling guns in the refueling station, the queuing time of the vehicle is:

$$T_n = \sum_{j \in O} \overbrace{t_c + t_c + \cdots + t_c}^{n-c} * x_j^x = \sum_{\mathrm{j} \in O} (n-c) * t_c * x_{\mathrm{j}}^v, \quad n \geq c \tag{6.13}$$

Replenishment time refers to the time when the vehicle enters the replenishment station and the staff time the vehicle replenishment until the end of replenishment and leaves the replenishment station. Note that the queuing time is not included. There are many factors that affect the vehicle replenishment time, such as: the number of refueling guns, refueling methods, the business proficiency of service personnel, the volume of vehicle fuel tank, etc., but it can be summarized as service efficiency. Through the field investigation of CNPC (5 km gas station, 7 km gas station, Qinglong Road gas station), Sinopec (junction station, Daojiao gas station), and Shell gas station in Nanan District of Chongqing, the average energy replenishment time of each vehicle is 3 ~ 4 min.

$$T_c = 3.5 \tag{6.14}$$

According to the formula (6.10), (6.13) and (6.14), it can be obtained that the total travel time $T_w$ of vehicles in the entire road network is:

$$\begin{aligned} T_w &= T_w^r + \zeta_v(T_\mathrm{n} + T_\mathrm{c}) \\ &= \sum_{a \in A} T_o \left[ 1 + 0.15 \left( \frac{v_{a,m_1} + v_{a,m_2}}{L_a + L * \zeta_v} \right)^4 \right] * x_j^v \\ &\quad + \zeta_v \left( \sum_{j \in O} (n-c) * t_c + T_c \right) * x_j^v \end{aligned} \tag{6.15}$$

In the process of driving, the travel cost $F_r$ mainly includes oil consumption $F_f$, carbon emission cost $F_e$, waiting cost of energy replenishment $F_n$, and refueling fee $F_o$.

Replenishment waiting cost includes labor waiting cost, vehicle waiting cost, that is

$$F_n = T_\mathrm{n}(\mu + \eta) \tag{6.16}$$

The cost of refueling is the product of the amount of refueling $q_v$ and the price of oil $p_o$ $p_o$

$$F_o = p_g q_o \tag{6.17}$$

The influencing factors of oil consumption and carbon emission cost include direct factors that can be quantified, such as driving speed, load, and driving distance; There are indirect factors that are difficult to quantify, such as road quality (road congestion, road grade), vehicle performance, driving style, etc. The load is not considered because the vehicles in the city queue to replenish energy are almost small cars, during the day, the main urban area trucks are restricted, and the possibility of queuing up at night is less. Focusing on the two most critical factors, driving distance and vehicle performance, the fuel consumption cost $F_f$, carbon emission $E_v$ and carbon emission cost $F_e$ in the process of vehicle replenishment on urban roads are respectively as follows:

$$F_f = \sum_{a \in A} (d_a + L * \zeta_v) * f_v * x_a^v * \lambda_f \tag{6.18}$$

$$E_v = \sum_{a \in A} (d_a + L * \zeta_v) * c_v * x_a^v \tag{6.19}$$

$$F_e = E_v * \lambda_e \tag{6.20}$$

The travel cost is expressed as:

$$F_r = F_f + F_e + F_n + F_o \tag{6.21}$$

Therefore, the impedance of the replenishing vehicle at a single replenishing station is:

$$c_\omega^r = f_\omega^t(T_\omega^a + T_n + T_c) + f_\omega^c(F_f + F_e + F_n + F_o) \tag{6.22}$$

When there is $G$ terminal energy supply station on the path $a$, because the replenishment time and replenishment waiting time of the replenishment vehicle at each supply station are different, the path impedance driven by the replenisher when selecting the replenishment station is also different. In order to calculate the replenishment waiting time $T_n$ and average replenishment time $T_c$ of the supply station on the path $a$, the probability of the replenishment vehicle on the path choosing the supply $j$ is expressed as $p_w^j$, then

$$p_w^j(v) = \frac{\exp[-\theta c_w^j(v)]}{\sum_{j=1}^N \exp[-\theta c_w^j(v)]} \tag{6.23}$$

Therefore, the impedance of the recharging vehicle on the path $a$ is:

$$c_w^r = \left[f_w^t T_w^a + P_j[f_w^t(T_n + T_c) + f_w^c(F_n + F_o)] + f_w^c(F_f + F)\right] \tag{6.24}$$

Therefore, the impedance of all replenished vehicles on all paths in the objective function road network is:

$$\min c_\omega^r = \sum_{r=1}^{R}\sum_{v=1}^{V}\left(f_\omega^t \mathrm{T}_\omega^a + \zeta_v \sum_{j=1}^{O} P_j[f_\omega^t(T_\mathrm{n} + T_\mathrm{c}) + f_w^c(F_n + F_o)] + f_w^c(F_f + F_e)\right) \tag{6.25}$$

## 6.4.3 Design of Random Vehicle Energy Balance Algorithm

The model is a two-layer structure NP (Nondeterministic Polynomially problem), and it is difficult to find an exact solution. Therefore, this paper adopts the two-layer model single-layer processing to find a "good" solution in the imprecise solution, so we design two-stage algorithms to solve the model, namely Method of Successive Average (MSA) and genetic algorithm.

### 6.4.3.1 Complementary Energy Distribution Model Solving Algorithm

The key point of this algorithm is to introduce traveler's regret theory and use Logit diversion model to continuously iterate on the selection of enabler gas stations. The algorithm steps are as follows:

Step 1: Under the initial condition, the replenishment vehicle randomly selects the travel path;
Step 2: Update the path. After completing a trip, the replenishment vehicle updating path and replenishment station are selected, without the replenishment vehicle updating path selection. The path selection probability is $P_w$: After the vehicle is selected, the number of vehicles with alternate energy such as updating the path and supplementing stations in the road network; For the case that the replenishing vehicle chooses the replenishing station, it is described as the change of the number of replenishing vehicles in the replenishing station. After the first diversion, the average waiting time of all replenishment stations on each path $r$ in the network is sorted.

$$T' = \sum_{j=1}^{N} P_j(T_n + T_c) \tag{6.26}$$

When the replenishment waiting time of a certain replenishment station is long, some replenishment vehicles will be diverted to other paths with a certain probability, and the probability of diversion is:

$$P = \frac{\sum_{j=1}^{N} P_j(T_n + T_c)}{\sum_{r=1}^{R} \sum_{j=1}^{N} P_j(T_n + T_c)} \tag{6.27}$$

Step 3: Update the next iteration based on MSA. Return to Step 2.
Step 4: Check for convergence.

#### 6.4.3.2 Optimization Model Solving Algorithm

Aiming at the upper road network planning model in this paper, genetic algorithm is adopted to solve it. The algorithm steps are as follows:

Step 1: Path search. All simple acyclic paths in OD pairs in the road network are taken as the set of effective travel replenishing paths $R_w$.
Step 2: The initial population is formed by coding. The supplementary energy station is digitally coded, the coding interval $[S_{\min}, S_{\max}]$, and the distribution of different supplementary energy stations.
Step 3: Initialize the proportion of the total flow of the replenishing vehicle in the road network and the selection probability of the replenishing station of the replenishing vehicle. Replenishing vehicle flow set: $\{v_{a,m}^1 | a \in A\}$ path set in the road network $\{f_w^{rc,1} | \forall r \in R_w\}$, calculate the replenishing impedance of each section and each replenishing station $\{c_w^r | \forall r \in R_w\}$, $\{c_w^n | \forall n \in N_w\}$.
Step 4: Through MSA algorithm, logit model is adopted to iteratively select the path of the enabler and the replenishment station to obtain the partial equilibrium of vehicle replenishment in the road network. See Sect. 6.4.1 for details. The modifier updates the selection probability, the updating path, and the updating road section $\{f_w^{rc,1} | \forall r \in R_w\}$ and the replenishing impedance of the replenishing station $\{c_w^r | \forall r \in R_w\}$, $\{c_w^n | \forall n \in N_w\}$.
Step 5: Integer coding is performed for each supplementary energy station, $i$ indicating that the node has supplementary energy stations, and the number of refueling guns is $i$, 0 indicating that the node has no supplementary energy stations. Each chromosome represents a layout scheme of the recharge station. In the process of cross-mutation, chromosomes are selected in the form of a roulette wheel for cross-mutation and mutation, with crossover probability $p_c = 0.8$ and mutation probability $p_m = 0.01$, $i \in \{0, 4, 6, 8, 18\}$ and the resulting new population is generated, as shown in Fig. 6.6.
Step 6: make $t = t + 1$, go back to Step 4.

When the number of evolution reaches the preset number of iterations, the genetic operation is terminated and the result is output.

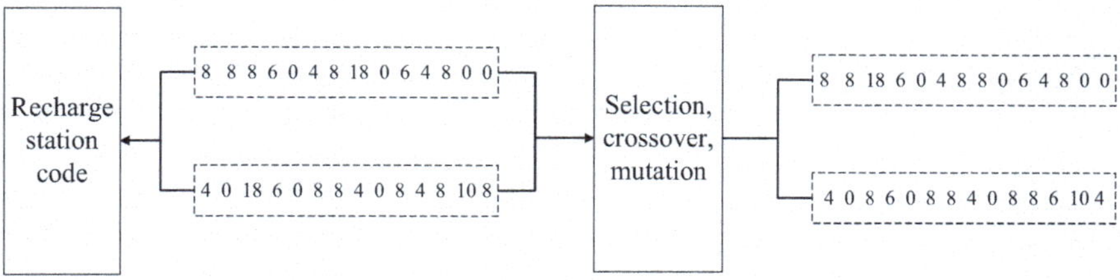


**Fig. 6.6** Schematic diagram of genetic process

### 6.4.4 *Analysis of Random Vehicle Energy Balance Calculation Examples*

The correctness and effectiveness of the model and algorithm are verified by using the Nguyen-Dupuis road network, which includes 500 random replenishment vehicles, 4 O-pairs {(1,2), (1,3), (4, 2), (4,3)}, 13 replenishment stations (numbered A-I), 19 road sections and 25 travel routes, as shown in Fig. 6.7. The algorithm is implemented in Python and the relevant parameters are as follows.

Table 6.3 shows the parameter Settings in the road network, assuming the number of refueling guns; Person and vehicle unit waiting cost reference; The road free driving speed is derived from the literature on urban road average driving speed, carbon emission, and fuel consumption.

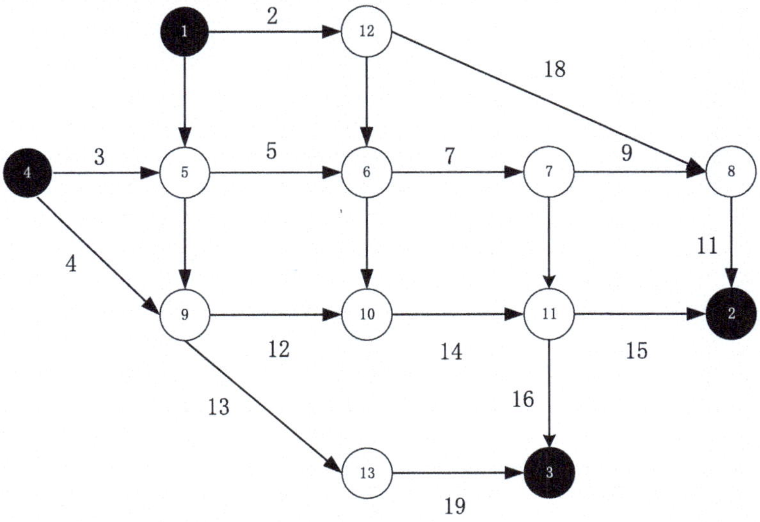

**Fig. 6.7** Nguyen-Dupuis network map

**Table 6.3** Model parameter settings

| Argument | Implication | Value | Argument | Implication | Value |
|---|---|---|---|---|---|
| $i$ | Number of refueling guns (only) | {0, 4, 6, 8, 18} | $c_v$ | Vehicle carbon emission rate per kilometer (kg/km) | 0.0414 |
| $\mu$ | Vehicle waiting unit cost (Yuan/hour) | 90 | $\lambda_e$ | Unit carbon cost (Yuan/kg) | 0.0528 |
| $\eta$ | Person waiting unit cost (Yuan/hour) | 24 | $f$ | Fuel consumption per kilometer (liter/kilometer) | 0.0108 |
| $v$ | Road free travel speed (km/h) | 40 | $\lambda_f$ | Unit fuel consumption cost (Yuan/liter) | 7.5 |
| $p_c$ | Cross variation | 0.8 | $p_m$ | Genetic variation | 0.01 |

#### 6.4.4.1 Data Description

In order to realize the solution of the model and algorithm, the basic attributes of urban gas stations, roads, and paths are described, as shown in Table 6.4.

Table 6.4 shows the number of vehicles that are queuing up to replenish energy at gas stations based on real-time traffic information. Formula (6.13) is used to calculate the waiting time of each station. Among them, gas station 5 has the longest queuing time, which should be avoided as far as possible, and gas station 8 has the least vehicles, which can be preferred.

Table 6.5 lists the relevant parameters of 19 sections, including section capacity, section flow, and section length.

Figure 6.8 shows the real-time traffic flow on 19 road sections, and the traffic flow on each road section is not equal. Among them, the actual traffic flow of road sections 1, 3, 9 and 7 exceeds the road capacity constraint and has formed

**Table 6.4** Queues at gas stations

| Node | Number of queued vehicles/vehicles | Wait time/minute | Recharge time/minute |
|---|---|---|---|
| 5 | 10 | 35 | 38.5 |
| 6 | 7 | 24.5 | 28 |
| 7 | 6 | 21 | 24.5 |
| 8 | 0 | 0 | 3.5 |
| 9 | 4 | 14 | 17.5 |
| 10 | 8 | 28 | 31.5 |
| 11 | 7 | 24.5 | 28 |
| 12 | 3 | 10.5 | 14 |
| 13 | 2 | 7 | 10.5 |

**Table 6.5** Basic attributes of 19 road sections

| A section of a road | Starting point | End point | Road capacity | Road length | Flow rate | T0 |
|---|---|---|---|---|---|---|
| 1 | 1 | 5 | 250 | 12 | 264 | 21.3575 |
| 2 | 1 | 12 | 150 | 9 | 131 | 6.5236 |
| 3 | 4 | 5 | 250 | 13 | 363 | 14.4451 |
| 4 | 4 | 9 | 250 | 8 | 84 | 5.3435 |
| 5 | 5 | 6 | 250 | 2 | 138 | 1.3519 |
| 6 | 5 | 9 | 250 | 4 | 0 | 2.6667 |
| 7 | 6 | 7 | 250 | 13 | 17 | 8.6667 |
| 8 | 6 | 10 | 250 | 5 | 49 | 3.3341 |
| 9 | 7 | 8 | 150 | 6 | 179 | 5.2167 |
| 10 | 7 | 11 | 250 | 5 | 27 | 3.3334 |
| 11 | 8 | 2 | 250 | 9 | 27 | 6.0001 |
| 12 | 9 | 10 | 250 | 16 | 246 | 12.1667 |
| 13 | 9 | 13 | 250 | 3 | 90 | 2.0050 |
| 14 | 10 | 11 | 250 | 8 | 199 | 5.6545 |
| 15 | 11 | 2 | 250 | 5 | 32 | 3.3335 |
| 16 | 11 | 3 | 250 | 8 | 245 | 6.0712 |
| 17 | 12 | 6 | 150 | 8 | 222 | 9.1716 |
| 18 | 12 | 8 | 250 | 4 | 127 | 2.6933 |
| 19 | 13 | 3 | 250 | 10 | 89 | 6.6827 |

traffic congestion. In the process of energy replenishment, this road section should be avoided as much as possible. Sections 2, 12 and 16 are close to saturation, but there is no traffic congestion; The remaining 12 roads have less traffic flow, and the vehicle speed can be increased. In the process of replenishment, these sections can be prefered. Among them, the traffic flow of section 6 is 0, and this section is the free driving speed.

Table 6.6 shows the number of paths in the network and the nodes through which each path passes.

#### 6.4.4.2 Distribution Result Analysis

The algorithm in Section 4 is used to solve the model in this paper. Where the value $\theta$ is 0.1, the vehicle allocation results and the supplementary impedance borne by each path are obtained, as shown in Table 6.7.

It can be found from Table 6.7 that 500 replenishment vehicles are allocated to different replenishment stations on 25 routes, among which gas station 8 is the most allocated vehicle. For example, for vehicles taking route 1 as the travel route, gas

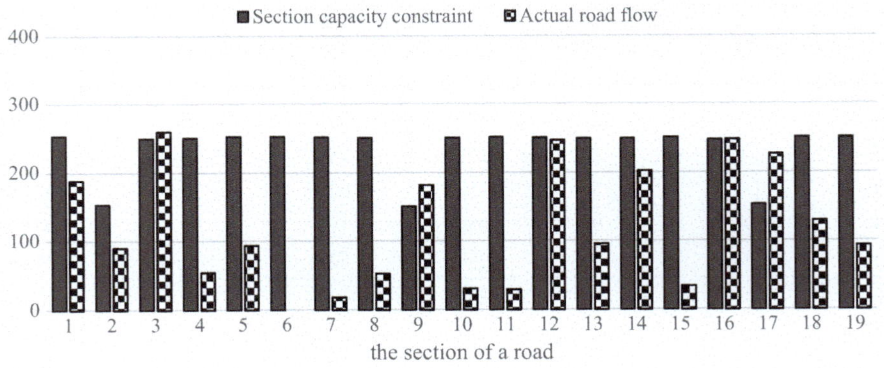


**Fig. 6.8** Comparison between road section capacity and actual flow

**Table 6.6** Path design

| OD | Path | Passing node | OD | Path | Passing node |
|---|---|---|---|---|---|
| 1 ↓ 2 | 1 | 1 → 12 → 8 → 2 | 4 ↓ 2 | | |
| | 2 | 1 → 12 → 6 → 7 → 8 → 2 | | 15 | 4 → 5 → 6 → 7 → 8 → 2 |
| | 3 | 1 → 12 → 6 → 7 → 11 → 2 | | 16 | 4 → 5 → 6 → 7 → 11 → 2 |
| | 4 | 1 → 12 → 6 → 10 → 11 → 2 | | 17 | 4 → 5 → 6 → 10 → 11 → 2 |
| | 5 | 1 → 5 → 6 → 7 → 8 → 2 | | 18 | 4 → 5 → 9 → 10 → 11 → 2 |
| | 6 | 1 → 5 → 6 → 7 → 11 → 2 | | | |
| | 7 | 1 → 5 → 6 → 10 → 11 → 2 | 4 ↓ 3 | 19 | 4 → 9 → 10 → 11 → 2 |
| | 8 | 1 → 5 → 9 → 10 → 11 → 2 | | 20 | 4 → 9 → 13 → 3 |
| 1 ↓ 3 | 9 | 1 → 5 → 9 → 13 → 3 | | 21 | 4 → 9 → 10 → 11 → 3 |
| | 10 | 1 → 5 → 9 → 13 → 3 | | 22 | 4 → 5 → 9 → 13 → 3 |
| | 11 | 1 → 5 → 6 → 10 → 11 → 3 | | 23 | 4 → 5 → 9 → 10 → 11 → 3 |
| | 12 | 1 → 5 → 6 → 7 → 11 → 3 | | 24 | 4 → 5 → 6 → 10 → 11 → 3 |
| | 13 | 1 → 12 → 6 → 10 → 11 → 3 | | 25 | 4 → 5 → 6 → 7 → 11 → 3 |
| | 14 | 1 → 12 → 6 → 7 → 11 → 3 | | | |

station 8 and gas station 12 can be selected for energy supplement, and the minimum energy impedance of gas station 8 is 1.86. Choose route 8 as the travel path and choose 9. The maximum impedance faced by the gas station is 7.2. In addition, there are four paths for gas station 8, one for gas station 7, six for gas station 9, two for gas station 11, four for gas station 12, three for gas station 13, and two for gas station 6 and 1.

After 20 iterations, the iteration diagram shown in Fig. 6.9 is obtained. The supplementary energy impedance of a single vehicle with 25 routes and the number of vehicles assigned to each route are obtained. It can be found from Fig. 6.10 that the number of vehicles assigned is inversely related to the supplementary energy

**Table 6.7** 25 path recharge impedance and vehicle allocation results

| OD | Path | Select site | Single vehicle recharge impedance | Vehicle allocation | Total impedance | OD | Path | Select site | Single vehicle recharge impedance | Vehicle allocation | Total impedance |
|---|---|---|---|---|---|---|---|---|---|---|---|
| 1 → 2 | 1 | 8 | 1.86 | 26 | 48.36 | 4 → 2 | | | | | |
| | 2 | 8 | 4.43 | 20 | 88.60 | | 15 | 8 | 4.37 | 20 | 87.4 |
| | 3 | 12 | 5.47 | 19 | 103.86 | | 16 | 7 | 6.25 | 19 | 118.83 |
| | 4 | 12 | 5.52 | 19 | 104.83 | | 17 | 11 | 6.6 | 19 | 125.4 |
| | 5 | 8 | 4.69 | 19 | 89.11 | | 18 | 9 | 6.88 | 19 | 130.72 |
| | 6 | 7 | 6.57 | 19 | 124.91 | 4 → 3 | | | | | |
| | 7 | 6\11 | 6.92 | 19 | 131.48 | | 19 | 9 | 5.51 | 22 | 121.22 |
| | 8 | 9 | 7.20 | 18 | 129.60 | | 20 | 13 | 1.981 | 27 | 53.50 |
| 1 → 3 | 9 | 13 | 3.68 | 23 | 84.67 | | 21 | 9 | 4.80 | 21 | 100.65 |
| | 10 | 9 | 6.48 | 18 | 116.69 | | 22 | 13 | 3.35 | 23 | 77.08 |
| | 11 | 11 | 7.06 | 18 | 127.08 | | 23 | 9 | 6.16 | 18 | 110.93 |
| | 12 | 7 | 6.71 | 18 | 120.86 | | 24 | 6\11 | 6.74 | 19 | 128.06 |
| | 13 | 12 | 5.66 | 19 | 107.49 | | 25 | 7 | 6.68 | 19 | 126.92 |
| | 14 | 12 | 5.61 | 19 | 106.54 | | | | 合计 | 500 | 2664.80 |

**Fig. 6.9** Iterative diagram of algorithm convergence

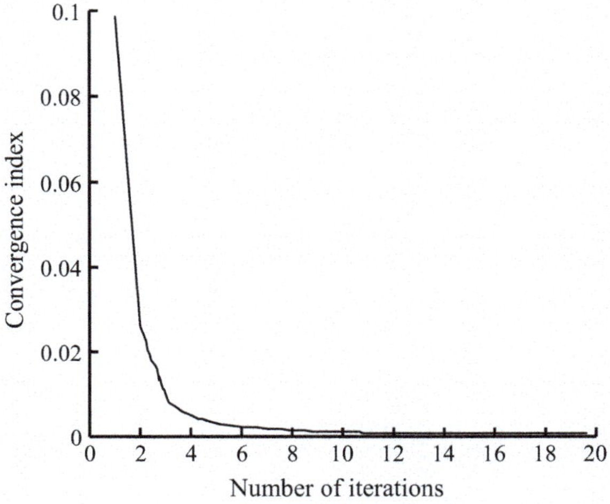


**Fig. 6.10** Replenishment impedance and vehicle distribution of 25 paths

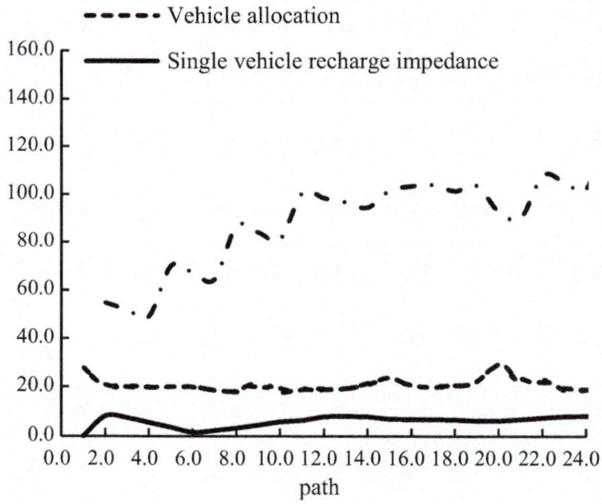


impedance. The larger the supplementary energy impedance, the fewer supplementary energy vehicles assigned at the station; conversely, the smaller the supplementary energy impedance, the more vehicles assigned. According to the congestion of the route and the station, the vehicle has the priority to choose the gas station according to the driving path.

#### 6.4.4.3 Validity Test

In order to verify the validity of the scheduling complement selection mechanism and model, the algorithm in Section 4 is applied to solve the problem from the two latitudes of "with scheduling mechanism" and "without scheduling mechanism".

Under the "congestion" scheduling mechanism, through real-time understanding of the number of vehicles waiting in line at gas stations, gas stations with low queues are preferentially selected for replenishment. The results are as follows: From the perspective of the travel cost of the enabler, it can be found from Fig. 6.11 that the energy cost of all travel routes in the road network with "congestion scheduling mechanism" is significantly better than that of "non-congestion scheduling mechanism", especially for routes 2, 5, 9, 15, 21 and 23, and the cost savings are more significant. The cost savings of other routes are not obvious, because the use of this model to allocate and select supplementary energy will increase the detour distance of supplementary energy vehicles, resulting in an increase in driving costs. In Fig. 6.12, it is found that 500 randomly replenishing vehicles in the road network can save 11,069.828 yuan by using this model for replenishing energy allocation. Therefore, when the enabler and the minimum cost are replenishment oriented, it is recommended to use the refueling option with "congestion scheduling mechanism".

From the perspective of the travel time of the enabler, it can be found in Fig. 6.13 and 6.14 that the use of "congestion scheduling mechanism" to allocate supplementary energy can save the total travel time, but it is not obvious. The reason is that the detour of the replenishment vehicle after dispatch, although the waiting time of the vehicle is saved by nearly 84.6%, the driving time is increased by 51%. The use of "congestion scheduling mechanism" can effectively alleviate traffic congestion and

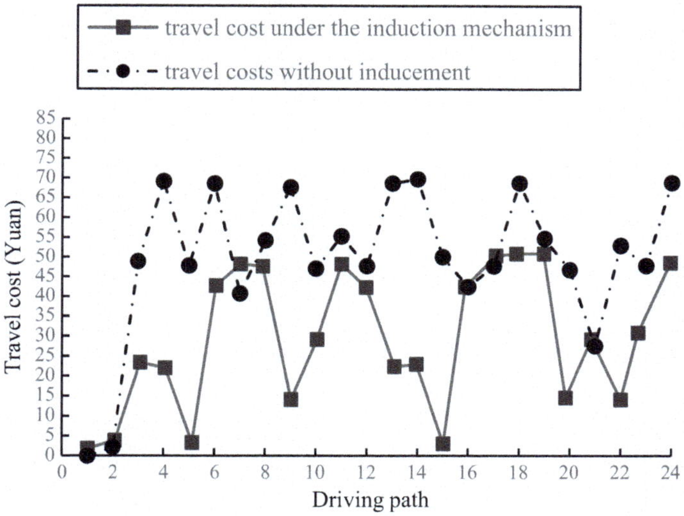


**Fig. 6.11** Travel cost analysis of "with or without scheduling" for each route

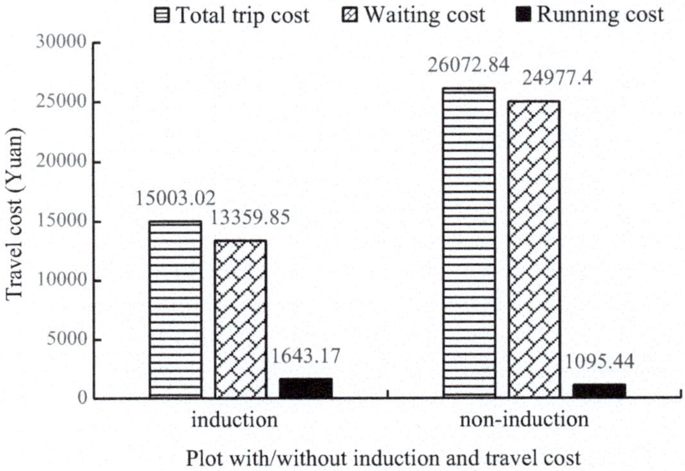


**Fig. 6.12** Comparison and analysis of various travel costs

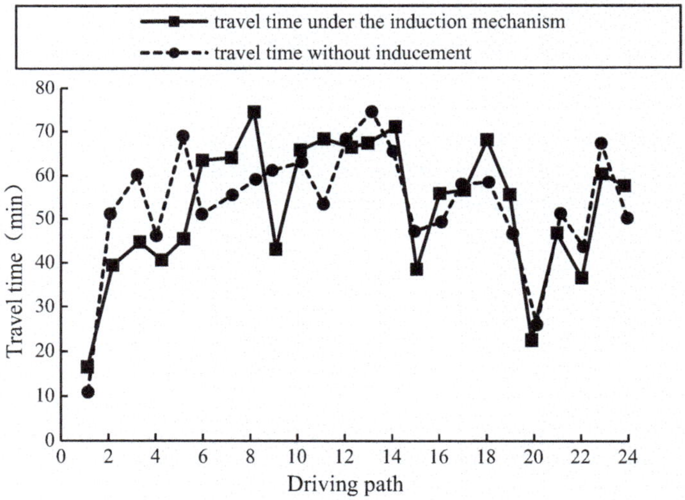


**Fig. 6.13** Travel cost analysis of "with or without scheduling" for each route

reduce the negative effect of urban energy congestion. Considering the optimum of road network system, the distribution strategy is effective.

Under the "price" scheduling mechanism, the government or enterprises can reasonably adjust the unit price of refueling during the peak time or provide additional services, so as to make the reasonable scheduling of vehicle replenishment and reduce the negative effect of replenishment congestion. For stations with frequent refueling congestion, the government gives subsidies to the stations to increase the

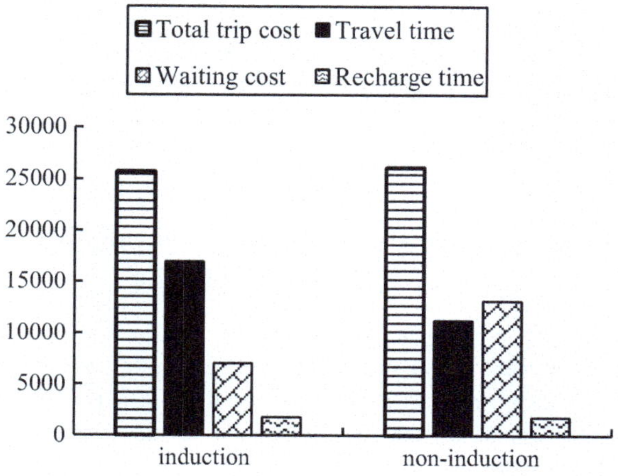


**Fig. 6.14** Comparison and analysis of various travel times

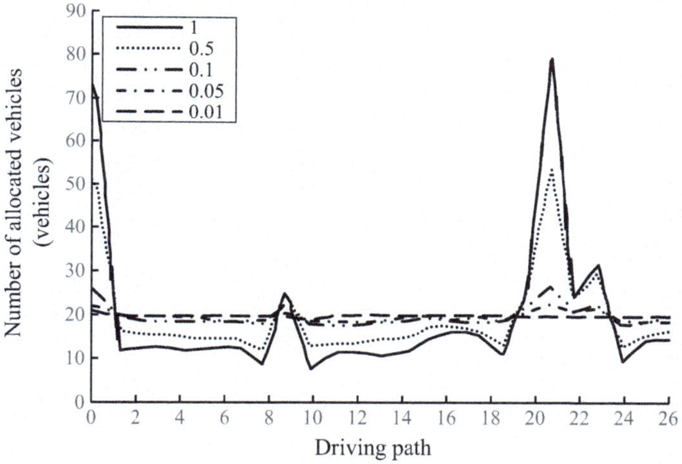


**Fig. 6.15** Vehicle allocation for each path with different values

oil price reasonably and reduce the number of vehicles to supplement the energy. For stations with few refueling vehicles, reasonably reduce the oil price or introduce additional services to attract vehicles to this service, and the government will appropriately raise taxes on the gas stations to adjust the market balance.

In order to investigate the influence of the driver's sensitivity to the recharge impedance on the random recharge vehicle allocation, Fig. 6.15 shows the vehicle allocation of the driver's joint choice of travel route and gas station under different perception levels. In this regard, the sensitivity parameters of drivers' perceived impedance difference $\theta$ were designed to be 0.01, 0.05, 0.1, 0.5, and 1, respectively.

When the parameter approaches 0, it means that the sensitivity is very low, and the driver does not care much about the "supplementary energy impedance", and the driver chooses the refueling path and gas station equally probability. On the contrary, when the parameter is infinite, it means that the driver attaches importance to the negative effect brought by the refueling process and is willing to choose the small impedance recharge scheme.

### *6.4.5 Research Conclusion of Random Vehicle Replenishment Equilibrium*

Considering the information sharing among vehicles, roads, and gas stations under the condition of complete information, the problem of random vehicle recharge allocation based on double-layer programming model is studied. The problem is solved by genetic algorithm and MSA algorithm. The example results show that the number of vehicles assigned is inversely proportional to the supplementary energy impedance. The larger the supplementary energy impedance, the fewer vehicles assigned to the station, the smaller the impedance, and the more vehicles assigned. At the same time, the travel cost is obviously reduced under the model and scheduling mechanism, and the travel time is saved slightly. The travel cost of 500 supplementary vehicles was reduced by 11,000 yuan, and the waiting time was saved by 6,114.5 min; The use of scheduling mechanism to allocate vehicles can avoid vehicle agglomeration and effectively reduce the negative effect caused by energy replenishment congestion. In the individual decision model, the driver's perception of the scheduling mechanism will affect the network allocation results. The more the driver pays attention to the impedance, the more reasonable the allocation results and the more significant the reduction of travel time and cost. Therefore, this allocation strategy can effectively alleviate traffic congestion and reduce the negative effect of urban energy congestion. The follow-up study will further explore the mechanism of network replenishment and burden reduction from the perspective of dynamic game.

## 6.5 Research on Replenishment Scheduling of Electric Vehicles Considering the Minimization of Negative Effects

### *6.5.1 Analysis of the Replenishment Scheduling Problem of Electric Vehicles*

The replenishment of electric vehicle energy is a necessary condition to ensure that the vehicle can complete the travel plan when the battery life is low. Usually, when the charging state of the electric vehicle is below a certain threshold or it cannot complete the remaining journey, the replenishment demand is generated. At this time, the driver needs to go to the charging station for replenishment and then drive to the destination. As shown in Fig. 6.16, there are two charging stations in a certain area, in which there are four charging piles remaining in charging station 1 and one charging pile remaining in charging station 2, and there are multiple replenishing vehicles in the road network, namely vehicle 1 and vehicle 2. The driver's choice of charging station is egoistic and random. In the case of no replenishment vehicle scheduling, on account of $t_2 = 10\,\text{min} < t_1 = 15\,\text{min}$, vehicle 1 will select charging station 2 for recharge; At the same time, vehicle 2 will also choose a charging station 2 that is close to it for recharge. As a result, replenishment vehicles gather at the two charging stations, resulting in high load at local charging stations. In addition, the replenishment time of electric vehicles is long, and the service capacity of charging stations is limited, resulting in a queuing phenomenon in the station, which increases the travel time of drivers. In addition, the distance of the vehicle to the charging station is not in the travel plan, which increases the mileage of the driver, and the driver is prone to mileage anxiety with the continuous reduction of the range. To this end, multiple vehicles in the network are scheduled to minimize a series of negative effects caused by energy replenishment behavior.

Consider the traffic network $G = (V, A)$, where $V$ is the node set, $V=\{\upsilon|\upsilon \in V\}$; $A$ is the section set, $A=\{a|a=1, 2, ..., m\}$; Multiple OD pairs exist on the network. The OD pairs are numbered in a unified manner $OD = \{i|i = 1, 2, ...N\}$. $g_i$ represents the $i$ travel demand between OD pairs, and $\beta_i$ is the penetration rate of $i$

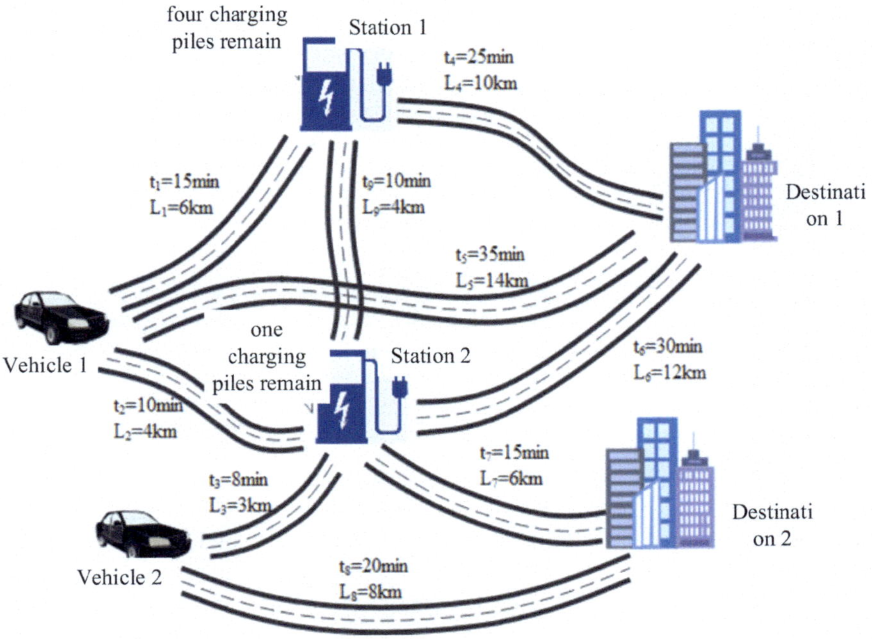


**Fig. 6.16** Example of electric vehicle replenishment

supplementing electric vehicles between OD pairs; The charging stations are all located on the network nodes, and the charging stations are uniformly numbered $J = \{j|j = 1, 2, ...r\}$, where $r$ is the number of charging stations; The vehicle takes the node that generates the energy demand as the starting point, finds the energy supplement station according to the station and road section information, and continues to drive to the destination after the station energy supplement. The position information of the vehicle at the starting point and the remaining battery life are known. The charging station is located on the road node, and the location information of the charging station and the queue length can be known.

## 6.5.2 *Dual-Layer Model of Energy Replenishment Scheduling for Electric Vehicles*

### 6.5.2.1 The Symbols Are Described in Table 6.8

### 6.5.2.2 Electric Vehicle Replenishment

Electric Vehicle Initial State of Charge Measurement

According to the statistical relationship between the state of Charge (SOC) of electric vehicles and the probability of charging behavior, the charging behavior of electric vehicles usually occurs when the state of charge is lower than 60%. In order to consider the gradual change of the initial state of charge, therefore, it is assumed that the initial charging state of EV recharge $B_o(i)$ in each OD pair follows a triangular distribution, and its probability density function is as follows:

$$f(x|a, b, c) = \begin{cases} \frac{2(x-a)}{(b-a)(c-a)} & \text{for } a \le x \le c \\ \frac{2(b-x)}{(b-a)(b-c)} & \text{for c } < x \le b \end{cases} \tag{6.28}$$

In the formula, $a$ is the lowest level of the initial charging state of the replenished vehicle, $b$ is the highest level of the initial charging state of the replenished vehicle, and $c$ is the mode of the initial charging state of the replenished vehicle; In this book $a = 0.2, c = 0.4, b = 0.6$.

Power Consumption Measurement of Electric Vehicles

The remaining power $B_{re}^n(i)$ of the recharge $n$ vehicle between OD pairs to $i$ when it arrives at the charging station is directly related to the distance traveled from the starting point $O(i)$ to the recommended charging station $j$, which is calculated as follows:

**Table 6.8** Symbol description table

| Symbol | Symbol specification |
|---|---|
| $V$ | The set of locations of all nodes in the network |
| $A$ | The collection of all sections of the road network |
| OD | The set of OD pairs of the replenishing vehicle |
| $q_i$ | Number of vehicles replenished by OD to $i$ |
| $J$ | The collection of charging station locations on the network, subscript $j$ |
| $L_a$ | Length of section $a$ |
| $v_a$ | Traffic flow on section $a$ |
| $t_a^0$ | Free travel time for section $a$ |
| $C_a$ | The capacity of the road section $a$ |
| $t_a$ | Travel time of the road section $a$ |
| $B_o^n(i)$ | ODpair $i$ indicates the initial charging state of the nth replenishing vehicle |
| $B_{re}^n(i)$ | ODpair $i$ indicates the amount of charge left at the charging station of the nth replenishing vehicle |
| $d_j^0$ | Queuing time coefficient of charging station $j$ |
| $u_j$ | Charge flow of charging station $j$ |
| $c_j$ | Service capability of charging station $j$ |
| $c_{ij}^n$ | OD pair $i$ indicates Range anxiety cost of the nth replenishing vehicle |
| $\omega$ | The unit price of incentive compensation for supplementary vehicles |
| $Z$ | Scheduling incentive compensation matrix of replenishing vehicle |
| $k_i$ | The path set for replenishing vehicles form $O(i)$ to $D(i)$ |
| $k_{i,j'}$ | A set of paths for recharging vehicles from the charging station form $O(i)$ to station $j$ |
| $k''_{j,i}$ | The set of paths that recharge the vehicle from the charging station $j$ to $D(i)$ |
| $\delta_{ak'_{i,j}}^{OJ}(i)$ | From $O(i)$ to charging station $j$, if section $a$ is on path $k_{i,j'}$, it is 1 |
| $\delta_{ak''_{j,i}}^{JD}(i)$ | From $O(i)$ to charging station $j$, if section $a$ is on path $k''_{i,j}$, it is 1 |
| $\delta_{ak_i}^{OD}(i)$ | From $O(i)$ to charging station $j$, if section $a$ is on path $k_i$, it is 1 |
| $X_{i,j}$ | The number of replenished vehicles in OD pair $i$ that are dispatched to charging station $j$ |

$$B_{re}^n(i)=\eta\left(B_c^n(i)-L(i)\cdot E_c\right) \tag{6.29}$$

$$L_{ij}=\sum_{a=1}^{m} L_a \delta_{ak_{i,j'}}^{OJ}(i) \tag{6.30}$$

$$\delta_{ak'_{i,j}}^{OJ}(i)=\begin{cases}1, & a \in k_i\\ 0, & a \notin k_i\end{cases} \tag{6.31}$$

In the formula, $\eta$ is the energy consumption coefficient; $L_{ij}$ is the distance of the vehicle from the starting point $O(i)$ to the recommended charging station $j$.

### Section Travel Time Measurement

In the road traffic network, the vehicle speed is constrained by the real-time traffic flow and road capacity, which will affect the driving time of electric vehicles on the road, and then affect the spatio-temporal distribution of the charging demand of supplementary vehicles. Therefore, this book assumes that the road section travel time adopts the resistance function form of BPR Bureau of Public Roads (1964), such as Eq. (6.32).

$$t_a(v_a)=t_a^0\left[1+0.15\left(\frac{v_a}{C_a}\right)^4\right] \tag{6.32}$$

Therefore, the travel time of the replenished vehicle from $O(i)$ arriving at the recommended charging station $j$ is:

$$T_{O,J}^{\text{travel}}(i)=\sum_{a=1}^{m}\delta_{ak_{i,j'}}^{OJ}(i)t_a(v_a) \tag{6.33}$$

### Charging Station Queue Time Measurement

In addition to the time spent on the road, drivers of electric vehicles also need to consider the queuing time for replenishment. According to the observation, both the replenishment queuing time and travel time increase with the increase of traffic flow. This book brings these two costs into a unified framework by establishing an extended transportation network, as shown in Fig. 6.17. First, the pseudo arc is added to indicate the replenishment waiting time, as shown by the virtual arc {3} in Fig. 6.17b. Then, expand the arc {2} shown in Fig. 6.17a to the arc opposite {(2′, 2″)} in Fig. 6.17b. For example, for a route through a charging station, {1,3,2″} is a charging route and {1,2′} is a non-charging route.

In the $M/M/1$ queuing model Baek and Moon (2014), the expression of the driver's average waiting time in the queue is

$$T_j^{\text{wait}}(\mu_j,\lambda_j)=\frac{\lambda_j}{\mu_j(\mu_j-\lambda_j)} \tag{6.34}$$

where $\lambda_j$ is the arrival rate per unit time of the charging station $j$, $\mu_j$ is the service rate for charging stations $j$, (In the $M/M/1$ queuing model, $\lambda_j \leq \mu_j$ to ensure the stability of the queuing system). By performing Taylor expansion on formula (6.34),

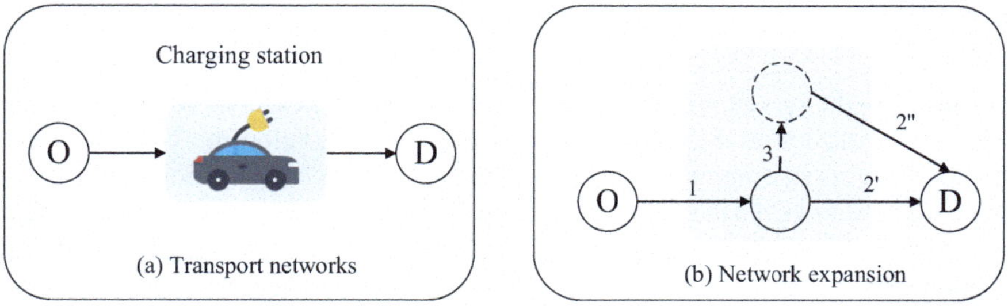


**Fig. 6.17** Transport Network (**a**) and extension network (**b**)

$$
\begin{aligned}
T_j^{\text{wait}}(\mu_j, \lambda_j) &= \frac{1}{\mu_j} \cdot \frac{1}{1 - \lambda_j/\mu_j} - \frac{1}{\mu_j} \approx \frac{1}{\mu_j}(1 + \frac{\lambda_j}{\mu_j} + \frac{\lambda_j^2}{\mu_j^2}) - \frac{1}{\mu_j} \\
&= \frac{1}{\mu_j}(\frac{\lambda_j}{\mu_j} + \frac{\lambda_j^2}{\mu_j^2})
\end{aligned}
\tag{6.35}
$$

Through the idea of the joint formula (6.35) and the BPR function, the relationship between the queuing time of the charging station and the recharge flow can be deduced:

$$
T_j^{\text{wait}} = d_j^0 \left[ \frac{u_j}{c_j} + \frac{u_j^2}{c_j^2} \right] \tag{6.36}
$$

In the formula, $d_j^0$ is the queueing time coefficient of the free flow replenishing vehicle, $u_j$ is the replenishing flow of the charging station $j$, and $c_j$ is the service capability of the charging station.

#### 6.5.2.3 Measurement of Negative Effect of Electric Vehicle Energy Supplement

After the demand for supplementary energy is generated, the electric vehicle needs to go to the charging station for supplementary energy, and then drive to the destination. Therefore, the replenishment behavior of the vehicle causes the driver to generate extra mileage and extra travel time, and when going to the charging station, the driver will have range anxiety as the range is reduced. In addition, when there is no replenishment vehicle scheduling, the driver's choice of charging station is self-interested and random. In the peak period of replenishment, vehicles in local charging station are prone to gather, which not only causes traffic congestion on the surrounding road, but also increases the driver's replenishment time. Therefore, aiming at the range anxiety, extra mileage, and extra travel time caused by EV replenishment, a double-layer optimization model of replenishment scheduling was established to minimize the negative effects of replenishment behavior.

Mileage Anxiety Cost Measurement

According to the survey, when the SOC of electric vehicles exceeds 20% to 25%, the driver will feel comfortable; However, when the SOC falls below this comfort range, drivers worry about running out of power. If the recharge vehicle has not yet arrived at the charging station, but the remaining charge drops below the comfort point $B_{ra}$, With the reduction of electricity, the driver will become more and more anxious, resulting in psychological costs; If the replenished vehicle arrives at the charging station, its SOC is above the comfort point $B_m$ and no range anxiety costs are incurred Hu et al. (2019). Therefore, the average mileage anxiety cost of replenishing vehicles $O(i)$ from the recommended charging station $j$ is as follows:

$$C_1(i,j)=\frac{\sum_{n=1}^{q_{ij}} c_{ij}^n}{q_{ij}} \tag{6.37}$$

$$c_{ij}^n=\begin{cases} 0, & B_{re}^n(i) \geq B_{ra} \\ \frac{B_{ra}-B_{re}^n(i)}{B_{ra}} \times c_p, & 0 \leq B_{re}^n(i) \leq B_{ra} \end{cases} \tag{6.38}$$

In the formula, $q_{ij}$ is the number of supplementary vehicles selected to charging stations $j$ for OD pairs to $i$, $c_p$ is the national average vehicle traction cost.

Extra Mileage Measurement

Considering that the direction of the replenishing vehicle to the selected charging station should be the same as the direction of the destination, that is, it does not take the "backtrack", the sum of the driving distance of the replenishing vehicle from the starting point $O$ to the charging station $J$ to the final destination $D$ is different from the driving distance of the vehicle from the starting point $O$ to the destination $D$ the vehicle does not trigger the replenishing demand, and this difference represents the additional driving distance generated by the replenishing behavior; Therefore, the additional mileage of the replenished vehicle from $O(i)$ to the selected charging station $j$ is shown in Formula (6.39). And because the driver needs to reach the charging station before the battery runs out, in the negative effect of extra mileage, the most critical influence on the driver is the distance from the starting point $O$ to the charging station $J$.

$$C_2(i,j)=\phi_1\left[\left(\sum_{a=1}^{m} L_a\delta_{ak_{i,j}'}^{OJ}(i)+\sum_{a=1}^{m} L_a\delta_{ak_{j,i}''}^{JD}(i)\right)-\sum_{a=1}^{m} L_a\delta_{ak_i}^{OD}(i)\right] \tag{6.39}$$

In the formula $\phi_1$ is the coefficient that converts miles traveled into costs.

Additional Travel Time Measurement

When the electric vehicle triggers the replenishment demand, the time used to complete the trip includes the driving time $T_{O,J}^{\text{travel}}(i)$ of the electric vehicle to reach the charging station, the waiting time $T_j^{wait}$ at the charging station, the charging time $T_{\text{charge}}$ and the time $T_{J,D}^{\text{travel}}(i)$ from the charging station to the destination. Therefore, the additional travel time caused by the replenishment behavior is the total travel time after triggering the replenishment demand minus the travel time without triggering the replenishment demand, as shown in Eq. (6.40). In addition, due to the randomness of the driver's selection of charging stations, the arrival rate of charging stations is different, and the replenishment time of electric vehicles is long, so the queuing time of charging stations is quite different. Therefore, in the negative effect of extra travel time, the most important effect on drivers is the queuing time at the charging station.

$$C_3(i,j)=\phi_2\left[\left(T_{O,J}^{\text{travel}}(i)+T_j^{\text{wait}}+T_{\text{charge}}+T_{J,D}^{\text{travel}}(i)\right)-T_{O,D}^{\text{travel}}(i)\right] \tag{6.40}$$

In the formula $\phi_2$ is the factor that converts the extra miles driven into the cost.

#### 6.5.2.4 Replenishment Scheduling Incentive Strategy

When the electric vehicle goes to the charging station to replenish energy, it needs a long drive or a long queue time, the driver's willingness to replenish energy at the charging station will be greatly reduced. Therefore, in order to improve the driver's response to the replenishment scheduling, the incentive mechanism is formulated according to the driving distance from the electric vehicle to the charging station and the replenishment queuing time. Firstly, the driver's driving distance and replenishment queuing time are divided into three levels: low (L), medium (M) and high (H) according to formula (6.41) to Formula (6.43), and formula (6.41) and formula (6.42) are linear. Then, the driving distance level is combined with the replenishment queuing time level, and the incentive price is divided into 6 levels: very low (VL), low (L), medium (M), high (RH), high (H) and very high (EH), The value of each level is $\gamma_{ij} = 1, 2, 3, 4, 5, 6$. As shown in Table 6.9, the driving distance scheduling levels in cases 1–3 are all low level (L), and the queuing time scheduling levels increase step by step, so the incentive level also increases from very low (VL) to medium (M). In case 4–6, the driving distance scheduling level is medium (M), and the queuing time scheduling level increases successively, so the incentive level increases from low (L) to high (RH). In case 6.9, the driving distance scheduling level is high level (H), and the queuing time scheduling level increases step by step, so the incentive level also increases from very low (M) to medium (EH). Therefore, the longer the distance of the charging station selected by scheduling and the longer the queuing time, that is, the more compensation incentives for the driver, as shown in Eq. (6.44).

**Table 6.9** Classification of incentive levels for supplementary scheduling

| Scheduling situation | Travel distance scheduling level | Queue time scheduling level | Incentive level |
|---|---|---|---|
| Case 1 | L | L | VL |
| Case 2 | L | M | L |
| Case3 | Case | Case | Case |
| Case 4 | M | L | L |
| Case 5 | M | M | M |
| Case 6 | M | H | RH |
| Case 7 | H | L | M |
| Case 8 | H | M | RH |
| Case 9 | H | H | EH |

$$\zeta_{ij}' = \frac{L_{ij}}{L_{ij}^{\max}} \tag{6.41}$$

$$\zeta_{ij}'' = \frac{T_{ij}^{\text{wait}}}{T_{\max}^{\text{wait}}} \tag{6.42}$$

$$\psi_{ij} = \begin{cases} L, & 0 < \zeta_{ij} \le 0.35 \\ M, & 0.35 < \zeta_{ij} \le 0.7 \; \zeta_{ij}', \zeta_{ij}'' \in \zeta_{ij} \\ H, & 0.7 < \zeta_{ij} \le 1 \end{cases} \tag{6.43}$$

$$Z_{ij} = \omega \gamma_{ij} \tag{6.44}$$

In the formula, $L_{ij}$ is the distance from the replenishing vehicle $O(i)$ to the charging station $j$, $L_{ij}^{\max}$ is the farthest distance from the replenishing vehicle $O(i)$ to all charging stations $j$; $T_{ij}^{wait}$ is the queue waiting time of the charging station, and $T_{\max}^{wait}$ is the maximum queue waiting time that the replenishment vehicle can bear; $\psi_{ij}$ is the scheduling level of the driving distance and queuing time of the vehicle from the charging station $j$ to the OD pair for $i$ is supplemented. $Z_{ij}$ is the incentive cost of compensating vehicle to charging station $j$ for OD alignment to $i$, $Z$ is $Z_{ij}$'s in matrix form; $\omega$ is unit scheduling incentive value; $\gamma_{ij}$ is the replenishment dispatching incentive value of the vehicle to the charging station $j$ in the OD pair for $i$.

#### 6.5.2.5 Two-layer Scheduling Model of Electric Vehicle Energy Supplement

Due to the diversity of charging station distribution, it is not suitable to schedule only with the goal of minimizing the negative effect of all the supplementary vehicles in the system. Therefore, this book also considers the interests of the intermediate planner and the balance of the negative effect of the driver's supplementary energy, and establishes a double-layer optimization model of supplementary vehicle scheduling, which makes the model more widely applicable. The upper layer model establishes the global optimal model from the point of view of the middle planner, which aims to minimize the negative effect of all supplementary vehicles and the supplementary incentive given by the planner to the driver.

The objective function of the upper model is:

$$F_1(Z) = \min \sum_{i=1}^{N} \sum_{j=1}^{r} X_{i,j}\left[C_1(i,j) + C_2(i,j) + C_3(i,j) + Z\right] \tag{6.45}$$

The constraint conditions are:

$$X_{i,j} \in \arg\min F_1(Z) \tag{6.46}$$

$$Z_{ij} > 0, \quad \forall Z_{ij} \in Z \tag{6.47}$$

$$0 < \zeta'_{ij} \leq 1 \tag{6.48}$$

$$0 < \zeta''_{ij} \leq 1 \tag{6.49}$$

$$T_{ij}^{\text{wait}} \leq T_{\max}^{\text{wait}} \tag{6.50}$$

$$B_{re}(i) \geq 0 \tag{6.51}$$

$$\delta_{ak_{i,j}'}^{OJ}(i) = \{1, 0\} \tag{6.52}$$

$$\delta_{ak_{j,i}''}^{JD}(i) = \{1, 0\} \tag{6.53}$$

$$\delta_{ak_i}^{OD}(i) = \{1, 0\} \tag{6.54}$$

Formula (6.45) is the objective function of the upper layer model, where $Z$ is the decision variable of the layer model, that is, the incentive matrix of the OD pair of complement energy. The constraints (6.46) indicate that the lower variable $X_{i,j}$ is the solution that minimizes the upper model; $X_{i,j}$ represents the number of vehicles selected to the charging station $j$ for supplementing OD to $i$; The constraint conditions (6.47) are the constraints of the excitation matrix; The constraints (6.48) and (6.49) are the constraints of distance scheduling contribution and queue waiting time scheduling contribution of replenishment vehicles. The constraint condition (6.50) is the driver's waiting time constraint. If the queuing time is greater than the driver's maximum patience time, the charging station will be re-selected. The constraint condition (6.51) is the remaining power constraint of the replenishing vehicle, so that the state of charge of the replenishing vehicle is greater than or equal to zero when it arrives at the charging station. Constraints (6.52)–(6.54) are 0–1 variable constraints.

The lower model constructs the user equilibrium model of the negative effect of energy replenishment according to the key factors of minimizing the negative effect of energy replenishment. When the vehicle is replenished, the two most critical factors affecting the negative effect are the driving distance from the starting point $O$ to the charging station $J$ and the queuing time at the charging station $J$. First of all, when the electric vehicle is in a state of low endurance, the driver is more concerned about whether it can safely reach the charging station before the battery runs out, and when the SOC is lower than the comfortable point $B_{ra}$, the driver will also have range anxiety cost, and because the range anxiety cost is directly related to the driving distance of the vehicle to the charging station, in order to minimize the negative effect of the driver's energy supplement, the driver will have to pay more attention to the cost. The driving distance from the starting point $O$ to the charging station $J$ is regarded as one of the key factors affecting the negative effect. Moreover, because the number of charging piles and the number of replenishing vehicles in each charging station are different, the queuing time of the charging station is also different, and even may be quite different; Therefore, the queue waiting time is regarded as another key factor affecting the negative effect. In addition to the above two factors, the intermediate planner's incentive to the driver's supplementary energy scheduling is also an important factor affecting the driver's choice of charging station. Therefore, the lower level model is built from the perspective of the negative effect equilibrium of driver's energy compensation.

The lower model is:

$$F_2(X_{i,j}) = \min\left[\sum_{i=1}^{N}\sum_{j=1}^{r} X_{i,j}\left(\phi_1 \sum_{a=1}^{m} L_a \delta_{ak_i}^{OJ}(i,j) - Z\right) + \phi_2 \sum_{j=1}^{r}\int_0^{u_j} T_j^{\text{wait}}(y)\mathrm{d}y\right] \tag{6.55}$$

The constraint conditions are:

$$0 \le X_{i,j} \le q_i,\ X_{i,j} \in \mathbf{Z}, \forall\, i \in N, \forall j \in J \tag{6.56}$$

$$q_i = \sum_{j=1}^{r} X_{i,j} \tag{6.57}$$

$$u_j = \sum_{i=1}^{N} X_{i,j} \tag{6.58}$$

$$\phi_1 \sum_{a=1}^{m} L_a \delta_{ak_i}^{OJ}(i,j) + \phi_2 T_j^{\text{wait}} - Z_{ij} = \phi_1 \sum_{a=1}^{m} L_a \delta_{ak_i}^{OJ}(i,s) + \phi_2 T_s^{\text{wait}} - Z_{is} \tag{6.59}$$

$$0 < \zeta'_{ij} \leq 1 \tag{6.60}$$

$$0 < \zeta''_{ij} \leq 1 \tag{6.61}$$

$$T_{ij}^{\text{wait}} \leq T_{\max}^{\text{wait}} \tag{6.62}$$

$$B_{re}(i) \geq 0 \tag{6.63}$$

$$\delta_{ak'_{i,j}}^{OJ}(i) = \{1, 0\} \tag{6.64}$$

$$\delta_{ak''_{j,i}}^{JD}(i) = \{1, 0\} \tag{6.65}$$

$$\delta_{ak_i}^{OD}(i) = \{1, 0\} \tag{6.66}$$

Formula (6.55) is the objective function of the lower model, where $X_{i,j}$ is the decision variable, that is, the number of vehicles selected to recharge at the charging station $j$ for supplementing OD to $i$. Constraints (6.56) are $X_{i,j}$'s value constraints. The constraint condition (6.57) is the flow conservation constraint of replenished vehicle in OD pair to $i$. The constraint condition (6.58) is the constraint of recharge flow of the charging station $j$, that is, the recharge flow of the charging station $j$ is equal to the sum of the flow of each recharge OD to the charging station $j$. The constraint conditions (6.59) represent the negative effect equilibrium of the same OD pair of medium replenishment vehicles. Constraints (6.60) to constraints (6.66) are part of the constraints in the upper model, and are also constraints in the lower model. Constraints (6.60) and (6.61) are constraints on the distance scheduling contribution and queuing waiting time scheduling contribution of replenishment vehicles. Constraints (6.62) are the driver's waiting time constraints. If the queuing time is greater than the driver's maximum patience time, the charging station will be re-selected. The constraint condition (6.63) is the remaining power constraint of the replenishing vehicle, so that the state of charge of the replenishing vehicle is greater than or equal to zero when it arrives at the charging station. The constraints (6.64)–(6.66) are 0–1 variable constraints.

### 6.5.3 Design of GA-FW Algorithm for Energy Replenishing Scheduling of Electric Vehicles

In order to solve the two-layer optimization model of complement scheduling proposed in the previous section, genetic algorithm is used to solve the upper model and search the complement excitation matrix with minimal negative effect. The improved Frank-Wolfe algorithm is used to solve the lower model, and the distribution scheme of the replenishment vehicle is obtained and the path is planned. The specific steps are as follows:

Step 1: Initialize. Input path network, OD travel demand, charging station, initial travel time of each arc, section capacity, related parameter values, algorithm end conditions, and other data.
Step 2: Code and generate the initial population. The chromosome is encoded in a discrete integer way, and the encoding interval is as follows $[S_{\min}, S_{\max}]$, the incentive scheme is digitally encoded, and the initial population is formed through the encoding, that is, the incentive scheme for replenishment vehicles to go to different charging stations.
Step 3: Replace the upper energy scheduling incentive scheme into the lower model, and use the Frank-Wolfe algorithm to allocate the replenished vehicles and non-replenished vehicles in the OD pair, and update the path flow and road section travel time. Check whether the conditions are met. If yes, go to Step 4. If no, go to Step 2.
Step 4: Calculate the fitness function. The vehicle allocation scheme obtained in step 3 is returned to the upper model, and the fitness function is calculated and evaluated. According to the evaluation results, the chromosomes were crossed and mutated to produce a new population.
Step 5: Check whether the end condition of the algorithm is met. If yes, output the replenishment scheduling scheme. Otherwise, return to Step 3.

### 6.5.4 *Analysis of Calculation Examples of Energy Replenishment Scheduling for Electric Vehicles*

To verify the proposed model and algorithm, the Nguyen-Dupius network is taken as an example, as shown in Fig. 6.18. This small network consists of 13 nodes, 19 sections, and 4 OD pairs, and the numbers in brackets at the arrows represent section length, free flow time, and section capacity respectively. There are three fast charging stations in the network, located at nodes 6, 9, and 11. It is assumed that the models of electric vehicles in the model are the same, the charging stations of 6, 9, and 11 are all homogeneous charging stations, and the vehicles that recharge energy on the way can recharge energy at most once. The model parameters are shown in Tables 6.10 and 6.11.

#### 6.5.4.1 Analysis of Numerical Results

The algorithm proposed in this book is used to solve the calculation example, and the optimal incentive level scheme is obtained, as shown in Table 6.12.

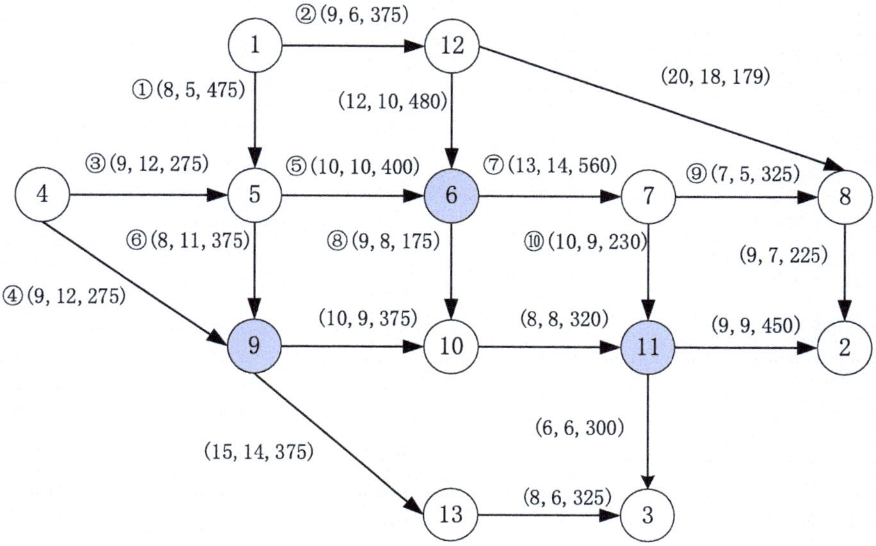


**Fig. 6.18** Nguyen-Dupius network

**Table 6.10** Model parameters

| Argument | Significance | Value |
|---|---|---|
| $E_c$ | Vehicle electricity consumption per kilometer | 0.18 KWH/km |
| $B_f$ | Battery capacity | 54 KWh |
| $\eta$ | Energy consumption coefficient | 0.9 |
| $B_{ra}$ | Psychological threshold of mileage anxiety | 20% |
| $\phi_1$ | The cost of electricity consumed per unit length | 1.5 yuan/km |
| $\phi_2$ | Time value conversion Coefficient | 0.6 yuan/minute |
| $c_p$ | National average vehicle towing cost | 150yuan |
| $d_j^0$ | Queueing time coefficient | 30 min |
| $\beta_i$ | Replenish the permeability of the vehicle | 20% |
| $\omega$ | Unit scheduling incentive value | 1yuan |

**Table 6.11** OD travel demand

| Serial number | OD pair | OD pair indicates the travel requirement |
|---|---|---|
| 1 | 1–2 | 400 |
| 2 | 1–3 | 300 |
| 3 | 4–2 | 300 |
| 4 | 4–3 | 800 |

**Table 6.12** Optimal incentive level scheme

| ODpair | Station 6 | Station 9 | Station 11 |
|---|---|---|---|
| 1–2 | 3 | 1 | 6 |
| 1–3 | 3 | 5 | 3 |
| 4–2 | 3 | 2 | 4 |
| 4–3 | 3 | 3 | 3 |

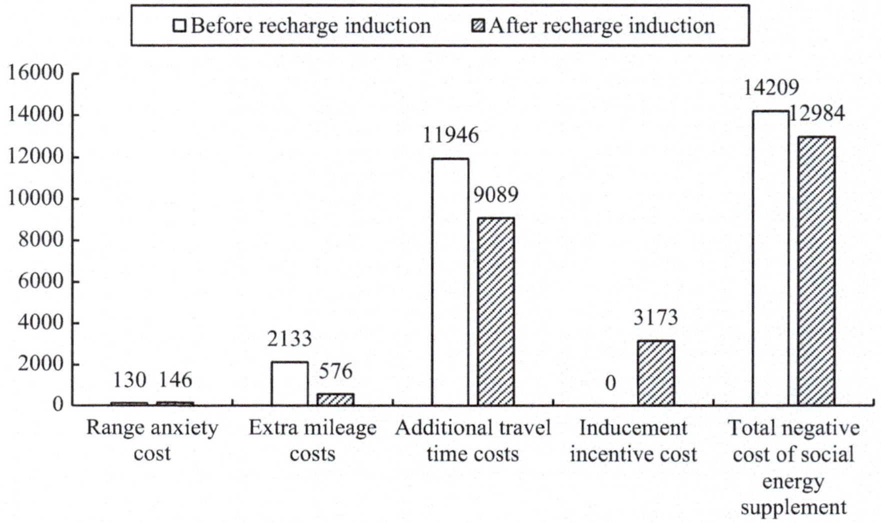


**Fig. 6.19** Comparison of negative effects before and after supplementary energy scheduling

As shown in Fig. 6.19, the total negative cost of social energy supplement for all electric vehicles in the road network after energy supplement scheduling is 12,984 yuan, among which, the cost of driver range anxiety is 146 yuan, the cost of extra mileage is 576 yuan, and the cost of extra travel time is 9089 yuan. The scheduling incentive cost of the planner is 3173 yuan. In contrast, when there is no replenishment incentive, all vehicles are replenished by the nearest charging station. By calculation, the total negative effect of energy replenishment in the road network is 14,209 yuan, in which the driver's mileage anxiety cost is 130 yuan, the extra mileage cost is 2133 yuan, and the extra travel time cost is 11946 yuan.

Through the analysis of the above calculation results, we can see that although the supplementary energy scheduling model proposed in this book increases the incentive cost of planners, the negative effect cost of total social supplementary energy decreases, which proves the effectiveness of supplementary energy scheduling. Among them, in this calculation example, the proportion of mileage anxiety cost and extra mileage cost is relatively small because the Nguyen-Dupius network is relatively small and the location of charging stations is more balanced.

**Table 6.13** Comparison of queuing time at charging stations before and after scheduling

| Station | Total station queuing time after scheduling (minutes) | Total queuing time before scheduling (minutes) | Optimization effect (%) |
|---|---|---|---|
| Station 6 | 2364 | 2982 | 20.7 |
| Station 9 | 2364 | 3931 | 39.9 |
| Station 11 | 2364 | 1416 | −66.9 |

The queuing time of recharge vehicles at each charging station before and after dispatch is shown in Table 6.13. The results show that after adopting the optimization model and algorithm of replenishment scheduling proposed in this book, reasonable charging stations can be allocated to drivers according to the remaining power of replenishment vehicles, road network information, and queuing time of charging stations, which avoids the phenomenon that replenishment vehicles are concentrated in charging station 6 or 9 before scheduling, and makes the utilization rate of various charging stations in the region more balanced. The negative effect of centralized recharge on the charging station is reduced, and the driver's queuing time of recharge is also reduced as a whole.

#### 6.5.4.2 Sensitivity Analysis

Unit Price Sensitivity Analysis of Supplementary Scheduling Excitation

In order to explore the influence of different incentive unit price on the negative effect of energy replenishing vehicles, sensitivity analysis of incentive unit price of energy replenishing scheduling is carried out below, and specific results are shown in Table 6.14. When the incentive unit price is 0.5, the incentive cost given by the planner is less, and the willingness of the vehicle to go to the charging station is lower, so the optimization effect of the negative effect of energy supplement is not obvious. When the incentive unit price is 1, the incentive cost of supplementary energy given by the planner is relatively large, and the willingness of supplementary vehicles to go to the scheduled charging station is relatively high, so the optimization effect of the negative effect of supplementary energy is relatively obvious. When the incentive unit price is 2, the planner gives more incentive cost of supplementary energy, and the supplementary vehicle has a higher willingness to go to the scheduled charging station, but the total cost of additional travel time of supplementary vehicle increases, among which the main influencing factor of the additional travel time cost is the queuing time of the charging station. The main reason for this result is that drivers may choose charging stations with long waiting times in order to obtain higher incentive costs. According to the sensitivity analysis of incentive unit price, when the incentive of supplementary energy scheduling is low, the willingness of supplementary vehicles to go to the charging station is low, and the optimization effect of the negative effect of supplementary energy is poor. When the incentive

**Table 6.14** excitation unit price sensitivity analysis

| $\omega$ | Negative cost change of supplementary energy | | | | |
|---|---|---|---|---|---|
| | Range anxiety cost | Extra mileage costs | Additional travel time costs | Incentive cost | Total social cost |
| 0.5 | 162 | 633 | 10,671 | 1473 | 12,939 |
| 1 | 146 | 576 | 9089 | 3173 | 12,984 |
| 2 | 132 | 301 | 11,946 | 4464 | 16,843 |

**Table 6.15** Permeability sensitivity analysis of electric vehicles

| $\beta$ (%) | Total cost of negative effect of energy replenishment before scheduling | Total cost of negative effect of energy replenishment after dispatch | Optimization effect (%) |
|---|---|---|---|
| 20 | 14,209 | 12,984 | 8.63 |
| 40 | 17,017 | 15,279 | 10.21 |
| 60 | 31,055 | 29,557 | 4.82 |
| 70 | 47,859 | 46,802 | 2.21 |

of replenishment is high, replenishment vehicles may choose charging stations with longer queuing time in order to obtain higher compensation costs, which aggravates the negative effect of replenishment.

Permeability Sensitivity Analysis of Electric Vehicles

Previous analyses have been based on a fixed penetration rate of 20% for electric vehicles. In order to explore the influence of different electric vehicle share on the supplementary energy scheduling proposed in this book, sensitivity analysis was carried out on the penetration rate of electric vehicles while keeping OD internal travel demand unchanged. The specific results are shown in Table 6.15. When the penetration rate of the EV in the OD pair increases from 20 to 40%, the optimization effect of the EV in the OD pair increases from 8.63 to 10.21%, but when the penetration rate of the EV in the OD pair increases from 40 to 60%.

Or 70%, the optimization effect is reduced. The reason is that there are too many electric vehicles in the road network, and the demand for supplementary energy surges, far exceeding the supply capacity of charging stations in the road network. At this time, it is no longer the imbalance in the time and space of supplementary energy demand, but the insufficient number of charging station facilities in the road network, and additional charging stations should be considered.

According to the permeability sensitivity analysis of electric vehicles, under the condition that OD internal travel demand remains unchanged, appropriately increasing the permeability of electric vehicles can improve the optimization effect of energy replenishing scheduling. However, when the penetration rate of electric vehicles continues to rise, if the number of charging station facilities in the road network is not increased, the negative effect of replenishing vehicles in the road network will be aggravated.

## 6.6 Suggestions on Optimization Strategy of Urban Transportation Energy Supply and Demand Matching

Since China pledged to reach the peak of carbon in 2030 and carbon neutrality in 2060 at the international conference in 2020, the country has introduced a number of energy conservation and emission reduction policies, among which easing urban traffic congestion is an important starting point to promote the "dual carbon" goal. In the urban transportation hub, improving the capacity of key nodes is an important countermeasure to ease the traffic. The efficient and reasonable operation of urban transportation energy supply network is the core of transportation development. However, due to the complex structure of urban transportation energy supply network and large supply and demand volume; Replenishment station brands, industry competition is prominent; The unreasonable layout of some stations and the randomness of energy demand lead to the imbalance of energy station layout and energy demand in the city, resulting in a negative effect when the vehicle is replenishing energy. Specifically, due to the random spatio-temporal nature of vehicles in urban TESN and the unreasonable layout of energy replenishment stations, vehicles gather and queue up for energy replenishment at some stations during vehicle replenishment, leading to traffic congestion at stations, increasing energy consumption and carbon emissions, and causing environmental pollution. At the same time, for the individual driver, a long time waiting in line causes a waste of time, resulting in negative emotions, and these negative emotions will be further aggravated and bring more serious impact. In addition, there are also fewer vehicles arriving at the station to replenish energy, resulting in idle waste of resources, causing losses to the urban economy. The development of urban transportation causes adverse effects and affects the implementation of national strategies. Therefore, according to the above problems, the following countermeasures and suggestions are put forward to reduce the negative effect of energy supplement:

### *6.6.1 Accelerate the Construction of Vehicle Energy Replenishment Scheduling Information Platform*

Due to the phenomenon of non-sharing of information among replenishing vehicles, replenishing stations, and transportation networks, drivers cannot fully grasp the current traffic information and the queuing situation in replenishing stations when they need replenishing energy. Therefore, large-scale replenishing behaviors in the

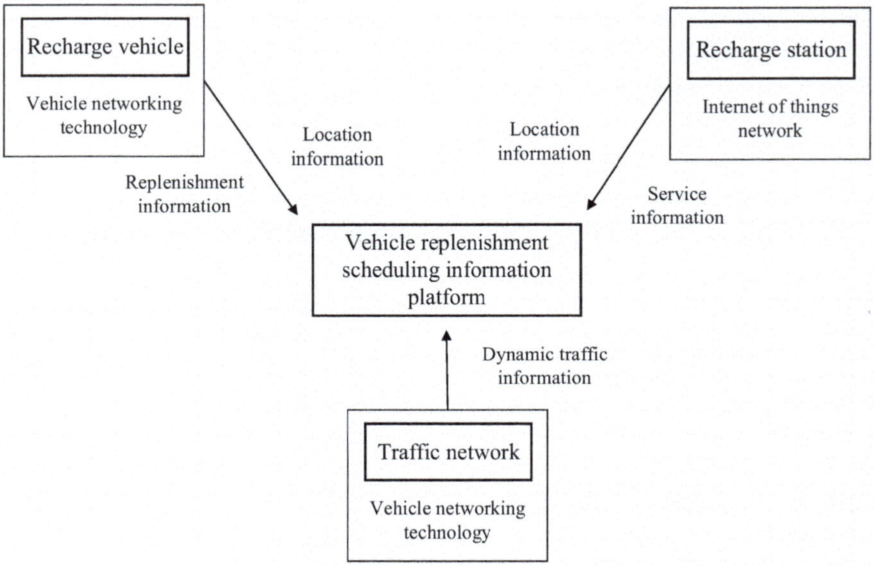


**Fig. 6.20** Vehicle scheduling information platform

road network not only increase the travel time and replenishing cost of drivers, but also increase the travel time and replenishing cost of drivers. Moreover, it is easy to cause mileage anxiety of drivers and congestion of local recharge stations, and even affect traffic and cause traffic jams. Therefore, an information platform for vehicle replenishment scheduling should be built by combining the location information and replenishment information of the replenishment vehicle, the location information and service information of the replenishment station, and the dynamic information of the traffic network, as shown in Fig. 6.20.

Among them, the replenishment vehicle information is mainly location information and replenishment information, and the replenishment information includes driving range, mileage anxiety threshold, driver's replenishment requirements, etc. The supplementary energy station information includes information and service information, in which the service information is the number of supplementary energy equipment of the supplementary energy station, the service time of a single user, the energy unit price, the number of available equipment in the station, the number of queuing vehicles and other information. Traffic network dynamic traffic information refers to the real-time traffic situation of each road. The location information of replenishment vehicles and road information of traffic network can be obtained in real time through vehicle networking technology, and the service information of replenishment stations can be connected to the Internet of Things through wireless communication network, and the information can be transmitted to the vehicle scheduling information platform in real time. The vehicle replenishment scheduling information platform will analyze and calculate the obtained information, and provide necessary information for subsequent replenishment vehicle scheduling.

The construction of vehicle scheduling information platform is the first step to realize intelligent vehicle scheduling. The government or other intermediate planners should actively assume the leading role in the construction of vehicle scheduling information platform, accelerate the construction of vehicle scheduling information platform through the power of the government or other intermediate planners, and provide a solid data foundation for subsequent vehicle scheduling.

### *6.6.2 Enabling the Site to Respond Positively to the Scheduling Platform*

As for the replenishment station, during the replenishment peak period, high-flow replenishment vehicles rush into the replenishment station within a short period of time, and the replenishment equipment in the station is in short supply. Especially for electric vehicles, electric vehicles take a long time to charge, so it is easy to cause the phenomenon of queuing in local charging stations during the replenishment peak period, which reduces the driver's satisfaction with the replenishment station. On the contrary, most of the replenishment equipment is idle during the peak period of replenishment, which results in the decrease of the unit time income of the replenishment station. Therefore, in view of the imbalance of the above energy replenishment demand in time and space, the energy replenishment station should actively respond to the vehicle energy replenishment scheduling platform, and achieve the balance of supply and demand in time and space through the vehicle scheduling information platform. For example, in the time dimension, the charging station can dynamically adjust the unit price or service fee of energy to form a time-sharing price, and appropriately increase the unit energy cost during the peak period of energy replenishment. At this time, the vehicle dispatching platform will pass the price of the replenishment station to the driver, and the driver will obtain the replenishment price in different periods of time, and induce the driver to replenish energy in the off-peak period through the price. To achieve the effect of peak cutting and valley filling, but the fixed time-sharing price sometimes causes new energy peak, so that the burden reduction effect is not obvious. Therefore, the dynamic time-sharing price should be set according to the real-time service situation in the station or the number of energy replenishing vehicles to minimize the time accumulation of energy replenishing demand. In terms of spatial dimension, the driver's choice of supplementary energy station is egoistic and random, which easily leads to the congestion of some supplementary energy stations and the excess of some supplementary energy stations. Therefore, if the supplementary stations actively respond to the scheduling platform, the supplementary stations in the region will form an alliance, cooperate with multiple agents, implement the benefit distribution mechanism, and balance the supplementary energy demand among the supplementary stations. The benefit distribution method is used to divide the benefit, so as to achieve the balanced effect of equipment utilization rate of each replenishing station in the region, and improve the customer satisfaction and the unit time benefit of replenishing station.

In order to further encourage the replenishment stations to respond positively to the scheduling platform, first of all, the intermediate planners can provide incentives to the replenishment stations with high enthusiasm to respond to the platform by formulating relevant incentive policies. Secondly, a reasonable benefit distribution mechanism should be developed to avoid the site refusing to respond to the dispatching platform because of unfair benefit distribution. Finally, a reasonable restraint and punishment system should be formulated to standardize the operation and management of the supplementary energy station under the scheduling platform.

### *6.6.3 Guide Replenishment Vehicles to Actively Participate in Platform Scheduling*

For the driver, on the one hand, how to choose the best replenishment station and replenishment path to avoid the phenomenon of mileage emergency during the driving will be the main problem that users pay attention to; On the other hand, how to avoid the negative impact of overloading and congestion on the energy supply network caused by large-scale replenishment vehicles will be a key issue for planners to consider. Therefore, it is necessary to guide the replenishing vehicle to actively participate in the platform scheduling, rationally select the replenishing station and carry out orderly replenishing, and reduce a series of negative effects caused by the replenishing behavior.

The driver's personal factors are the key factors influencing the uncertainty of replenishment demand. The driver decides when, where, and with how much remaining range the vehicle can recharge. Due to the heterogeneity of drivers, different gender, age, education level, driving age, risk attitude, and income will have an important impact on drivers' choice of recharge stations and driving paths. However, for drivers, the most reluctant to see the complement queue, if it is often encountered the complement queue, the ordinary complement is not prepared in advance, some drivers will have mood swings, especially for their own mood is not good, there is an emergency, in a hurry for people, there may be aggressive behavior. Because the long queue is easy to cause psychological irritability of the driver, the formation of a "road rage" family, further stimulate contradictions, resulting in worse effects. According to research, people's psychological state in the face of queuing is divided into three stages: the first stage, calm stage, this stage is relatively short, people are still in the rational stage; In the second stage, impatience and anxiety stage, people's patience reaches the limit, they are disgusted with waiting, and they are easy to lose control of emotions and drastic actions; Stage three: Return to calm. Therefore, in order to meet the replenishment needs of different drivers, the vehicle replenishment scheduling platform provides drivers with the queuing congestion degree of all nearby replenishment stations based on real-time traffic information and real-time service information of replenishment stations. Drivers can choose whether to arrive at the station or not according to the congestion degree of replenishment energy, saving drivers' replenishment time. In addition, the driver can also obtain the energy price of the supplementary energy station in real time through the vehicle dispatching platform, and respond to the preferential policies of the supplementary energy station, such as compensating energy in the flat peak period of low energy prices, which can not only reduce the supplementary energy cost, save the queuing time of supplementary energy, but also balance the equipment utilization rate.

In order to better guide energy replenishment vehicles to participate in platform scheduling, it is necessary to encourage energy replenishment vehicles. Although the government or other intermediate planners are not the core subjects of negative effect of energy replenishment to reduce the burden, their participation will make the burden reduction effect of energy replenishment-induced scheduling work double with half the effort. From the perspective of minimizing the negative effect of energy replenishment on the whole road network, intermediate planners give appropriate incentives to drivers who contribute more to reducing the negative effect of energy replenishment. With the improvement of incentive effect, the planners' incentive cost of inducing energy replenishment vehicles increases, but the negative effect of energy replenishment vehicles decreases first and then increases. Therefore, the planner should choose the appropriate supplementary incentive unit price according to the price sensitivity of the driver, which not only effectively reduces the cost of the planner, but also reduces the negative effect of supplementary vehicles in the road network. The results show that although the proposed supplementary incentive mechanism increases the incentive cost of planners, the total negative social supplementary cost of supplementary vehicles decreases. Therefore, the government or other intermediate planners should actively participate in the work of reducing the negative effects of supplementary energy and formulate relevant rules to play a role of management and restriction in the work of reducing the negative effects.

## 6.7 Summary of This Chapter

This chapter first analyzes the theory of energy replenishment vehicle scheduling, and analyzes the technical basis of vehicle energy replenishment scheduling, and puts forward a double-layer random energy replenishment scheduling plan for fuel vehicles. In order to balance the demand and supply of replenishment energy in the road network, the incentive scheduling method is adopted to set several incentive levels. Genetic algorithm and improved Frank-Wolfe algorithm are designed to solve the double-layer energy replenishment scheduling model, and the effectiveness of the model and algorithm is verified by a road network example. The numerical results show that:

(1) Although the dual-layer optimization model of supplementary energy scheduling proposed in this book increases the scheduling incentive cost of planners, the total negative social supplementary energy cost of all supplementary vehicles decreases, which proves the effectiveness of supplementary energy scheduling.
(2) With the improvement of incentive effect, the scheduling incentive cost of the replenishment vehicle increases, but the negative effect of the replenishment vehicle decreases first and then increases. Therefore, selecting the appropriate unit price of the replenishment dispatching incentive can effectively reduce the cost expenditure of the planner and reduce the negative effect of the replenishment vehicle in the road network.
(3) With the increase of the penetration rate of replenishment OD to medium electric vehicles, the replenishment demand in the road network far exceeds the service capacity of the station, and the scheduling incentive is provided at this time.

## References

Baek JW, Moon SK (2014) The M/M/1 queue with a production-inventory system and lost sales. Appl Mathematics Comput 233(8):534–544

Bureau of Public Roads (1964) Traffic assignment manual. Urban Planning Division, US Department of Commerce, Washington, DC

Hu L, Dong J, Lin Z (2019) Modeling charging behavior of battery electric vehicle drivers: a cumulative prospect theory based approach. Transp Res Part C: Emerg Technol 102(2):474–489

# Chapter 7
# Simulation Research of Urban Transportation Energy Supply Network Based on Anylogic

## 7.1 Analysis of Simulation Principle Based on Anylogic

### *7.1.1 Analysis of Simulation Principle*

AnyLogic is one of the early tools to introduce multi-method simulation modeling, and can use visual modeling languages such as flowcharts, state diagrams, and action diagrams. It supports multi-agent simulation, discrete event simulation, and system dynamics simulation in the fields of control, dynamic systems, computer systems, production, machinery, military, transportation, and logistics. In addition, it is an object-oriented modeling method based on UML language, based on flowchart modeling method, state diagram, differential and algebraic equations, and based on 100% Java language modeling. The process of designing AnyLogic model is actually to design the class of active objects, which can interact and transmit messages through ports, and define the relationship between them. Information is then transmitted from the real world to simulate the movement and transmission of information in real time. The use of AnyLogic technology can provide strong support for the next step of the design of data, reasonable simulation of the real scene, and optimization. Agent-based modeling is an individual-centered modeling method, which is essentially discrete. Agent-based modeling is applicable to multi-level and complex models, which can not only realize simple physical object detail modeling, but also suitable for more complex modeling such as roads and buildings. Agent modeling first needs to determine the simulated objects, such as roads, cars, traffic lights, signs, etc. Then it is necessary to define the behavior of different vehicles, such as driving force, reaction, state, etc. Finally, the vehicle agents with behaviors are put into the simulation model and the relationship between them is established. Any Logic provides modules that can describe almost all Agent behavior, and if you need to add some module libraries, users can write Java code to achieve it.



X. Ge, *Urban Transportation Energy Supply Network*, Industrial Ecology and Environmental Management 5, https://doi.org/10.1007/978-3-031-64619-5_7

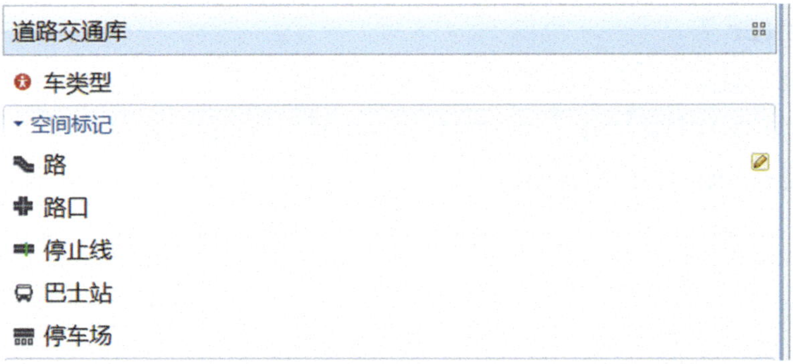


Fig. 7.1 AnyLogic road traffic Library

Fig. 7.2 AnyLogic road traffic module

Any Logic road traffic simulation module is mainly used for road traffic library, which is developed on the basis of process modeling library, including two parts: spatial marker element and road traffic module, as shown in Figs. 7.1 and 7.2.

The spatial marker elements are mainly used in road network environment base map modeling, such as roads, intersections, stations, parking lots, etc. The vehicle behavior module is mainly used to simulate the flow of vehicles on the road, such as replenishment behavior, normal driving, intersection turning, path selection, etc.

### 7.1.2 Simulation Modeling Example Based on AnyLogic

The simulation of a single site is mainly based on the construction of the internal lane of the site, the replenishment equipment, and the energy storage device. The data statistics of a single replenishing device and energy reserve in a station under different number of replenishing vehicles in multiple time periods were carried out, and the rules of vehicle queuing and replenishing at the station were sorted out. Multiple stations in the replenishment network analyzed the replenishment law in the energy supply network by statistical data of the queue leader of each station and the number of replenishment vehicles accumulated in the unit cycle time through the starting point traffic setting. The traffic environment simulation of AnyLogic software is mainly for the driving simulation of road vehicles, and the station is set to distinguish charging pile and refueling station through the time delay device. By adjusting the arrival rate of different stations to simulate the characteristics of the actual charging and refueling demand, so as to achieve the simulation of the actual recharge effect. Simulation environment modeling is mainly divided into urban transportation energy supply site environment modeling and urban transportation energy supply network environment modeling. The two models contain slightly different subjects.

#### 7.1.2.1 Main Analysis of Simulation Based on AnyLogi Software

The main body of the single station simulation model of urban transportation energy supply mainly includes four items. The first is the road infrastructure, including the main road of the road where the energy supply station is located and the road of the recharge vehicle inside the station. The second is the recharge server, and the refueling gun and charging pile set for the vehicle recharge. The third is dynamic vehicles, including vehicles with no supplementary energy demand and vehicles with supplementary energy demand. The fourth is an energy storage facility, mainly to observe the cumulative consumption of refined oil and cumulative consumption of electricity at the refueling station. The main body of urban transportation energy supply network simulation includes three main parts, namely, road infrastructure, recharge server, and dynamic vehicle. Therefore, according to the main body of the simulation, the urban transportation energy supply station and network environment model are built respectively, as shown in Fig. 7.3.

#### 7.1.2.2 Construction of Simulation Model of Urban Transportation Energy Supply Station

According to the survey and statistics of various stations in the main urban area of Chongqing, the internal four-lane gas station is selected according to the size of the layout, and the main road is three-lane, and the vehicles on the main road can change lanes to enter the gas station. The slip road into the petrol station is a two-lane road. An energy booster and a camera are placed on each of the four internal lanes to connect the four energy booster devices to the energy storage facility. Three energy storage facilities are set up for diesel, 92 gasoline, and 95 gasoline. The model construction is shown in Fig. 7.4.

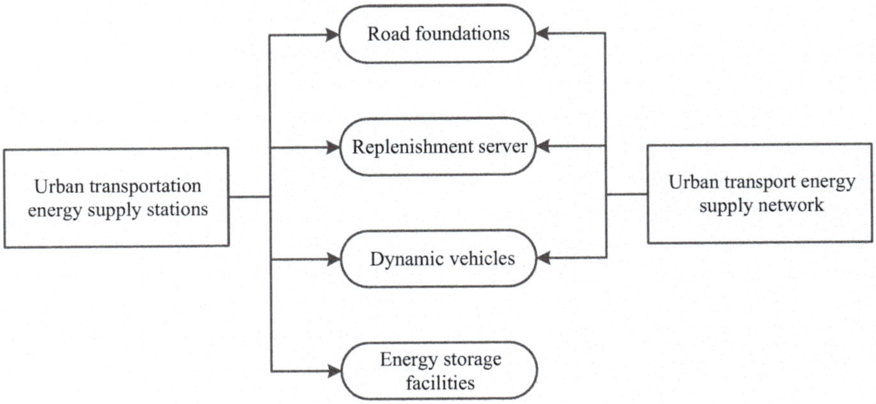


**Fig. 7.3** AnyLogic simulation model environment body

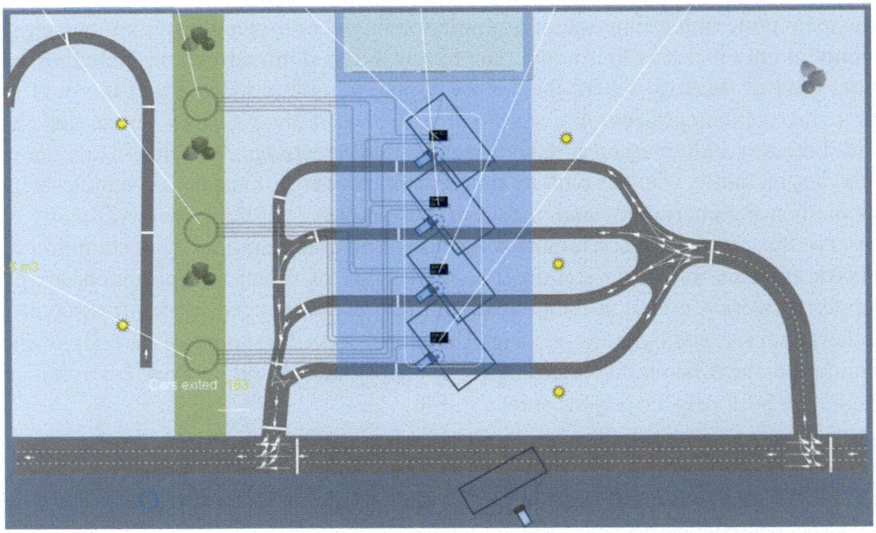

**Fig. 7.4** Effect of simulation model of point replenishment environment of urban transportation energy supply station

#### 7.1.2.3 Construction of Urban Transportation Energy Supply Network Simulation Model

According to the map of some areas of Jiulongpo in Chapter 4, the road feature model of this area is constructed. In order to be close to the reality of the road, according to the Baidu map road in the region, on the basis of the previous chapter to improve some regional roads. An energy supply station is set up at the corresponding location. The station distinguishes gas stations from charging stations by recharge time, arrival rate, and capacity. In addition, in order to make the model more simple and convenient, unnecessary reverse lanes on some roads are eliminated. Only the reverse lane road with the connected OD pair in the middle is left.

After building the basic environment model, there is still no dynamic vehicle input, so it is necessary to build the logical relationship of the model. It is not only necessary to divide the vehicle road selection and the road connection logic, but also to set the relevant parameters in the replenishing network according to the traffic flow and replenishing traffic flow.

## *7.1.3 Logical Relationship and Parameter Settings*

After building the environmental model, it is necessary to build the road network logic model of vehicle operation according to the connections between sections, so as to ensure that the vehicles input by the vehicle generator can select and drive between each section, complete the vehicle activities of all sections and stations required by the simulation, and obtain the corresponding simulation results.

### 7.1.3.1 Site Logical Relationship Building and Parameter Setting

The logic structure of the urban transportation energy supply station model mainly consists of two parts. The first part is the replenishment process of vehicles at gas stations, as shown in Fig. 7.5. The input makes a selection at the entrance intersection. The selection probability is 0.5, that is, half of the vehicles will enter the gas station to refuel. After entering the branch road, it will be divided into four refueling roads, and when there is no vehicle refueling, it will first enter the first refueling road and refuel at the No. 1 refueling machine. Subsequent refueling vehicles will be refueled according to the refueling distribution of the previous vehicle to the next refueling machine. According to Chapter 3, the time period is divided into 15 h from 7 a.m. to 10 p.m., and the refueling data is used as the number of vehicles arriving.92 Gasoline tank capacity is set to 30 cubic meters, the initial quantity is 20 cubic meters, and the maximum output rate is 10 L/meter; 95 Gasoline tank capacity is set to 30 cubic meters, the initial quantity is 20 cubic meters, and the maximum output rate is 10 L/meter; The diesel tank capacity is set at 30 cubic meters, the initial quantity is 15 cubic meters, and the maximum output rate is 10 L/meter.

The second part is the product replenishment process, as shown in Fig. 7.6. The fuel vehicle will make up the energy according to the remaining fuel amount of the gas station to meet the replenishment behavior of the refueling vehicle in this period of time. The length of the tanker is set to 7 m, the initial speed of the vehicle is 5 km/h, the maximum acceleration is 1.8 square meters/second, and the maximum deceleration is 4.28 square meters/second. The tank capacity is 50 L and the injection rate is 1 L/second. The storage facility is connected to the tanker through a pipeline at a rate of 25 L/m.

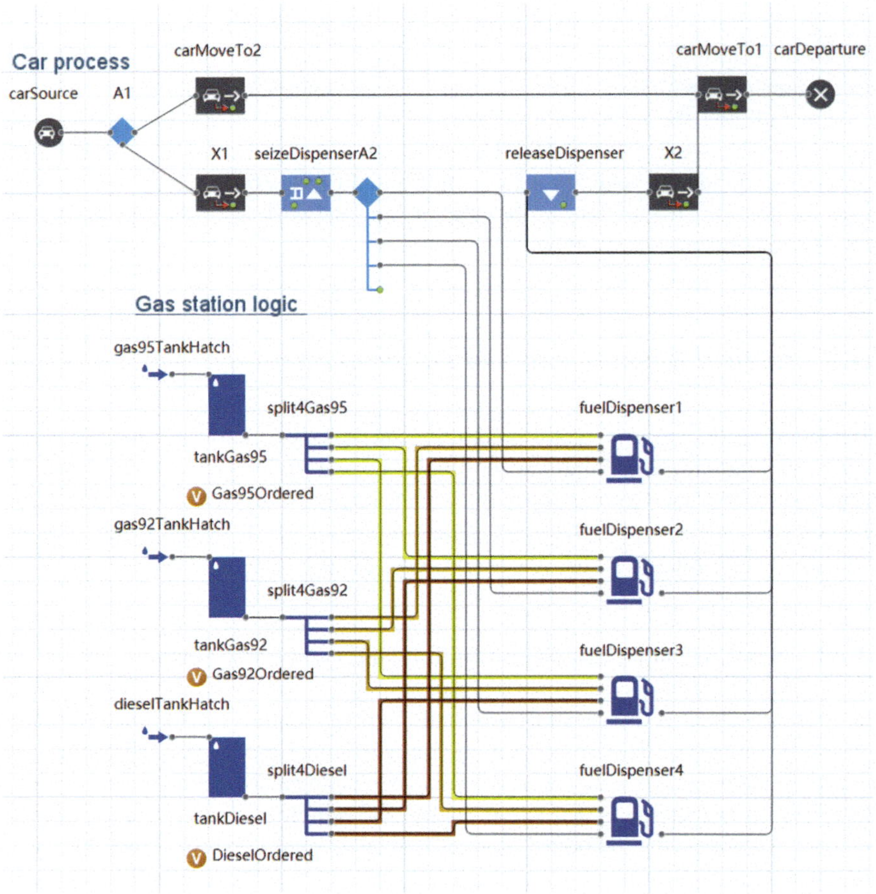


**Fig. 7.5** Logical structure model of vehicle replenishment at the station

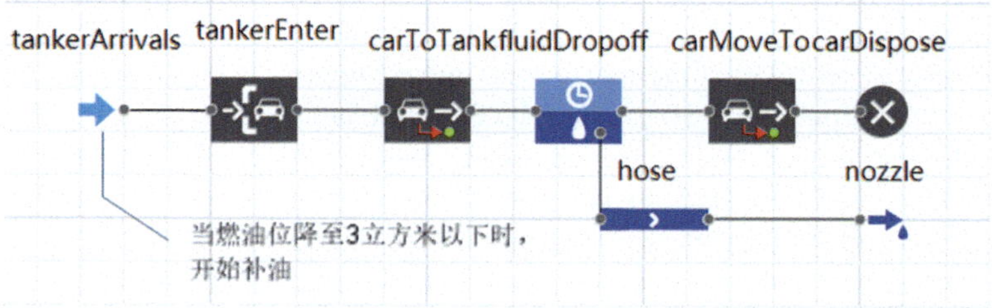


**Fig. 7.6** Logical structure model of replenishment energy of oil storage device

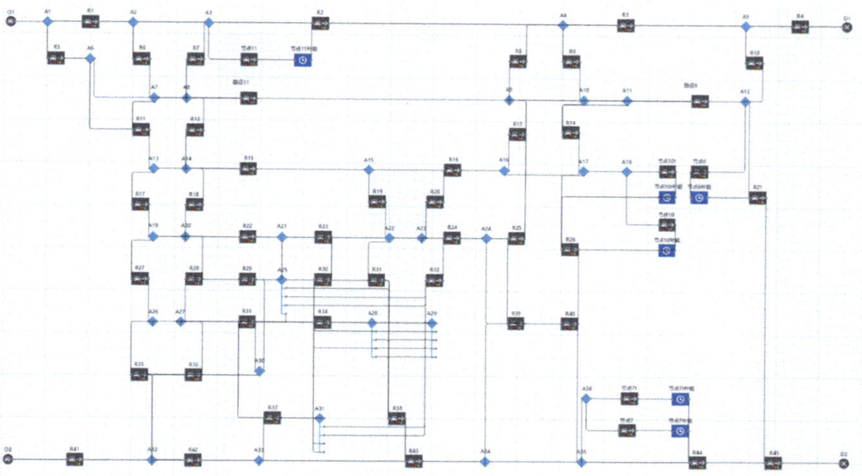

**Fig. 7.7** Logical structure model of road network

#### 7.1.3.2 Network Body Logical Relationship Construction and Parameter Setting

The main logic of the main body of the urban transportation energy supply network is the path selection and connection between various sections, as shown in Fig. 7.7. The number of two vehicle generators is divided into six time periods from 6 a.m. to 12 noon according to the peak period data, and the number of each time period is entered according to the vehicle flow data in Chapter 4. The length of the vehicle is 5 m, the length of the parking space at the recharge station is 6 m, the initial speed is 60 km/h, the maximum acceleration is 1.8 square meters/second, and the maximum deceleration is 4.28 square meters/second. The selection probability of the vehicle at the fork is average, the selection probability at the recharge station is 0.9 for the gas station and 0.1 for the charging station respectively. There are 8 refueling parking Spaces at gas stations, 30 parking Spaces at charging stations, and 20 parking Spaces at node 7 and node 11 respectively due to modeling limitations. The recharge time is 5 min delay at the refueling station and 15 min delay at the charging station.

#### 7.1.3.3 Logical Relationship and Parameter Setting of Replenishment in the Network

In order to obtain multidimensional replenishment data, the replenishment site selection model is built according to the logical structure in the path, as shown in Fig. 7.8. According to the different types of recharge stations, the selection probability of charging stations is 0.1, 0.3, 0.5, and 0.8 respectively. The charging station capacity means that the charger is set to 1, and the filling station capacity means that the

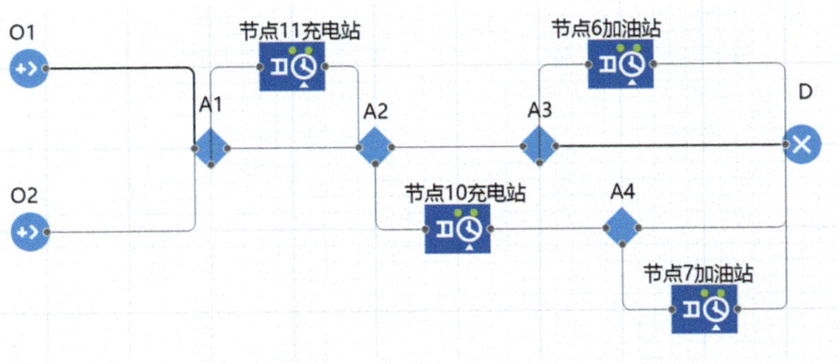


**Fig. 7.8** Logical structure model of the replenishment network

refueling gun is set to 4. The charging station delay time is set at 30 min and the gas station delay time is set at 10 min. The number of vehicle generator inputs is consistent with the road logic model. In addition, statistical charts are placed to calculate the queue and probability density histogram of each station.

#### 7.1.3.4 Logical Relation and Parameter Setting of Load Reduction Mechanism

The three load reduction mechanisms are realized by adjusting the setting parameters from the environment model and the two logical structure models of the network simulation model respectively.

##### Traffic Sections Are Restricted

As shown in Fig. 7.9, the selection probability coefficient of the section entering node 6 is taken as the traffic section restriction coefficient, and the section R47 input the section selection commands A9, A10, and A11, where A9 and A10 connect to the front road and are adjusted to 0.4, 0.3 and 0.2 respectively, while A11 is not adjusted because it connects to two stations of node 6 and node 10. By running a time period, the traffic of each road section and the traffic change of the station after one hour are calculated.

Similarly, the selection probability coefficient of the road section entering node 7 is taken as the traffic section restriction coefficient, as shown in Figs. 7.10 and 7.11. Section R44, the input section selection commands are A34 and A24, which are adjusted to 0.4, 0.3, and 0.2 respectively. By running a time period, the traffic of each road section and the traffic change of the station after one hour are calculated.

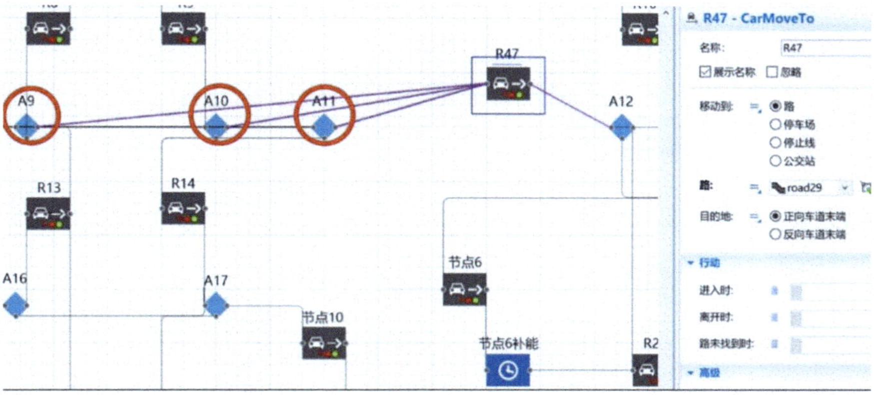


**Fig. 7.9** Link R47 Front-enters the selection command

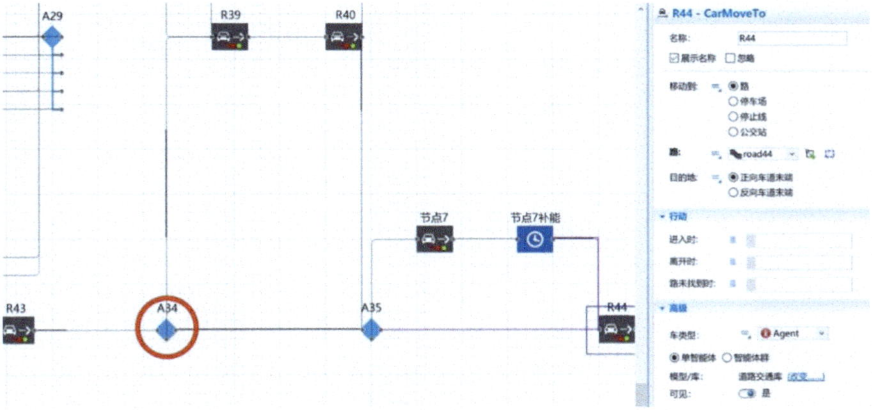


**Fig. 7.10** Section R44 front-enters the selection command A34

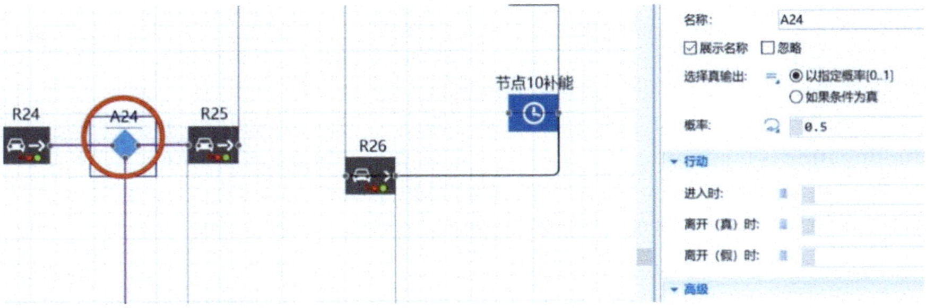


**Fig. 7.11** Section R44 front-enters the selection command A24

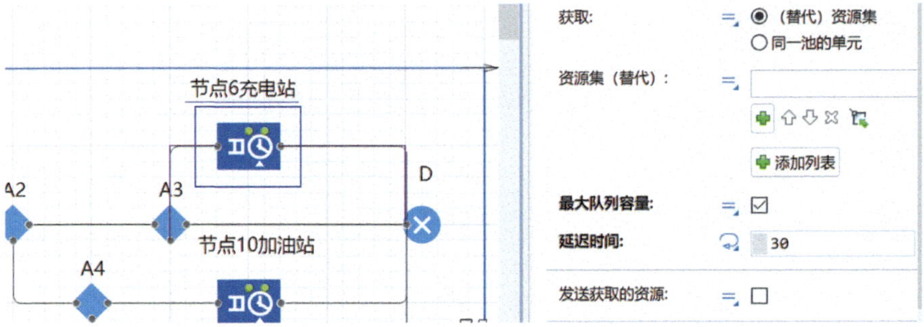


**Fig. 7.12** Node 6 replenishment delay time

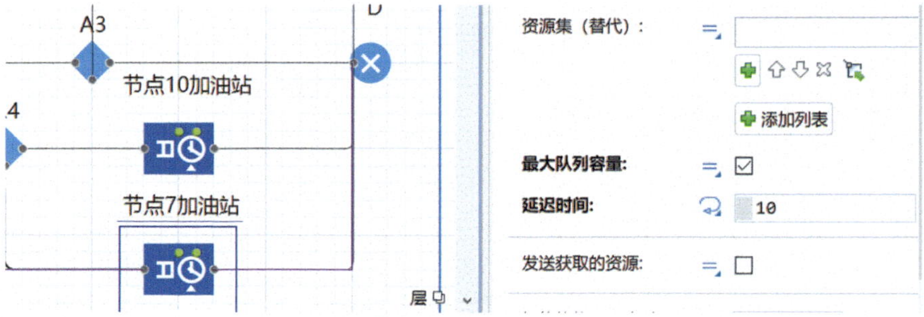


**Fig. 7.13** Node 7 Replenishment delay time

Site Service Rate Adjustment

In the simulation, the queuing delay time in node 6 and node 7 represents the site service time, as shown in Figs. 7.12 and 7.13. Node 6 is the charging station, and the service time of the charging station mainly depends on the charging speed. At present, the shortest fast charging time on the market is about 15 min, and the longest is about 30–40 min. With the development of fast charging technology for charging vehicles, the subsequent fast charging time may be shorter, and the delay time is set to 30 min, 20 min, and 10 min, respectively. Node 7 is a gas station, and the average refueling time is 6 min when there is no queue. When the vacancy rate is large or the vacancy rate of the gas station forecourt is small, it can reach 10 min, and the delay time is set as 10 min, 7 min, and 5 min, respectively.

Price Adjustment of Different Sites

Here the node selection probability represents the price adjustment, as shown in Figs. 7.14 and 7.15. The selection probabilities of node 11 are set to 0.95, 0.93, and

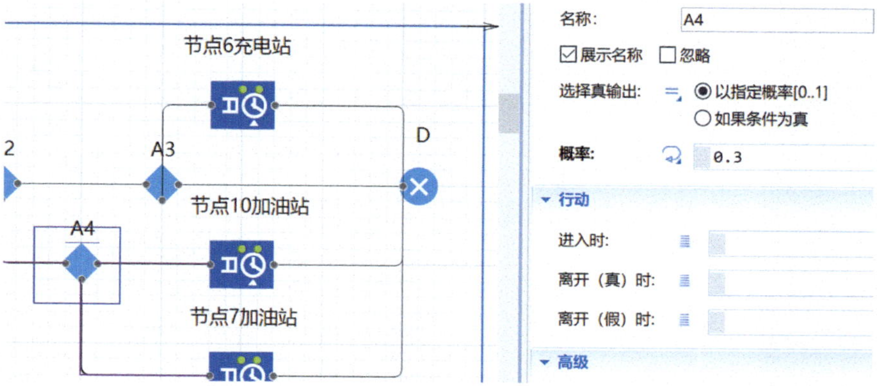


**Fig. 7.14** Node 7 Select commands

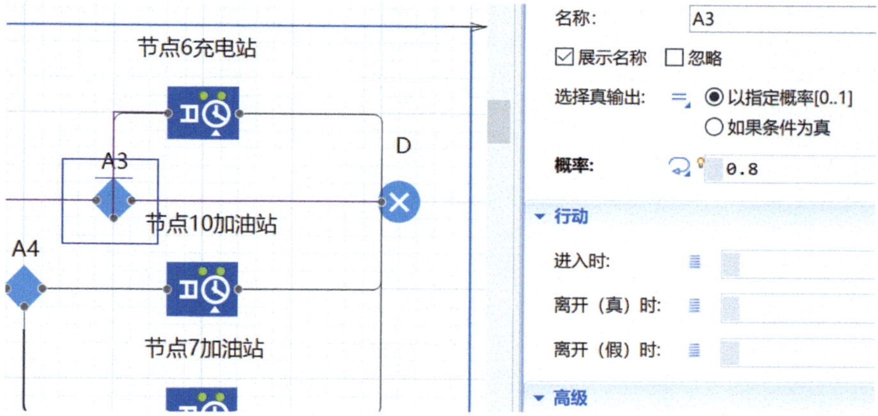


**Fig. 7.15** Node 6 Select commands

0.90 respectively, and the selection probabilities of node 10 are set to 0.3, 0.4, and 0.5 respectively.

### *7.1.4 Analysis of Simulation Results*

The running results of the above simulation model can be observed on the running interface. Figure 7.16 shows part of the simulation results. Data tags can be annotated with detailed data meaning by clicking the command. The operating models are single station, multi-station, and load-reducing models respectively. By analyzing the operation data of the model, the operation law of urban transportation energy supply is revealed, and the results of load reduction are compared and analyzed.

**Fig. 7.16** Data meaning remarks of simulation results

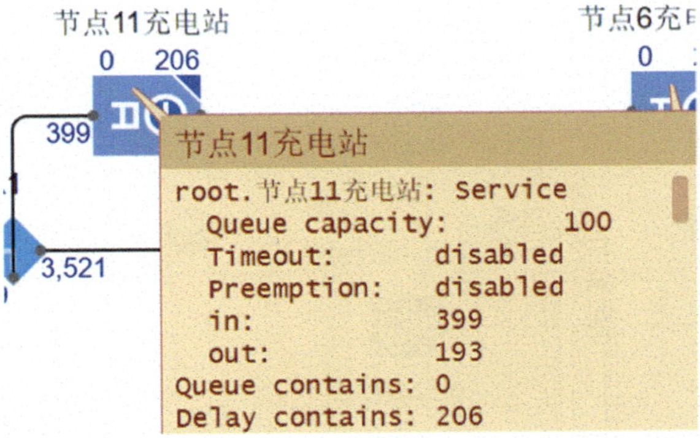


### 7.1.4.1 Analysis of Single-Site Replenishment Simulation Results

The point replenishment model of urban transportation energy supply station can obtain the operation results of each fuel tanker within the cycle time, as shown in Figs. 7.17, 7.18, 7.19, and 7.20. In this model, the sales volume of 95 oil is higher than that of 92 oil before 3 o'clock, and the opposite is true for the next three or four hours. Diesel sales are highest between 9 a.m. and 10 a.m., up to 900 L, and after 9 p.m. diesel sales are 0 L.

There were 11 h in which 92 oil sales were higher than 95 oil sales, and diesel sales were basically between 92 and 95 oil sales. Overall, the sales volume of the three refined oil products is relatively balanced.

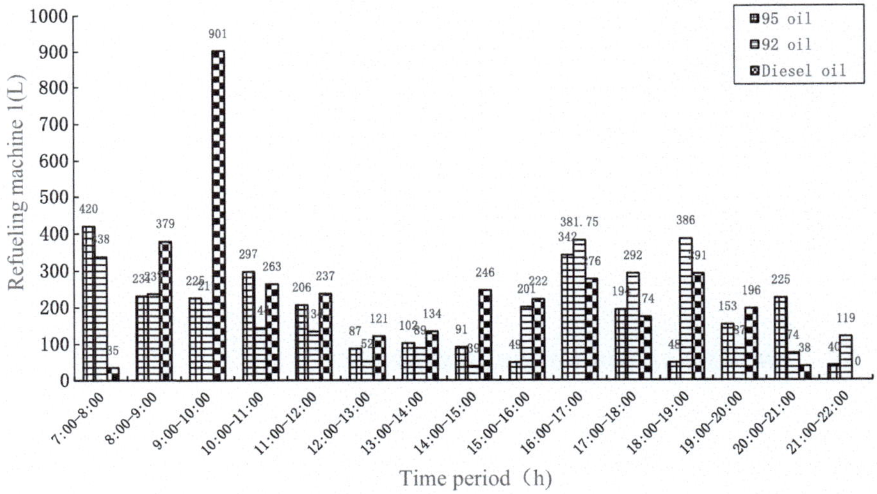


**Fig. 7.17** Unit time sales of three kinds of oil products of No. 1 tanker

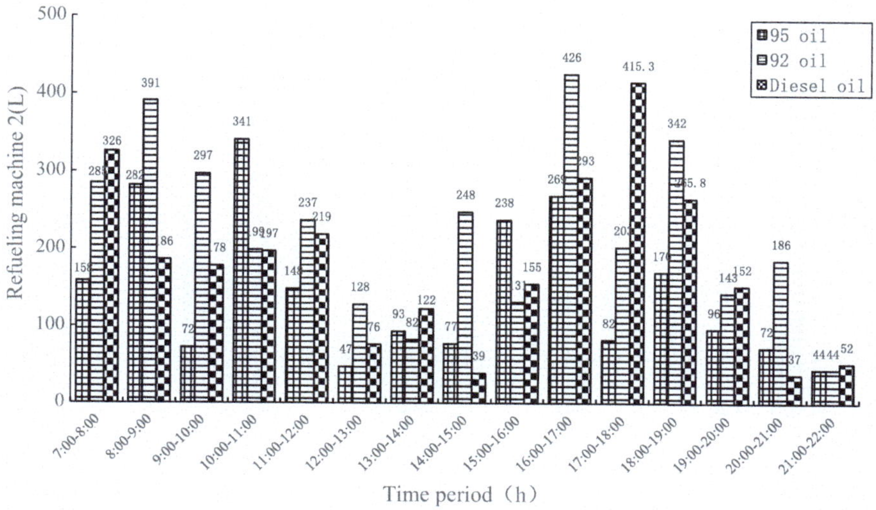


**Fig. 7.18** Unit period sales of three refined oil products of No. 2 tanker

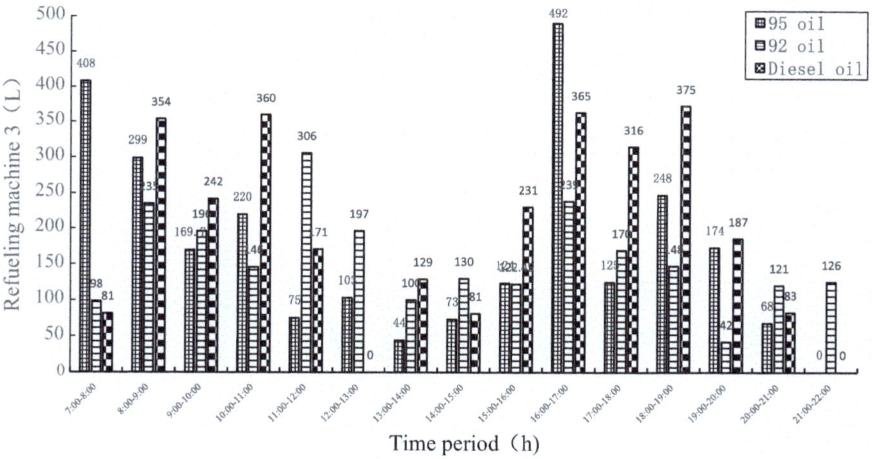


**Fig. 7.19** Unit time sales of three kinds of refined oil products of No. 3 tanker

No. 3 tanker 95 oil sales and diesel sales fluctuated greatly, while 92 oil sales fluctuated relatively small. 95 oil sales are highest between 7 a.m. and 8 a.m. and 4 p.m. to 5 p.m., and only 92 oil sales remain after 9 p.m.

No. 4 tanker diesel sales compared to higher, 92 oil sales compared to smaller. At 8 a.m. to 9.a.m diesel sales are much higher than 92 oil and 95 oil sales.95 oil sold more than 92 oil overall, and 92 oil sold 0 L after 9 p.m.

The number of vehicles served by all refueling machines in the time period is shown in Fig. 7.21. The average number of vehicles served by all tankers is equal.

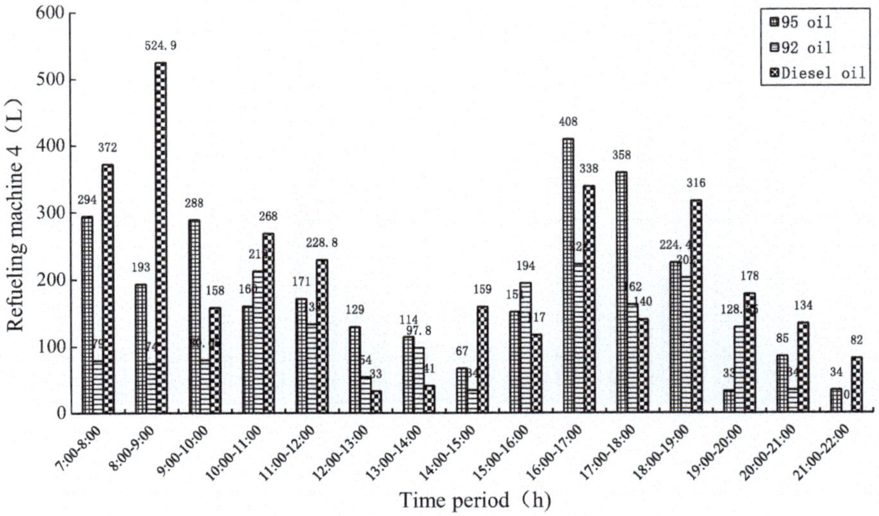


Fig. 7.20 Unit sales of three refined oil products of No. 4 tanker

The peak occurs between 4 p.m. and 5 p.m. and the lowest occurs between 9 p.m. and 10 p.m. At its peak, the number of No. 3 tanker service vehicles was up to 25; At the lowest point, No. 2, No. 3, and No. 4 tanker service vehicles were at least 3.

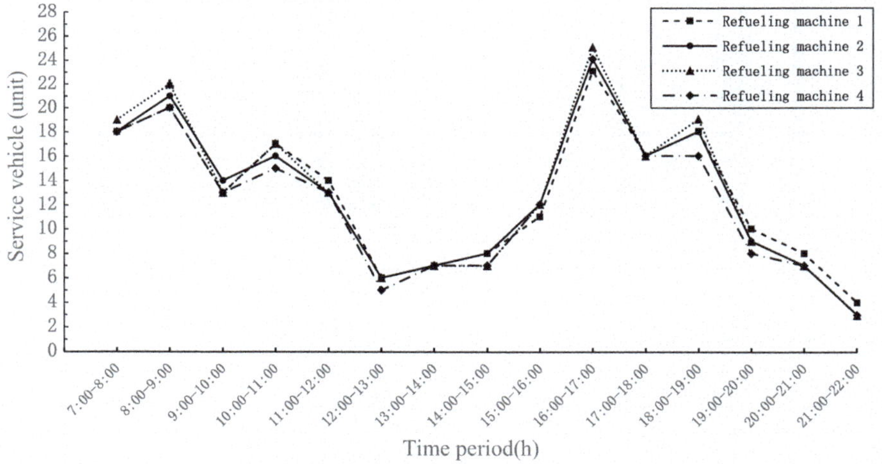


Fig. 7.21 Simulation results of the number of service vehicles per unit time period of each fuel tanker

**Table 7.1** Market share of electric vehicles 10% flow results of each section of the simulation run

| Sections | Traffic volume (units) | Sections | Traffic volume (units) | Sections | Traffic volume (units) | Sections | Traffic volume (units) | Sections | Traffic volume (units) |
|---|---|---|---|---|---|---|---|---|---|
| R1 | 531 | R11 | 366 | R21 | 967 | R31 | 406 | R41 | 1816 |
| R2 | 355 | R12 | 162 | R22 | 388 | R32 | 476 | R42 | 1008 |
| R3 | 585 | R13 | 770 | R23 | 666 | R33 | 226 | R43 | 925 |
| R4 | 300 | R14 | 310 | R24 | 663 | R34 | 534 | R44 | 880 |
| R5 | 522 | R15 | 311 | R25 | 578 | R35 | 134 | R45 | 1839 |
| R6 | 257 | R16 | 378 | R26 | 493 | R36 | 940 | R46 | 472 |
| R7 | 88 | R17 | 181 | R27 | 77 | R37 | 499 | R47 | 694 |
| R8 | 396 | R18 | 288 | R28 | 574 | R38 | 452 | | |
| R9 | 164 | R19 | 163 | R29 | 181 | R39 | 245 | | |
| R10 | 283 | R20 | 232 | R30 | 462 | R40 | 819 | | |

#### 7.1.4.2 Analysis of Multi-site Simulation Results

According to the simulation model of urban transportation energy supply network, the flow of each section can be obtained, as shown in Tables 7.1, 7.2, 7.3, and 7.4. O1 is the first starting point and the number of input vehicles is about 1100, O2 is the second starting point and the number of input vehicles is about 1800. As shown in Tables 7.1–7.4, the market share of charging vehicles is 10%, 30%, 50%, and 70%, respectively, under the flow rate of each road section within a cycle. R1–R47 means road sections, 47 in total. R41 is the O2 input section, in addition to the highest flow R45, the average flow of more than 1000 sections are R41, R42, R44, and R45. R7 and R27 had the smallest traffic flow, followed by R9, R12, R17, R19, and R35.

The four nodes in the road are four energy supply stations with different market share. The number of vehicles served at each station is shown in Fig. 7.22. Nodes 6 and 11 are charging stations, and nodes 7 and 10 are gas stations. It can be seen that the number of service vehicles at node 6 in the two charging stations is higher than that at node 11, and the number of service vehicles at node 7 in the two gas stations is higher than that at node 10. Electric vehicles accounted for 10% of the market, the number of 7 service vehicles was the highest, reaching nearly 600, and with the increase in the market proportion, the number of 6 service vehicles was the highest, reaching 470 in 70% of the market.

#### 7.1.4.3 Analysis of Network Load Reduction Simulation Results

The load reduction simulation of urban transportation energy supply network is set based on the actual market share of the number of fuel vehicles and the number of electric vehicles. Through the traffic section restriction strategy, station service

**Table 7.2** Electric vehicle market share 30% flow results of each section of the simulation run

| Sections | Traffic volume (units) | Sections | Traffic volume (units) | Sections | Traffic volume (units) | Sections | Traffic volume (units) | Sections | Traffic volume (units) |
|---|---|---|---|---|---|---|---|---|---|
| R1 | 552 | R11 | 357 | R21 | 1100 | R31 | 360 | R41 | 1811 |
| R2 | 346 | R12 | 148 | R22 | 370 | R32 | 472 | R42 | 1050 |
| R3 | 671 | R13 | 921 | R23 | 584 | R33 | 222 | R43 | 969 |
| R4 | 329 | R14 | 332 | R24 | 626 | R34 | 490 | R44 | 1226 |
| R5 | 527 | R15 | 326 | R25 | 711 | R35 | 148 | R45 | 2319 |
| R6 | 263 | R16 | 394 | R26 | 514 | R36 | 906 | R46 | 504 |
| R7 | 67 | R17 | 183 | R27 | 91 | R37 | 498 | R47 | 762 |
| R8 | 475 | R18 | 302 | R28 | 581 | R38 | 434 | | |
| R9 | 145 | R19 | 162 | R29 | 206 | R39 | 406 | | |
| R10 | 340 | R20 | 231 | R30 | 421 | R40 | 834 | | |

**Table 7.3** Market share of electric vehicles 50% flow results of each section of the simulation run

| Sections | Traffic volume (units) | Sections | Traffic volume (units) | Sections | Traffic volume (units) | Sections | Traffic volume (units) | Sections | Traffic volume (units) |
|---|---|---|---|---|---|---|---|---|---|
| R1 | 549 | R11 | 354 | R21 | 1200 | R31 | 393 | R41 | 1820 |
| R2 | 352 | R12 | 154 | R22 | 404 | R32 | 523 | R42 | 1029 |
| R3 | 676 | R13 | 1001 | R23 | 631 | R33 | 231 | R43 | 916 |
| R4 | 314 | R14 | 302 | R24 | 675 | R34 | 483 | R44 | 1288 |
| R5 | 507 | R15 | 330 | R25 | 809 | R35 | 148 | R45 | 2473 |
| R6 | 265 | R16 | 41 | R26 | 518 | R36 | 937 | R46 | 481 |
| R7 | 80 | R17 | 176 | R27 | 86 | R37 | 517 | R47 | 849 |
| R8 | 486 | R18 | 306 | R28 | 623 | R38 | 409 | | |
| R9 | 157 | R19 | 170 | R29 | 220 | R39 | 466 | | |
| R10 | 360 | R20 | 250 | R30 | 448 | R40 | 847 | | |

rate adjustment strategy, and different station price adjustment strategy, observe the operation of the replenishment network under each strategy.

### Traffic Section Restriction Strategy

The traffic section restriction strategy needs to adjust the selection probability of the pre-entry section of the station section, so it runs in the network main body logic model. The traffic flow of each road section under the road section restriction coefficient of 0.4, 0.3, and 0.2 is obtained respectively, as shown in Tables 7.5, 7.6, and 7.7. The front section of node 6 is R47, and the front section of node 7 is R44.

**Table 7.4** Market share of electric vehicles 80% flow results of each section of the simulation run

| Sections | Traffic volume (units) | Sections | Traffic volume (units) | Sections | Traffic volume (units) | Sections | Traffic volume (units) | Sections | Traffic volume (units) |
|---|---|---|---|---|---|---|---|---|---|
| R1 | 5443 | R11 | 415 | R21 | 1165 | R31 | 437 | R41 | 1838 |
| R2 | 335 | R12 | 130 | R22 | 348 | R32 | 519 | R42 | 1140 |
| R3 | 715 | R13 | 1067 | R23 | 684 | R33 | 225 | R43 | 998 |
| R4 | 362 | R14 | 290 | R24 | 664 | R34 | 513 | R44 | 1307 |
| R5 | 528 | R15 | 342 | R25 | 852 | R35 | 163 | R45 | 2463 |
| R6 | 265 | R16 | 436 | R26 | 507 | R36 | 859 | R46 | 429 |
| R7 | 65 | R17 | 211 | R27 | 104 | R37 | 571 | R47 | 819 |
| R8 | 543 | R18 | 268 | R28 | 512 | R38 | 435 | | |
| R9 | 160 | R19 | 172 | R29 | 161 | R39 | 524 | | |
| R10 | 351 | R20 | 268 | R30 | 498 | R40 | 839 | | |

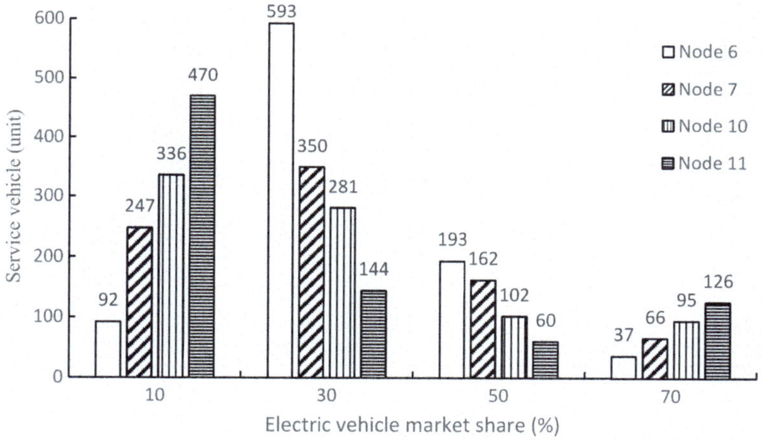


**Fig. 7.22** Simulation results of the number of service vehicles at different electric vehicle markets

The R44 dropped from 1,080 units to 879 units, down 18.61% year-on-year; The R47 dropped from 681 to 502 units, down 26.28% year-on-year.

According to the traffic restriction coefficient, the number of service vehicles at the four nodes is calculated, as shown in Fig. 7.23. The number of node 6 service vehicles decreased from 107 to 87, a year-on-year decrease of 18.69%; The number of Node 7 service vehicles decreased from 436 to 357, down 18.12% year-on-year; The number of Node-10 service vehicles decreased from 250 to 218, a year-on-year decrease of 12.80%; The number of node 11 service vehicles decreased from 40 to 27, a 32.50% decrease year-on-year.

**Table 7.5** Road restriction coefficient 0.4 traffic flow results of each section during simulation operation

| Sections | Traffic volume (units) | Sections | Traffic volume (units) | Sections | Traffic volume (units) | Sections | Traffic volume (units) | Sections | Traffic volume (units) |
|---|---|---|---|---|---|---|---|---|---|
| R1 | 561 | R11 | 410 | R21 | 1088 | R31 | 408 | R41 | 1817 |
| R2 | 337 | R12 | 141 | R22 | 414 | R32 | 489 | R42 | 1064 |
| R3 | 786 | R13 | 1059 | R23 | 661 | R33 | 242 | R43 | 959 |
| R4 | 374 | R14 | 393 | R24 | 647 | R34 | 509 | R44 | 1080 |
| R5 | 536 | R15 | 336 | R25 | 837 | R35 | 166 | R45 | 2159 |
| R6 | 283 | R16 | 428 | R26 | 596 | R36 | 916 | R46 | 470 |
| R7 | 69 | R17 | 199 | R27 | 99 | R37 | 523 | R47 | 681 |
| R8 | 631 | R18 | 267 | R28 | 582 | R38 | 454 | | |
| R9 | 179 | R19 | 157 | R29 | 216 | R39 | 443 | | |
| R10 | 411 | R20 | 250 | R30 | 465 | R40 | 844 | | |

**Table 7.6** Road restriction coefficient 0.3 flow results of each section during simulation operation

| Sections | Traffic volume (units) | Sections | Traffic volume (units) | Sections | Traffic volume (units) | Sections | Traffic volume (units) | Sections | Traffic volume (units) |
|---|---|---|---|---|---|---|---|---|---|
| R1 | 542 | R11 | 422 | R21 | 1009 | R31 | 445 | R41 | 1785 |
| R2 | 335 | R12 | 131 | R22 | 370 | R32 | 501 | R42 | 1066 |
| R3 | 980 | R13 | 1158 | R23 | 710 | R33 | 219 | R43 | 934 |
| R4 | 483 | R14 | 440 | R24 | 657 | R34 | 507 | **R44** | **1015** |
| R5 | 553 | R15 | 344 | R25 | 914 | R35 | 171 | R45 | 2006 |
| R6 | 262 | R16 | 448 | R26 | 638 | R36 | 887 | R46 | 447 |
| R7 | 64 | R17 | 218 | R27 | 116 | R37 | 540 | **R47** | **517** |
| R8 | 811 | R18 | 272 | R28 | 543 | R38 | 426 | | |
| R9 | 164 | R19 | 163 | R29 | 178 | R39 | 479 | | |
| R10 | 493 | R20 | 269 | R30 | 521 | R40 | 842 | | |

Site Service Rate Adjustment Strategy

Site service rate by adjusting the delay time of node 6 and node 7, node 6 is 30, 20, and 10. Nodes 7 are 10, 7, and 5, and the model operation results are shown in Figs. 7.24, 7.25, and 7.26. The number of input vehicles from O1 is 1022, and the number of input vehicles from O2 is 1014. The number of delayed waiting vehicles at node 6 decreased from 106 to 29, a year-on-year decrease of 72.64%; The number of delayed waiting vehicles at node 7 decreased from 86 to 36 vehicles, a decrease of 58.14% year-on-year. The probability densities were 6.82, 5.70, and 4.32, respectively. The

**Table 7.7** Road restriction coefficient 0.2 traffic flow results of each section during simulation operation

| Sections | Traffic volume (units) | Sections | Traffic volume (units) | Sections | Traffic volume (units) | Sections | Traffic volume (units) | Sections | Traffic volume (units) |
|---|---|---|---|---|---|---|---|---|---|
| R1 | 499 | R11 | 384 | R21 | 834 | R31 | 446 | R41 | 1848 |
| R2 | 309 | R12 | 149 | R22 | 394 | R32 | 492 | R42 | 1093 |
| R3 | 1341 | R13 | 1508 | R23 | 698 | R33 | 267 | R43 | 978 |
| R4 | 1077 | R14 | 340 | R24 | 678 | R34 | 511 | **R44** | **879** |
| R5 | 529 | R15 | 347 | R25 | 305 | R35 | 186 | R45 | 1708 |
| R6 | 251 | R16 | 411 | R26 | 541 | R36 | 939 | R46 | 465 |
| R7 | 71 | R17 | 191 | R27 | 88 | R37 | 574 | **R47** | **502** |
| R8 | 1191 | R18 | 304 | R28 | 596 | R38 | 464 | | |
| R9 | 161 | R19 | 185 | R29 | 208 | R39 | 769 | | |
| R10 | 256 | R20 | 249 | R30 | 514 | R40 | 677 | | |

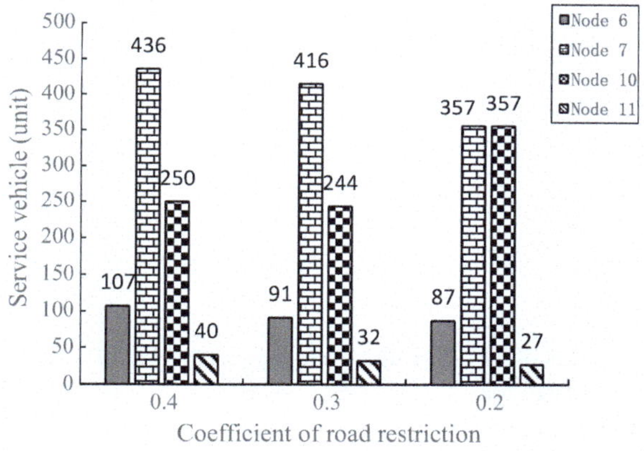


**Fig. 7.23** Number of service vehicles at road restricted sites

number of vehicles exported increased from 1,740 to 1,867, a year-on-year increase of 6.80%.

Price Adjustment Strategies for Different Sites

According to the adjustment of node selection probability, the selection command probability of node 11 is set to 0.95, 0.93, and 0.90 respectively, and the selection command probability of node 3 is set to 0.3, 0.4 and 0.5 respectively. The operation results are shown in Figs. 7.27, 7.28, and 7.29. The proportion of delayed service

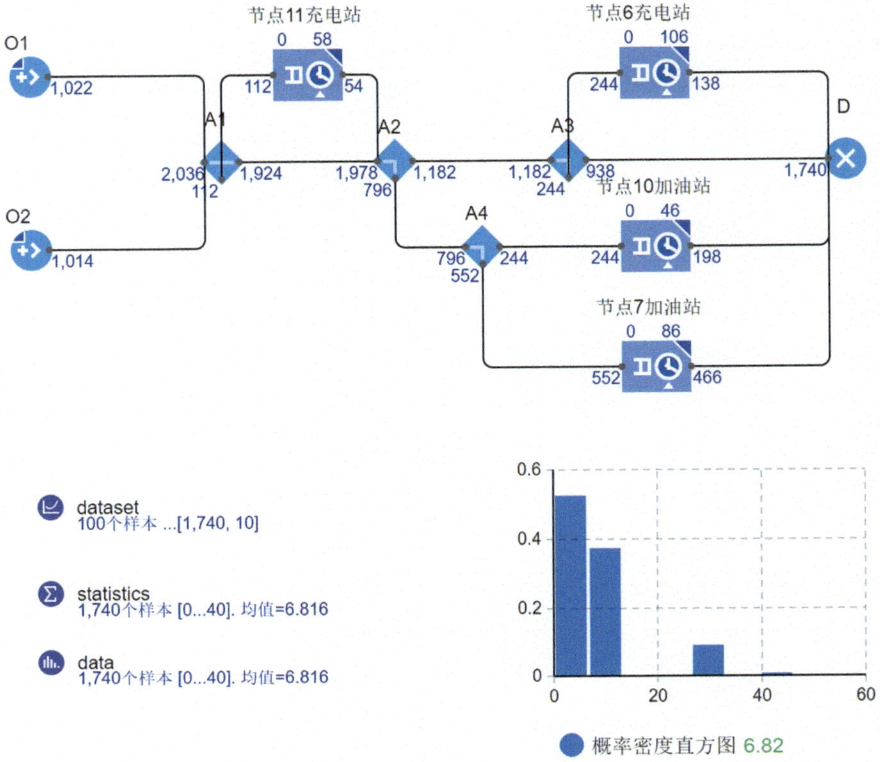


**Fig. 7.24** Simulation results of service rate adjustment by 10%

vehicles on node 11 and node 6 is 0.763, 1.059, and 1.719, respectively. The proportion of delayed service vehicles at node 10 and node 7 was 0.535, 0.781, and 0.859, respectively. The probability densities were 6.31, 6.24, and 6.37, respectively.

Based on the simulation results of the three load reduction strategies, it can be seen that charging stations are highly sensitive to the three load reduction strategies. In particular, improving charging efficiency will greatly reduce the accumulation of negative effects, followed by traffic section restriction strategies. The response of the three strategies is similar, and the sensitivity of the efficiency strategy is higher than the other two strategies. Therefore, the staff deployment of the site and the rational use of the refueling gun have the most significant impact on the negative effect of replenishment, and also suggest the importance of high-precision management of the site.

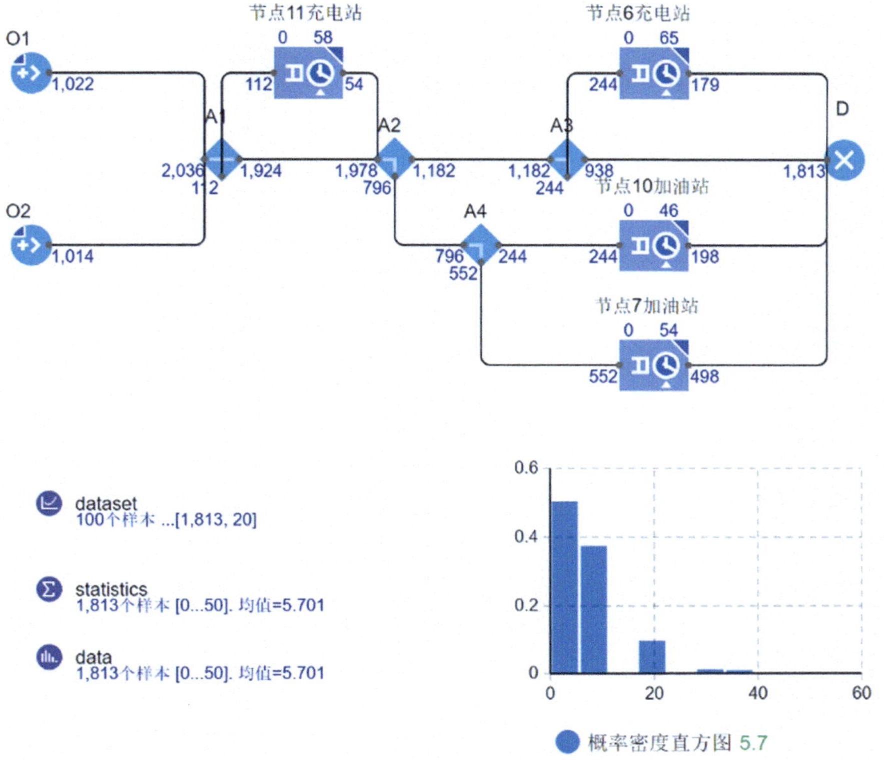


**Fig. 7.25** Simulation results of service rate adjustment by 30%

## 7.2 Simulation Study of Urban Transportation Energy Supply Network—A Case Study of Nanan District of Chongqing

### *7.2.1 Simulation Scenario Analysis*

Chongqing is located in the southwest of China, the upper reaches of the Yangtze River, Chongqing terrain is mainly hilly, low mountains, mountains, sloping land, known as "mountain city". According to the requirements of economic development and environmental protection in Chongqing's main urban areas, the main types of supply stations in Chongqing's main urban areas are: gas stations, filling stations, and charging stations. According to statistics, there are 219 charging stations, 347 gas stations, and 205 filling stations in Chongqing Ring Expressway. In order to facilitate the modeling and simulation of road network, part of the road network and gas station of Xuefu Avenue were selected as the research objects according to the terrain and road network condition of Chongqing. AnyLogic input the GIS map of the current

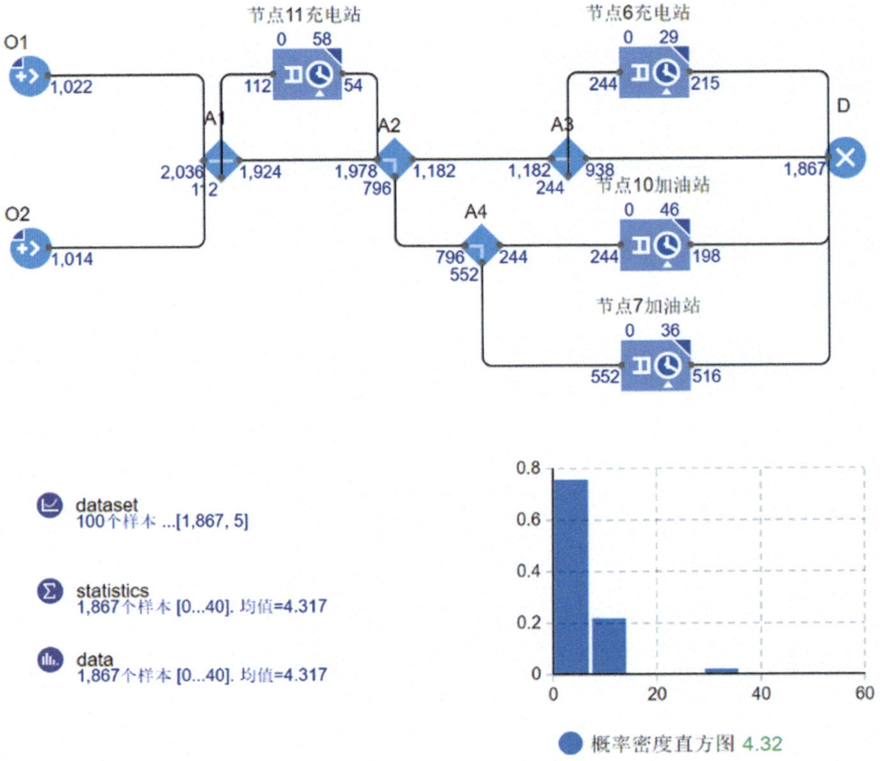


**Fig. 7.26** Service rate adjusted by 50% simulation results

situation of Chongqing, the location and service of some gas stations, and assume two starting places and two destinations (that is, four OD pairs), and assume that the starting place is the location of the vehicle with supplementary energy demand. The model is created by using the agent, and the logic structure between the agents is designed for simulation. Detailed data of gas stations and starting/destination points are shown in Tables 7.8 and 7.9.

## 7.2.2 AnyLogic-Based Simulation Model Construction

Based on some regional maps of Nanan District of Chongqing, the road feature model of this area is constructed. In order to be close to the reality of the road, according to the Baidu map road in the region, on the basis of the previous chapter to improve some regional roads. An energy supply station is set up at the corresponding location. In addition, in order to make the model more simple and convenient, unnecessary

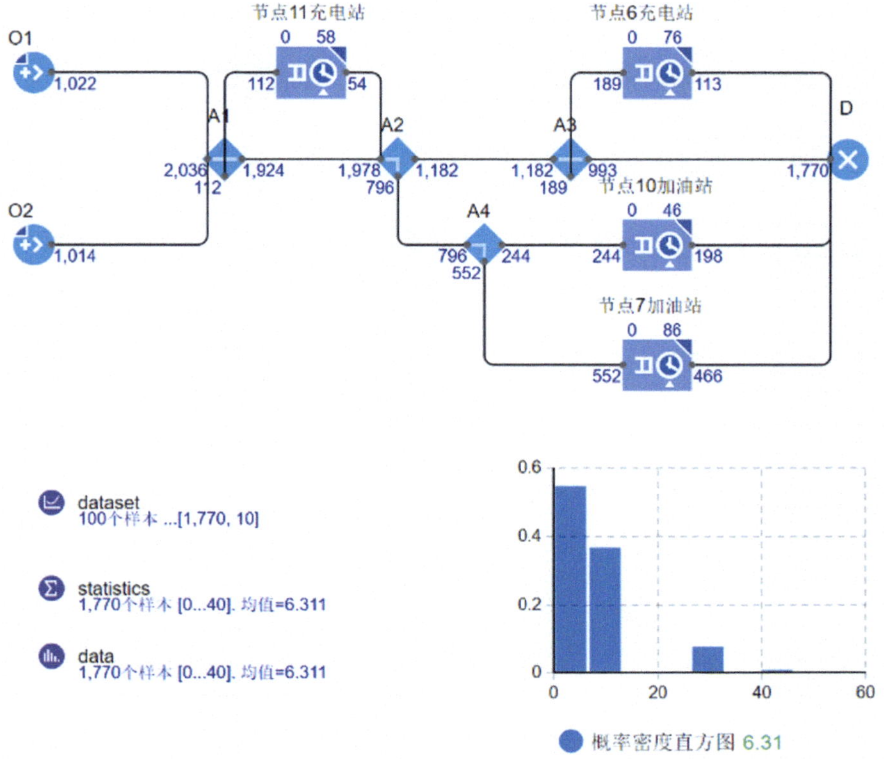


**Fig. 7.27** Simulation results of 10% price adjustment

reverse lanes on some roads are eliminated. Only the reverse lane road with the connected OD pair in the middle is left.

#### 7.2.2.1 Road Traffic Construction

Road button is used to build the whole road network structure.
is used to represent the recharge station.

#### 7.2.2.2 Construct the Logical Structure of the Road Network

: It is the starting point of the road traffic flow chart to generate vehicles.

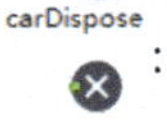

: It is the end of the road traffic flow chart to deal with vehicles.

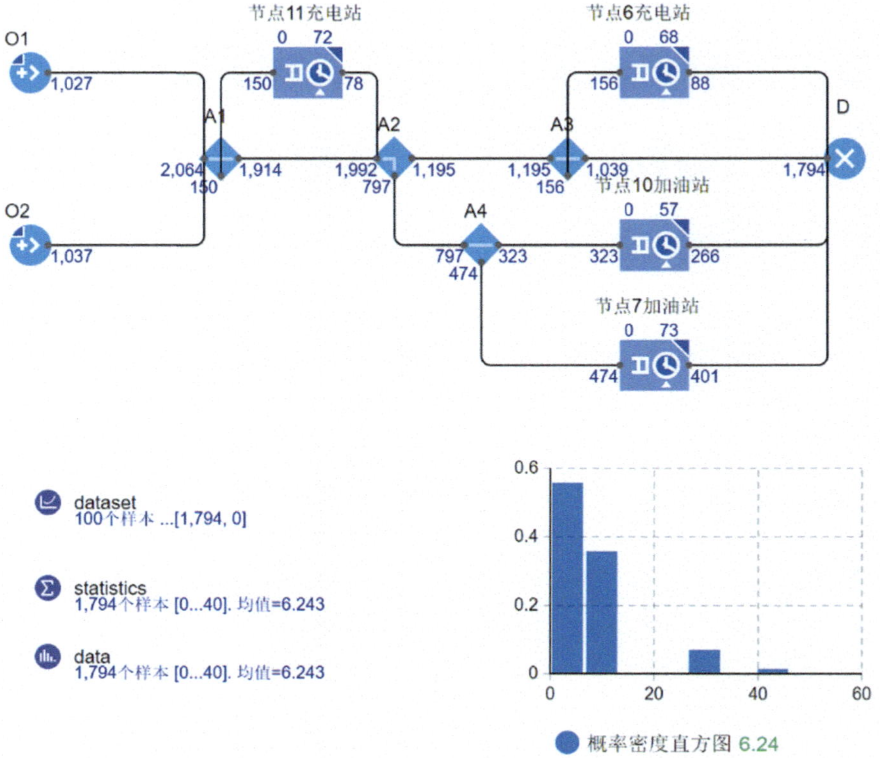


**Fig. 7.28** Simulation results of 30% price adjustment

carMoveTo

 : Get the car to go somewhere.

selectOutput51

 : Location judgment.

## *7.2.3 Logical Relationship and Parameter Settings of Network Agents*

### 7.2.3.1 Logical Relationships of Network Entities

After building the environmental model, it is necessary to build the road network logic model of vehicle operation according to the connections between sections, so as to ensure that the vehicles input by the vehicle generator can select and drive between each section, complete the vehicle activities of all sections and stations

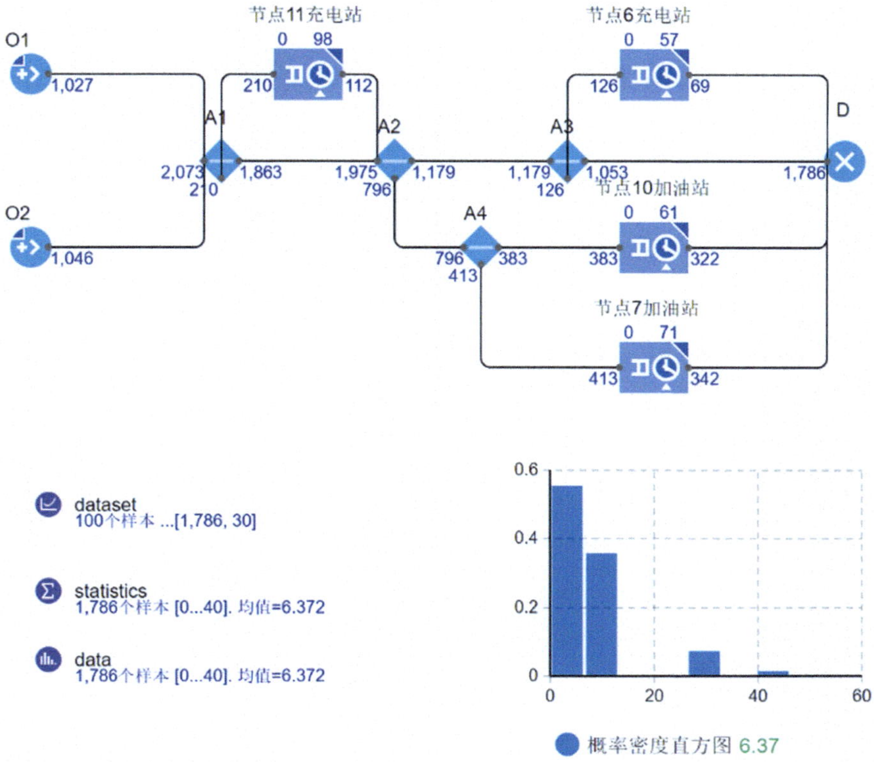


**Fig. 7.29** Simulation results of price adjustment by 50%

**Table 7.8** Latitude and longitude of origin/destination

| Category | Serial number | Latitude | Longitude |
|---|---|---|---|
| Starting point | 1 | 29.470425 | 106.555009 |
| | 2 | 29.495075 | 106.577201 |
| Point of purpose | 1 | 29.53694 | 106.56622 |
| | 2 | 29.587707 | 106.59784 |

required by the simulation, and obtain the corresponding simulation results. The main logic of urban transportation energy supply network is the path selection and connection between each section.

#### 7.2.3.2 Set Parameters

There are 5 starting points and 5 locations in the road network, and the vehicle generation rate of each starting point is set at 3 vehicles/min, as shown in Fig. 7.30.

**Table 7.9** Longitude and latitude of gas station

| Serial number | Gas station | Latitude | Longitude |
|---|---|---|---|
| 1 | Gas station 1 | 29.48666 | 106.56616 |
| 2 | Gas station 2 | 29.4749 | 106.55792 |
| 3 | Gas station 3 | 29.4934 | 106.56286 |
| 4 | Gas station 4 | 29.49928 | 106.56863 |
| 5 | Gas station 5 | 29.52117 | 106.57448 |
| 6 | Gas station 6 | 29.52081 | 106.55814 |
| 7 | Gas station 7 | 29.53653 | 106.57972 |
| 8 | Gas station 8 | 29.54024 | 106.55686 |
| 9 | Gas station 9 | 29.5462 | 106.58247 |

It is assumed that all models are the same, the vehicle length is 5 m, the initial speed of vehicles is 60 km/h, the preferred driving speed of vehicles is 60 km/h, and the maximum acceleration is 1.8 m/s. The maximum deceleration is 4.2 m/s; The selection probability of the input device is 0.5, that is, half of the vehicles will enter the gas station to refuel, and the probability of selecting the gas station is adjusted according to the utility of the station, as shown in Fig. 7.31.

There are 5 gas stations in the road network, and the average time for vehicles at each station is 3.5 min. There are 3 service stations in gas station 1 and 5 service stations in gas station 2, as shown in Fig. 7.32. Gas station 3 has 5 service desks; Gas station 4 has 3 help desks and gas station 5 has 3 help desks.

### *7.2.4 Analysis of Simulation Results*

According to the urban transportation energy supply network simulation model, the number of vehicles served by each of the five energy supply stations in the road network and the number of vehicles in the service station. Refueling stations 1 and 4 have served 11 and 16 vehicles respectively; Refueling station 2 serves 20 vehicles, and it can be seen that the resources are fully idle; At refueling station 3, there are still 3 vehicles replenishing energy, and the average number of vehicles in line for replenishing energy is 3. At refueling station 5, 39 vehicles have been served, but there are still 3 vehicles in the service for replenishing energy, and the average number of vehicles in line for replenishing energy is as high as 6. As a result, some sections of the road network are delayed, which shows that there is an extremely unbalanced distribution of traffic flow and replenishment vehicles in the road network.

**Fig. 7.30** Starting point vehicle information

**Fig. 7.31** Road intersection selection probability

## 7.3 Research on Optimization of Energy Supply Site Layout in Nanan District of Chongqing

### *7.3.1 Construction of Simulation Model for Layout Optimization of Energy Supply Sites*

Based on the starting point, current point and gas station location given in Sect. 7.1, the GIS distribution map is simplified into a road network, the road network structure is established, logical relations are created, and attributes are assigned to each agent.

Road traffic library is a traffic engineering and planning tool used to simulate and manage road traffic systems. In order to simplify the modeling and avoid the

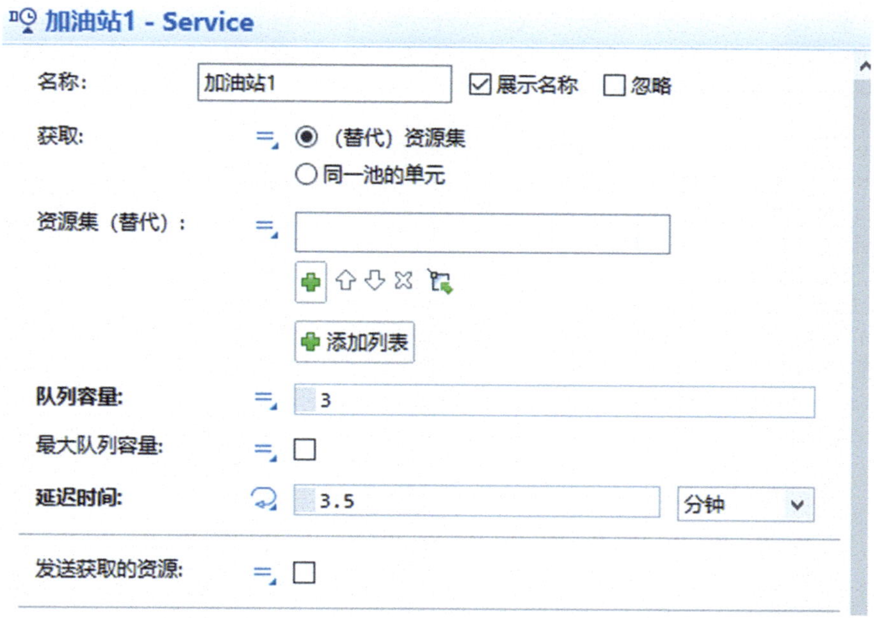


**Fig. 7.32** Gas station 1 information

influence of other factors on the simulation, only the network structure, gas stations, and vehicles are created when the network structure is created. This book constructs a road network topology based on the current urban road network in Chongqing, including 2 OD pairs, 16 one-way sections (named Link0-Link15), 8 paths, and 9 gas stations (named gas Station 1-Gas station 9). Vehicles enter the road network from the two O points at the left end. Vehicles are divided into vehicles that need to be replenished and vehicles that do not need to be replenished according to the amount of remaining oil. When the replenished vehicles are driving on the road, they will choose the nearest gas station for replenishment.

Agents in AnyLogic are the main building blocks of the logical model and can represent vehicles, devices, items, information, and more. Inter-agent communication is realized by defining variables, events, states, system dynamic inventory, and flow charts in the agent. The agent is used to replace several objects, such as gas station, starting place, destination, replenishing vehicle, and identifying vehicle needs replenishing energy, each of which involves its own unique attributes and behaviors. The details are as follows:

The location and number of gas stations are determined by creating gas station intelligence group. In addition, the gas station also needs to provide real-time information to the system in a timely manner, indicating that the traveler is in a "replenishment crowded state" or "replenishment idle state", in order to ensure that the replenishment vehicle can quickly replenish energy, avoid the vehicle queuing at the replenishment station, and reduce the replenishment time.

In Fig. 7.33, serial number 3 represents the state flow diagram, and a message appears when the site is idle. The status enters "statechart". "normal" indicates that the site is crowded and needs to queue up. "MakeOrder" indicates the idle state of the station, and vehicles do not need to queue. The number 1 indicates the body of the function, which informs the system of the site information. When there is no vehicle refueling at the station, the station sends "idle status of gas station" to the demand side (represented by "O point" in the simulation). At this time, the station will appear idle in combination with the process of serial number 3 (Fig. 7.34).

The agent is mainly used to determine the number of locations for both the starting point and the destination point. When creating vehicle intelligence, since the initial position of the vehicle is located at the starting point, the vehicle agent should be created in the starting point agent, as shown in Fig. 7.35: Parameter "place" is used to determine the starting point position; Serial number 1 indicates the vehicle agent group, named "trucks". Serial number 2 is the parameter, indicating the number

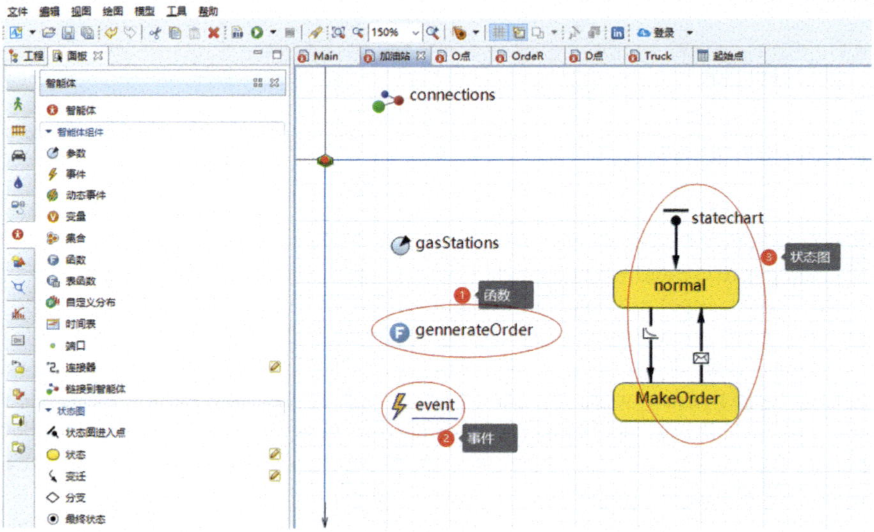


**Fig. 7.33** Simulation diagram of creating a gas station agent

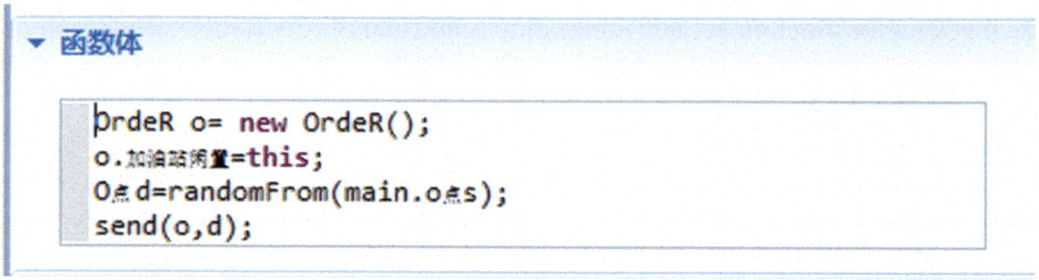

函数体

```
OrdeR o= new OrdeR();
o.加油站网点=this;
O点 d=randomFrom(main.o点s);
send(o,d);
```

**Fig. 7.34** Body code of the information prompt function

of vehicles with supplementary energy requirements in each demand area, named "num". Number 3 is the function body, which has three main functions. First, it is used to create a physical vehicle and determine the number of vehicles; Second, add the created vehicle to the vehicle agent "trucks"; Third, determine the starting position of the vehicle and place the vehicle at the starting point, as shown in Fig. 7.36. The number 4 is the body of the function, representing vehicles with supplementary energy requirements.

Vehicle agent is the core of the simulation. In Fig. 7.37, serial number 3 represents the vehicle status diagram, and the starting vehicle is located at the starting point (represented by "O point" in the simulation). According to "variable", the vehicle determines whether it has a replenishment demand and goes directly to the destination without replenishment demand (represented by "D point" in the simulation). For vehicles with supplementary energy needs, the information of surrounding gas stations is obtained according to "variable" and serial number 1, and the station is selected according to the road network. The selection basis is to preferentially select the "idle" gas station for supplementary energy according to the congestion of gas stations, as

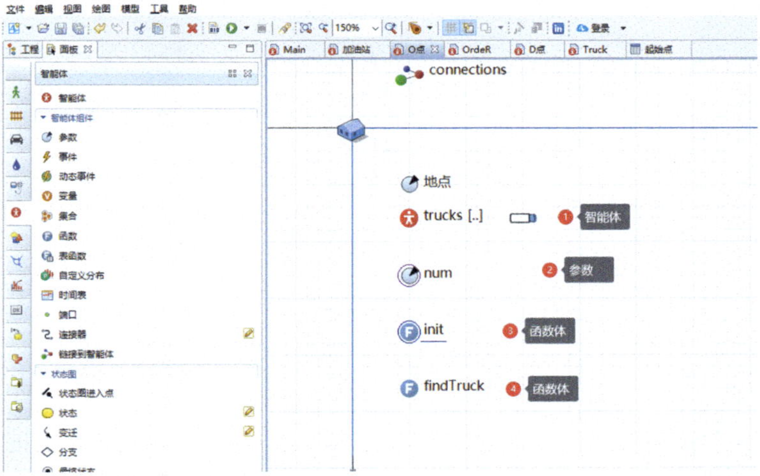


**Fig. 7.35** Create start-point agent simulation diagram

▾ 函数体

```
for(int i=0;i<num ;i++){
    Truck t=add_trucks();
    t.setLocation(this);
}
```

**Fig. 7.36** Vehicle function body code

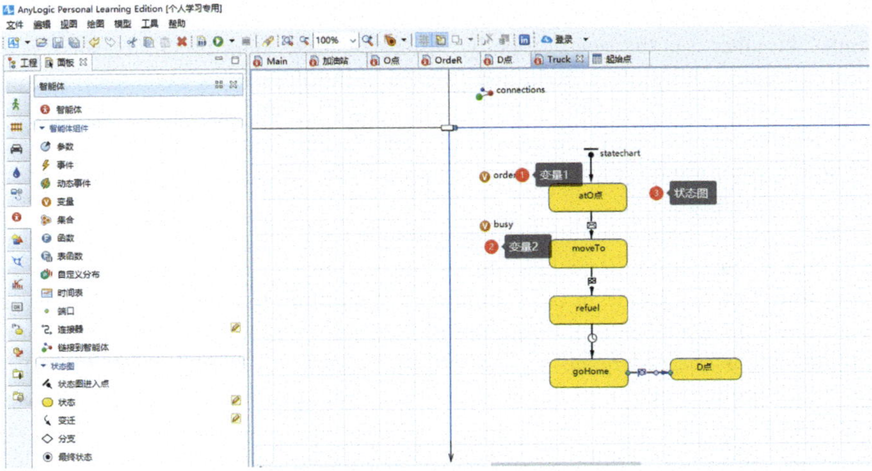


**Fig. 7.37** Creating a vehicle agent simulation diagram

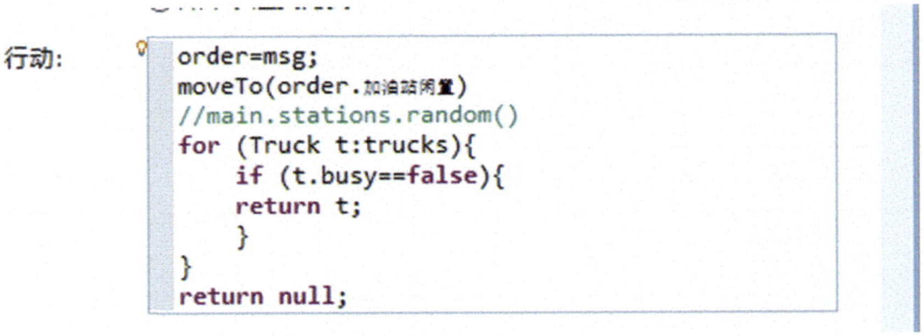


```
order=msg;
moveTo(order.加油站闲置)
//main.stations.random()
for (Truck t:trucks){
    if (t.busy==false){
    return t;
    }
}
return null;
```

**Fig. 7.38** The vehicle selects the gas station code

shown in Fig. 7.38. After completing the replenishment, the vehicle continues to drive to the destination point (represented by "D point" in the simulation).

#### 7.3.1.1 Create the Object of Judging Vehicle Replenishment Agent

Whether the vehicle needs replenishment is mainly judged according to the remaining fuel amount of the vehicle and the situation of the surrounding station. When the fuel amount is too low and the surrounding station has acceptable replenishment, the traveler will choose replenishment. In the simulation, it is mainly measured by the queueing situation of vehicles at the surrounding stations and the amount of vehicle fuel. The surrounding vehicle queue situation is constrained by creating a discriminating vehicle replenishment agent. When the station is in the "idle state" and the vehicle fuel is low, the vehicle is defined as the replenishment vehicle.

### 7.3.2 *Analysis of Simulation Results*

From the simulation of the original model, it can be found that the number of vehicles replenishing energy to stations 1–5 is 11, 20, 24, 16, and 45 respectively. Among them, the queuing situation of energy supplement stations 3 and 5 is more serious, especially at station 5, the queuing situation of energy supplement stations 1, 2, and 4 is not obvious, especially at station 2, which has serious idle resources. Therefore, by optimizing the layout of stations in the road network and expanding and compressing some station planning, the queuing time of vehicle energy supplement can be reduced and the efficiency of energy supplement of the road network can be improved.

On the main road, station 5 has a large traffic flow and a high arrival rate of energy replenishment vehicles, which is easy to cause vehicles to queue up for energy replenishment at station 5, resulting in traffic congestion on local sections. Therefore, station 5 is relocated to avoid energy replenishment queuing. For site 3, the size of the site should be expanded appropriately, the service capacity of the site should be improved, and the original 5 service desks should be expanded to 7. The arrival rate of vehicles at station 2 is low, which is easy to waste resources. Therefore, the service capacity can be properly compressed, and the original 5 service desks can be compressed to 3.

As can be seen from the optimized road network condition in Fig. 7.39, the number of vehicles replenishing energy to stations 1–5 is 16, 28, 20, 25, and 29 respectively, and there is no long-time queuing for replenishing energy, and there is no vehicle delay at nodes 16 and 33. Moreover, the average number of vehicles queuing at refueling stations 3 and 5 is reduced to 1. Therefore, by optimizing the site layout and some site planning in the road network, the queuing time of vehicle replenishment energy is effectively reduced, and the efficiency of road network replenishment energy and road network traffic efficiency are improved.

## 7.4 Multi-Agent Cooperation Game Simulation of Energy Supply Station in Nanan District of Chongqing

### 7.4.1 *Game Simulation of Cooperation Between the Two Sides of the Supplementary Energy Site*

#### 7.4.1.1 Gas Station 1, 3 Cooperation Game

According to the simulation of the original model, the vehicle replenishment service efficiency of refueling station 1 is high, and the resource utilization rate is not high; the queuing replenishment condition of refueling station 3 exists for a long time, and the average number of queuing vehicles is 3. Therefore, through the game cooperation between refueling station 3 and refueling station 1, it can be seen that the vehicle

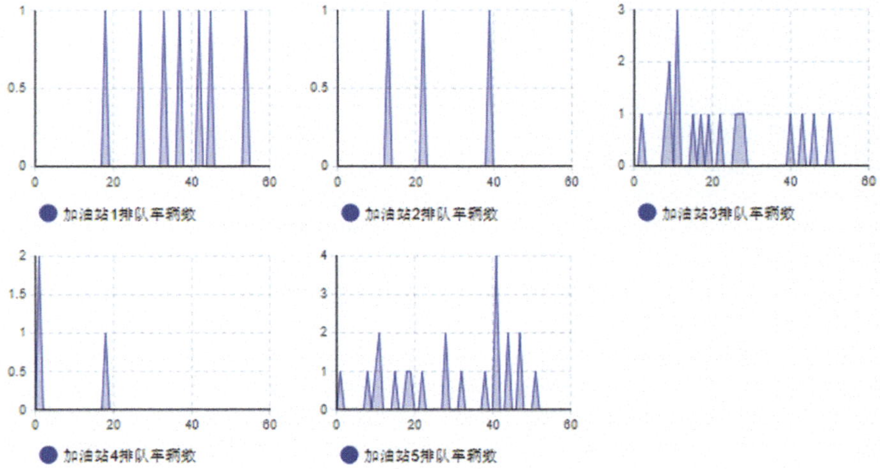


**Fig. 7.39** Optimized refueling station 1–5 queue situation

replenishment energy of refueling station 1 has increased. The number of queuing vehicles at refueling station 3 has decreased somewhat, but there is still a high number of queuing vehicles at refueling station 3, and there is still a large amount of idle resources for refueling stations 2 and 4, as shown in Figs. 7.40 and 7.41.

#### 7.4.1.2 Gas Station 4, 5 Cooperation Game

In the simulation of the original model, for refueling station 4, the average number of queuing vehicles is very small, and the station resources are idle; for refueling station 5, there is a very high average number of queuing vehicles, namely 6. Therefore, through the game cooperation between refueling station 4 and refueling station 5, it is obvious that the vehicle replenishment capacity of refueling station 4 has increased, and the average number of queuing vehicles reaches 2. The average number of queuing vehicles at refueling station 5 has decreased, but there are still queuing peaks in some periods, as shown in Figs. 7.42 and 7.43.

### *7.4.2 Game Simulation of Tripartite Cooperation in the Replenishment Site*

In the above original model, there is no queueing replenishment of vehicles at refueling station 2, while the queueing number of vehicles at refueling stations 3 and 5 is relatively high. Through the game cooperation between refueling stations 2, 3, and 5, the number of vehicle replenishment at refueling station 2 has a large increase, and

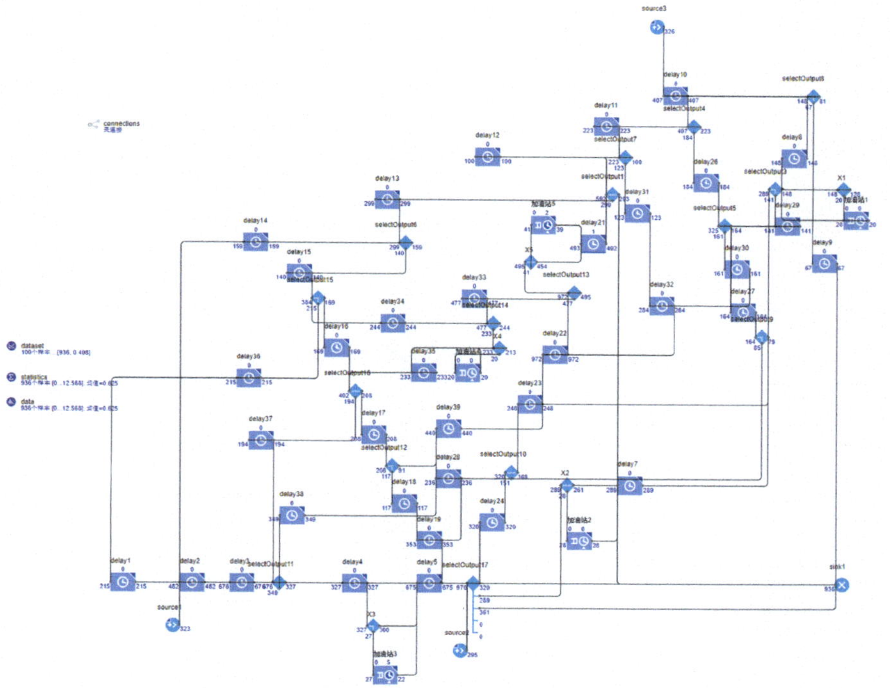

**Fig. 7.40** Network condition after game cooperation between refueling stations 1 and 3

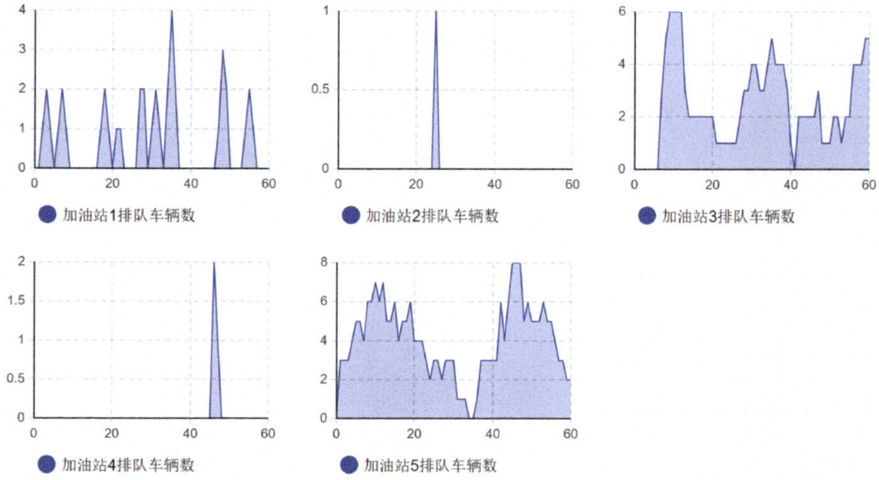


**Fig. 7.41** Queuing status of gas stations after game cooperation between refueling stations 1 and 3

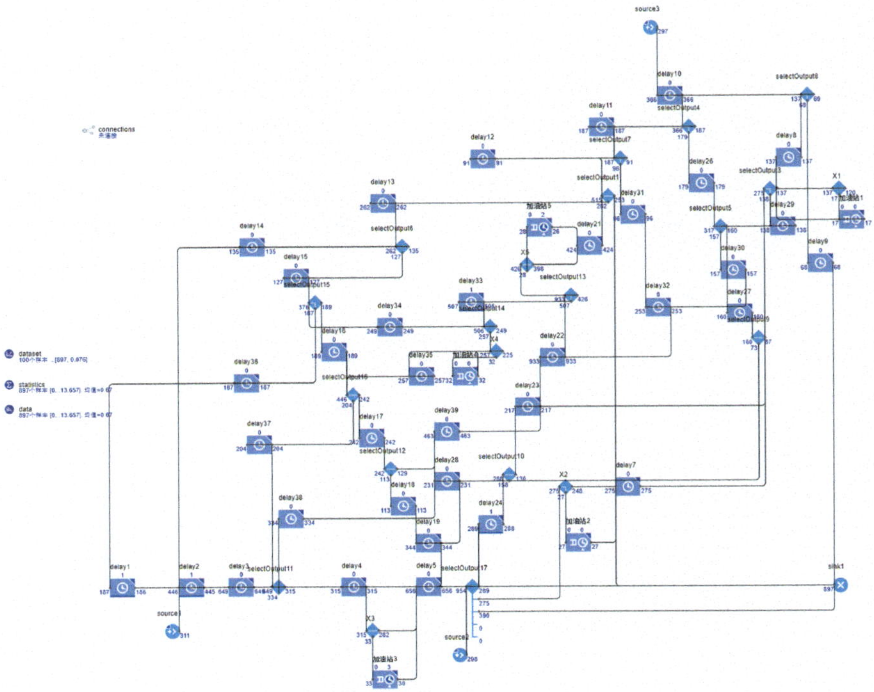

**Fig. 7.42** Network condition after game cooperation between refueling stations 4 and 5

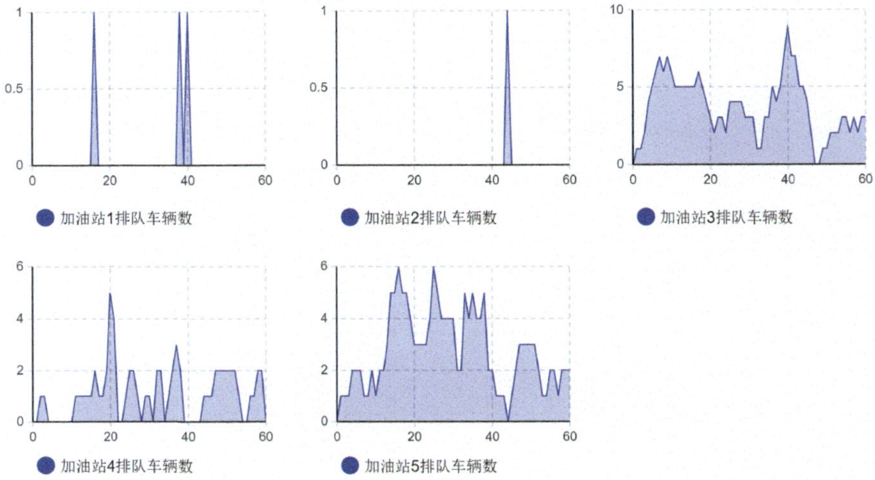


**Fig. 7.43** Queuing status of gas stations after game cooperation between refueling stations 4 and 5

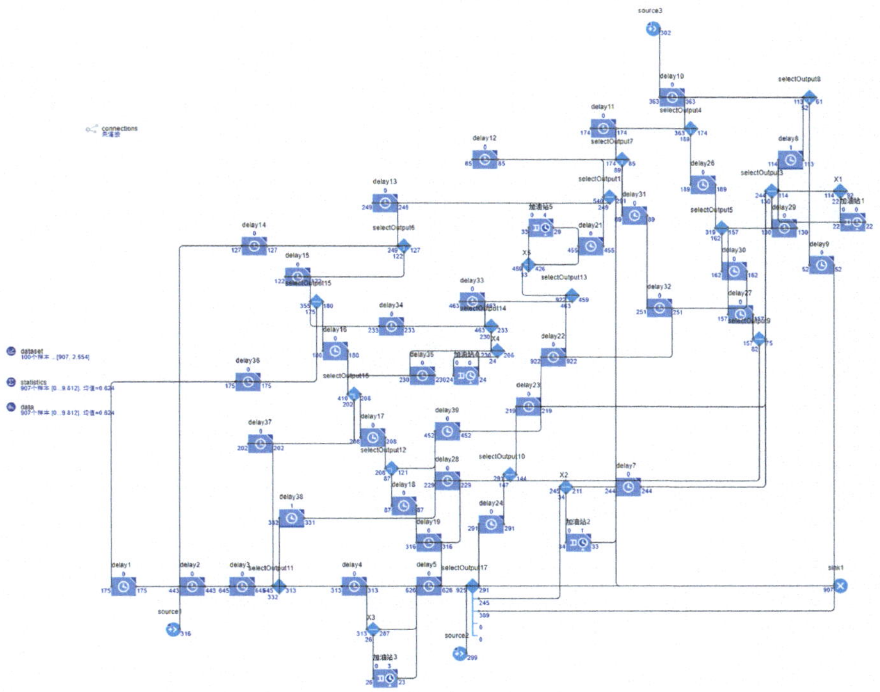

**Fig. 7.44** Network condition after game cooperation of refueling stations 2, 3 and 5

the average queueing number of vehicles at refueling stations 3 and 5 has a decrease. However, vehicle delay occurred at node 8, as shown in Figs. 7.44 and 7.45.

## 7.5 Simulation of Vehicle Energy Replenishment Scheduling in Nanan District of Chongqing

### *7.5.1 Simulation Scheme Design*

In the simulation of the original model, the number of vehicles to stations 1–5 were 11, 20, 24, 16, and 45 respectively. However, the average energy replenishment of vehicles queuing at energy replenishment stations 3 and 5 is relatively high, especially at station 5, where there are up to 6 vehicles queuing at energy replenishment stations 1, 2, and 4 are not obvious, and there is no vehicle replenishment queuing at station 2. Therefore, the induction strategy is adopted to induce part of the vehicles arriving at stations 3 and 5 to be allocated to stations 1, 2, and 4 in a balanced way. Reduce vehicle replenishment queues.

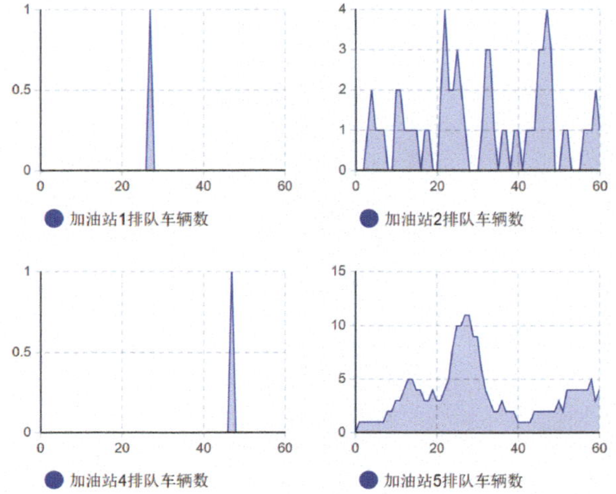


**Fig. 7.45** Queuing status of refueling stations 2, 3, and 5 after game cooperation

### *7.5.2 Analysis of Scheduling Simulation Results*

The optimized results are shown in Figs. 7.46 and 7.47. Compared with the simulation of the original model, the number of service vehicles at refueling stations 1–5 is 18, 20, 26, 29, and 23, respectively, through the optimization of energy replenishment scheduling. The number of vehicle replenishment at refueling stations is more balanced, and the average number of queuing vehicles at refueling stations 3 and 5 is reduced to 1, and the resource utilization rate of gas stations 1 and 4 is improved. Although the latency of nodes 16 and 33 no longer exists, node 8 has a delay.

## 7.6 Research on Negative Effect Reduction Strategy Based on Anylogic Simulation

In order to explore the negative effect reduction mechanism of urban transportation energy supply network and its application, simulation software is needed. AnyLogic simulation software has traffic pedestrian library, which can be used to simulate energy supply from a single station to the network and optimize the negative effect of the network, and analyze the operation result data.

The urban transport energy supply network is based on the road energy demand vehicles and energy supply stations. During the operation of the network system, due to the increase of vehicle flow, the demand for energy supplement increases, the overload of energy supply stations, and the unreasonable distribution of resource use at each station in the network, the spatio-temporal mismatch between vehicles and stations occurs, and the queuing of vehicles generates carbon dioxide, which

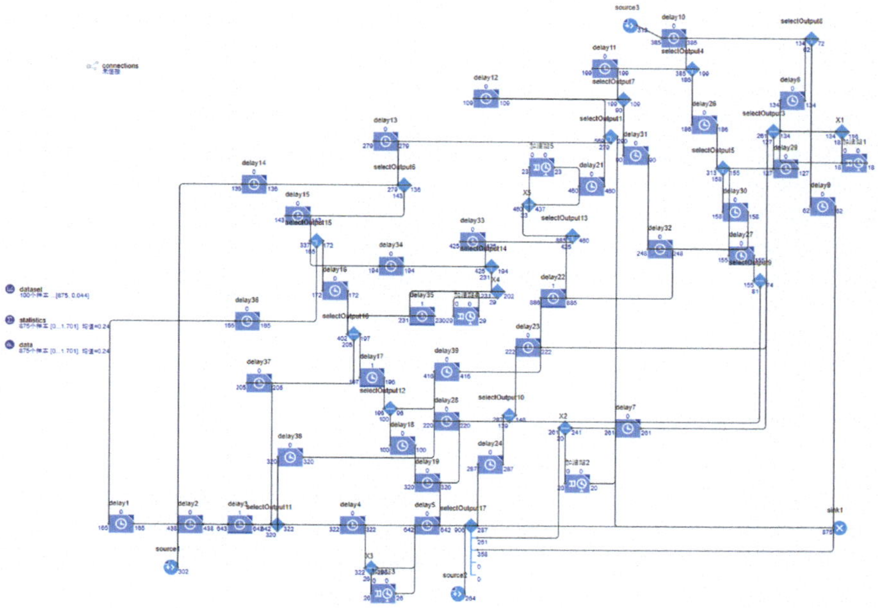

Fig. 7.46 Optimization of road network condition by energy replenishment scheduling

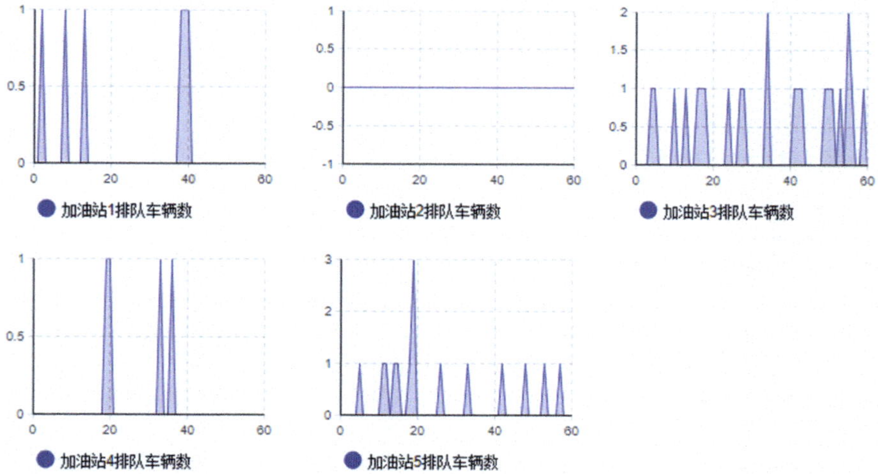


Fig. 7.47 Station queuing condition optimized by energy replenishment scheduling

increases environmental pollution and energy consumption. This runs counter to the current low-carbon policy of reducing the burden. Therefore, AnyLogic simulation software is used to build a simulation model for urban transportation energy supply network, simulate the actual running state of the energy supply network, analyze the energy recharge bottleneck points accumulated by negative effects, adjust different burden reduction mechanisms and analyze corresponding results, and provide ideas for energy supply network management.

The negative effect reduction of urban transportation energy supply network is to achieve the goal of reducing the overall negative effect by coordinating all aspects of road traffic and energy supply stations. This paper puts forward three strategies, which are traffic section restriction, station service rate adjustment, and different station price adjustment. By adjusting the correlation coefficients in the logic model of road network and station replenishment network in AnyLogic software, the purpose of optimization is achieved.

### *7.6.1 Traffic Section Restrictions*

Traffic restrictions originated in Argentina in the 1870s to control the number of vehicles on the road in order to minimize the impact on pedestrian travel. Compared with car purchase restriction, traffic restriction is currently the primary strategy of local governments to deal with urban traffic congestion. Among them, some cities only carry out road restrictions on some main roads and key areas, and some are time transition areas between the central city and the peripheral city. The restrictions range from morning and evening peak to all day, and the number of banned vehicles varies according to the scheme. Due to the specific situation of different cities, the effect of the restriction policy is not the same.

### *7.6.2 Adjust the Site Service Rate*

The station service rate has the greatest influence on the filling station, and the station efficiency mainly depends on the basic filling time and the filling gun vacancy rate. The position of the staff should be determined according to the number of personnel in the station, the site situation, the frequency of use of the refueling gun, and other factors.75% of vehicles enter through the station entrance and will first go to the location located near the entrance 1 tanker; 25% of the vehicles entering from the exit will go to the No. 2 tanker near the exit; Then the fuel truck will go to station 3 to refuel, and then down. During the peak period, it is necessary to strengthen cooperation between employees and increase the number of employees to reduce the vacancy rate of refueling guns.

### *7.6.3 Price Adjustment of Different Sites*

The pricing of domestic refined oil products is based on the crude oil price in the international market. The National Development and Reform Commission determines the price rise and fall of domestic refined oil products according to the law of the rise and fall of crude oil prices in the international market. But because refined oil is a product with a fixed price, there is a price curve that fluctuates around that price. As a marketing place for refined oil products, gas stations must of course face market competition and ensure that their prices can be profitable. Therefore, gas stations can independently set oil prices under the premise of not exceeding the maximum price proposed by the National Development and Reform Commission and not reducing costs, taking into account various factors such as operating costs, wholesale costs of refined oil products, and geographical location. The charging time and charging price of the State Grid charging station can be divided into three stages, corresponding to normal charging, premium pricing, and preferential pricing respectively. The charging pile has two main parts, namely the basic electricity price and the charging service fee. In order to adjust the electricity consumption at different times, the peak and trough of electricity consumption are not the same. In addition, the charging service operator will charge a certain charging service fee. According to different charging pile service providers, the service fee is also different, resulting in different prices at different times and different charging stations for different companies.

## 7.7 Summary

On the basis of fully investigating the operation data of the urban transportation energy supply network, the negative effects of the urban transportation energy supply network are evaluated and analyzed. AnyLogic simulation software is used to simulate and analyze the proposed burden reduction mechanism of the transportation energy supply network. According to the simulation results, the negative effects evaluation method and burden reduction mechanism of the urban transportation energy supply network are further improved. The typical transportation energy supply network sites were selected for empirical research, the theoretical research results were compared with practical data, and then the existing problems were discussed and targeted strategy suggestions were put forward to provide decision-making reference for urban transportation energy planning.

The manufacturer's authorised representative in the EU is Springer Nature Customer Service Centre GmbH, Europaplatz 3, 69115 Heidelberg, Germany. If you have any concerns regarding our products, please contact ProductSafety@springernature.com

Printed and bound by CPI Group (UK) Ltd, Croydon, CR0 4YY
01/10/2026
02225962-0006